AF684712

Volume 5

Ceramic Transactions

ADVANCED CHARACTERIZATION TECHNIQUES FOR CERAMICS

Edited by Wayne S. Young, IBM-San Jose,
Gary L. McVay, Battelle
Northwest, Gordon E. Pike,
Sandia Labs

The American Ceramic Society, Inc.
Westerville, Ohio

Proceedings of the 41st Pacific Coast Regional Meeting of the American Ceramic Society held in San Francisco, CA, on October 24–26, 1988.

LIBRARY OF CONGRESS
Cataloging-in-Publication Data

Pacific Coast Regional Meeting (41st: 1988: San Francisco, Calif.)
Advanced characterization techniques for ceramics/[edited by] Wayne S. Young, Gary L. McVay, Gordon E. Pike.
p. cm—(Ceramic transactions; v. 5)
"Proceedings of the 41st Pacific Coast Regional Meeting of the American Ceramic Society held in San Francisco, CA, on October 24–26, 1988"—Verso t.p.
"Sponsored by the Electronics Division of the American Ceramic Society and the Materials Research Society"—Foreword.
ISBN 0–944904–19–X
1. Ceramic materials—Testing—Congresses. I. Young. Wayne S. II. McVay, G.L. III. Pike, Gordon E. IV. American Ceramic Society. Electronics Division. V. Materials Research Society. VI. Title. VII. Series.
TA455.C43P33 1988 89–17924
620.1'4'0287—dc20 CIP

ISBN 0–944904–19–X

No part of this book may be reproduced, stored in a retrieval system, or transmitted in any form or by any means, electronic, mechanical, photocopying, microfilming, recording, or otherwise, without written permission from the publisher.

Printed in the United States of America

1 2 3 4 5—94 93 92 91 90 89

Foreword

As the applications of ceramic materials become more sophisticated and demanding, the need to characterize and understand their properties increases. However, advanced characterization techniques tend to be developed and first applied only within specialized research communities which may or may not be familiar to ceramists. The symposium on ''Advanced Characterization Techniques for Ceramics'' was organized primarily to focus on techniques developed in traditionally nonceramic fields—but for which ceramic applications are beginning to be found. While many techniques may qualify for such a designation, we tried to highlight those which have proven value and are realistically accessible (by collaboration, contract, or purchase) by inviting speakers who were practitioners of the techniques on ceramic materials. Results obtained on ceramic materials were used to illustrate the utility of the techniques, thus making the symposium primarily tutorial in nature. Supplementing the symposium was a literature exhibit by vendors of equipment, instrumentation, and services discussed in the sessions.

This proceedings volume is a collection of the invited and contributed papers presented at the symposium. Our intent in publishing this volume with its tutorial papers, subject index, and acronym glossary is to introduce the available techniques presented at the symposium to the wider ceramics community.

A special addition to this volume is the paper by Professor John B. Wachtman on ''Characterization of Ceramics,'' which was originally delivered as the Arthur L. Friedman Memorial Lecture at the 1989 Annual Meeting of the American Ceramic Society. Because of its broad and introductory nature, the editors have placed it first in the book.

The symposium on ''Advanced Characterization Techniques for Ceramics'' was held as part of the 41st Pacific Coast Regional Meeting of the American Ceramic Society in San Francisco, CA, October 24–26, 1988. This symposium was jointly sponsored by the Electronics Division of the American Ceramic Society and the Materials Research Society.

Wayne S. Young
Gary L. McVay
Gordon E. Pike

Ceramic Transactions is a new proceedings series designed to meet two needs: high quality content and rapid publication. Volumes in the series come from meetings, symposia, and forums. Each paper is reviewed by two peers, and final manuscripts are prepared by authors in a "camera-ready" format. The volumes in this series would not be possible without the hard work, dedication, and cooperation of editors, reviewers, and authors, who all deserve a great deal of thanks.

*Your comments, questions, and suggestions for future **Ceramic Transactions** volumes are welcomed and should be addressed to the Director of Publications, The American Ceramic Society, Inc.*

Acknowledgments

We wish to thank all the symposium contributors, particularly the tutorial and invited speakers, who provided excellent presentations and set the tone of the meeting. They are:

Tutorials: Richard J. Blattner, Charles Evans & Associates
Mehmet Sarikaya, University of Washington
Gareth Thomas, University of California/Berkeley

Invited Speakers:

Peter Arrowsmith
Albert J. Bevolo
Lynn A. Boatner
Dawn A. Bonnell
Gregory J. Exarhos
Sahba Ghaderi
Terry W. Gullion
H. Ernst Keller
Joan B. Pallix
Larry R. Pederson
Christopher H. Schilling
W. Kent Schubert
Carleton H. Seager
William J. Thomson

We are indebted to our session chairmen who directed the sessions and guided the discussions. In addition to the co-chairmen, these were:

Albert J. Bevelo
Lynn A. Boatner
Dawn A. Bonnell
Gregory J. Exarhos

This symposium was partially supported by the following companies who participated in the associated Literature Exhibit. We wish to express special appreciation to:

Buehler, Ltd.
Carl Zeiss, Inc.
Charles Evans & Associates
Instruments SA, Inc.
ISI, Inc.
JEOL USA, Inc.
Kratos Analytical
Leybold-Heraeus, Inc.
Perkin-Elmer Corporation
Surface Science Laboratories

Special thanks are also due the peer reviewers for their conscientious review of all the manuscripts in this proceedings. We are grateful to Marie Jimenez for coordinating the publication of this proceedings and Kathryn Demeter for her assistance at the meeting.

Contents

Overview

Section I. Near-Surface Techniques

*Tutorial

Section II. Bulk Techniques

Section III. Electronic Techniques

Section IV. Electron Microscopy and Microprobe Techniques

*Tutorial

Section V. Optical Techniques

Overview
Section I. Near-Surface Techniques

CHARACTERIZATION OF CERAMICS

by

John B. Wachtman

Rutgers, The State University of New Jersey

Friedberg Lecture for

The American Ceramic Society

Abstract

Modern instrumental techniques offer powerful means of characterizing ceramics. The importance of characterization is increasing as the composition and microstructure of ceramics and composites involving ceramics become more complex and sophisticated to produce higher performance. The large range of instrumental techniques applicable to ceramics is associated with other disciplines and has generally not been systematically covered in the education of ceramic engineers. In-depth coverage is not possible in a survey, but a sketch of the range and power of many of the principal instrumental characterization techniques as well as their limitations is given. A scheme based on the type of radiation or particles input and output is useful to classify these techniques and to understand their capabilities.

1. Introduction

Characterization is a critically important aspect of research on ceramics and is an increasingly critical aspect of production as more specialized properties and greater reliability are required. Every ceramist learns some characterization techniques. The rapid development of new instrumental techniques has made available a large variety of sophisticated characterization tools. No one can be expert in all of these techniques but it is important to have to have at least some

---

Editor's Note: This paper was presented as the 1989 Friedberg Lecture at the American Ceramic Society Meeting held in Indianapolis, Indiana, April 24, 1989. With the author's permission, the editors have included this paper because of its broad-ranging and tutorial nature.

qualitative overall knowledge of the range of techniques and what can be learned from each. Experts can then be consulted for details on selected techniques. This paper attempts to give a general picture even though it is surely incomplete.

Some twenty years ago a National Research Commission Committee on Characterization of Materials[1] laid down a definition as follows: Characterization describes those features of the composition and structure, including defects, of a material that are significant for a particular preparation, study of properties, or use, and suffice for reproduction of the material.

The committee went on to say that basically characterization is a description of the kinds and locations of the constituent atoms and ions, to the extent that such a description is needed for correlation with properties, with performance in use, and for reproducing and improving the material. Today they would probably add control of processing to their list. In practice, coarser features of structure, such as sizes, shapes, and distributions of grains and pores are often used as part of the characterization. We can classify characterization into three broad categories: elemental analysis, phase analysis, and microstructural analysis.

2. Classification by Input and Output

Many instrumental characterization techniques utilize some form of incident beam which produces some type of output which is analyzed to give information about the specimen. The input can be protons, electrons, ions or neutrons. Any one of these four can in principle produce any one of the four types of output. There are thus sixteen possible combinations of input and output. Some are not of particular value. On the other hand, there are other considerations such as the energy range of the input and of the output particles. Different energy ranges may provide different information and may require quite different experimental facilities so that the total number of characterization types is larger than one might at first think. This characterization schematic does not fit all characterization techniques but is very useful to provide an organized way of thinking about some of the most important characterization techniques and is the basis for how many of them are named, e.g. X-ray photoelectron spectroscopy refers to a technique in which X-ray photons are incident and electrons are ejected.

For each characterization technique which fits into this input and output scheme of classification, there are several things to consider. First, of course, is the input-to-output conversion process taking place in the material. In this way information about the material being studied is loaded onto the output beam. Understanding the physics and chemistry of the conversation process is the core of understanding the characterization technique in terms of both its value and its limits. The next essential feature is how the information is derived from the output beam. In many cases this is done by detecting the beam and converting the results into a spectrum in terms of mass, charge, energy or related quantities such as wave length or frequency. Another important aspect is that some techniques allow an image of the specimen to be formed. This raises questions of magnification and resolution. Each technique typically has a characteristic range in which analysis is carried out. Sometimes a single analysis system will cover the entire range and in other cases, more than one analysis system is needed. The issue of how much of the specimen can be sampled is important as greater emphasis is being put on ceramics and composites with extremely fine scale structures. The minimum size that can be sampled in terms of lateral dimensions and depth determines the scale of use of the technique. One would like to sample a volume smaller than the finest feature of the microstructure.

This paper concentrates primarily on techniques which allow micro sampling or in-process sampling. It begins with an overview and listing of the principal techniques and then gives a very terse summary of the basic facts about each and an example illustrating its use. Several older techniques are not discussed even though these are very important because most ceramists are already familiar with traditional techniques of analytical chemistry and X-ray diffraction.

An excellent summary of the essential characteristics of many techniques has been compiled by J. J. Friel of Princeton Gamma Technology (PGT)[2] and is summarized in Table 1. Most ceramists are quite familiar with the first few but may be less familiar with some of the others. It is useful to consider this table in terms of families of techniques and to become familiar with the principle techniques and their acronyms.

The first family of techniques uses an input electron beam to form a magnified image and to provide other information. The SEM, or scanning electron microscope, is of course a workhorse in any modern materials laboratory. When combined with EDS, energy dispersive spectroscopy, or WDS, wavelength dispersive spectroscopy, it provides information on elemental concentration. The electron probe microanalyzer, EPMA, is an electron beam device set up to optimize micro analysis in this backscattering technique. Another family of electron beam instruments uses thin samples and a transmission mode giving transmission electron microscopy or TEM. This can be adapted for analytical electron microscopy, AEM, by operating in the scanning mode as a scanning transmission electron microscope, STEM, with energy dispersive spectroscopy, EDS, and/or electron energy loss spectroscopy, EELS.

Another family of instruments is based on analysis of emitted electrons and is adapted for surface studies. These include Auger electron spectroscopy and X-ray photoelectron spectroscopy which also goes under the name of ESCA meaning electron spectroscopy for chemical analysis.

Still another family of techniques is based on using a mass spectrometer to analyze ions knocked out of the specimen itself by some form of incident beam and is used for surface analysis. These techniques include secondary ion mass spectrometry, SIMS, ion scattering spectrometry, ISS, and laser microprobe mass analysis.

Yet another family of ion beam instruments typically operates at much higher energies and often produces a composition profile extending some thousands of Angstroms into the sample. These include Rutherford backscattering spectroscopy, RBS, elastic recoil detection, ERD and particle induced X-ray emission, PIXE.

When both the incident and output particles are photons, we have micro-Raman spectroscopy and micro-Fourier transform infrared spectroscopy.

Additional important characterization techniques include scanning tunneling microscopy, STM, which provides atomic scale resolutions of surface features. Nuclear magnetic resonance and electron paramagnetic resonance allow both chemical analysis and a probing of details of chemical bonding in the right circumstances. Low energy electron diffraction, LEED, allows study of unique surface structures. Neutron diffraction and neutron scattering allow probing of the vibrational spectrum of solids and are especially sensitive probes for magnetic materials and for hydrogen containing materials.

Techniques used in non-destructive evaluation draw on some of the above types of characterization and use them in ways adapted to examination of specimens of substantial size. Important types include microfocus X-ray projection, X-ray tomography, magnetic resonance tomography and ultrasonics.

3. Electron Beam Instruments

Turning now to each technique in turn, a very brief summary and an example for each is given. Scanning electron microscopy uses electrons with energy typically up to 30,000 V as input and forms images either from the same electrons in a backscattered mode or secondary electrons which come off at a much lower energy or from X-rays. Energy analysis of the X-rays allows identification of specific elements down to about 1000 ppm with the important limitation that the lightest element which can be detected is sodium. This is a relatively routine technique with reasonably easy simple preparation and gives good imaging and analysis. Figure 1 by Oliver, Giszpenc and Klein of Rutgers University[3] shows a microporous silica gel.

The electron probe micro analyzer is a similar instrument optimized for chemical analysis. It uses wave length dispersive analysis of the X-rays and gives a resolution of 2 um by 2 um and a detection limit of 50 ppm. The lightest element detectable is beryllium. This EPMA instrument uses a bent crystal analyzer as shown in Figure 2. A recent development by Dale Newbury and colleagues at NIST[3] carries out a complete quantitative electron probe analysis at each point of a matrix scan. That is, the analysis including all the correction factors is carried out pixel by pixel before the image is formed. Figure 3 shows that this technique allows simultaneous identification of a low concentration boundary of zinc in poly-crystalline copper as well as very high concentration regions. Such a range of composition and structural detail is very difficult to show in a conventional gray scale picture.

In transmission electron microscopy thin samples are used. Input electrons at a constant energy which may be as high as 1 meV or even more in the larger machines are used to form an image with excellent lateral resolution and good depth resolution. This is the premier tool for very fine scale studies of materials. It can provide lattice images as well as selected area electron diffraction.

High resolution electron microscopy is capable of providing

very fine detail in the hands of people expert in both the experimental technique and the interpretation including complex simulation of electron images. This image by Ernst, Pirouz, and Heuer of Case-Western Reserve University[5] shows the interface between copper and aluminum oxide. Their analysis indicates that the alumina is oxygen-terminated rather than aluminum terminated against the copper. This is an exquisitely fine detail and shows the power of the technique in expert hands.

Scanning tunneling microscopy is an exciting new development in which an atomically sharp needle is moved with great precision across a surface. A minute electric current tunnels between the specimen and the needle through the intervening space. By using a feedback circuit, the distance between the needle and surface can be kept constant. The position of the needle is plotted for each scan. The result builds up a relief map of the surface. The image shown in Figure 5 is on silicon carbide and was obtained by Bonnell and Clarke of IBM.[6] Under favorable circumstances individual atoms can be imaged although the image relates to the density of electron states as well as surface profile. This new tool can be used in liquids and gases so that the progress of a reaction could perhaps be followed. Work is under way to expand this technique to non-conducting material such as ceramics. The scheme is to develop an atomic force microscope.

Analytical electron microscopy is done with a scanning transmission electron microscope equipped with an energy dispersive spectroscopy and electron energy loss spectroscopy by which secondary electrons and X-rays are detected. Recent work by Helen Chan and colleagues at Lehigh University[7] on calcium zirconate produced the electron energy loss spectrum shown in Figure 6. We note that the peaks for calcium and zinc ride on the shoulder of a peak for carbon and that the issue of background subtraction is important. Table 2 shows results for the calcium to oxygen ratio and zirconium to oxygen in w% determined both by the X-ray EDS and by EELS. We see that when a 50 electron volt window is used with EELS for background subtraction, the EELS results agree well with the X-ray results. Again we see that the result is dependent on operator skill in interpreting the spectrum. In this case, skill involves judgment in background subtraction.

4. Electron Analysis Techniques for Surfaces

A set of techniques for surface analysis is somewhat similar to the above techniques in using analysis of electrons but differs in detail. In Auger emission spectroscopy the input electrons causes the emission of Auger electrons. The energy is characteristic of the element. This can be used for quantitative surface analysis and for good imaging. Figure 7 by R. Caracciolo of Rutgers University[8] shows an Auger electron map of a portion of a printed circuit. The left portion is an aluminum map and the right portion is a silicon map showing a clear distinction between the coating and the substrate.

In X-ray photo electron spectroscopy incident X-rays cause the emission of electrons. The energy identifies the atom from which the electron came. Small shifts in the energy provide information on the chemical bonding state. Figure 8 from Hegde and Caracciolo[8] of Rutgers University shows the oxygen spectrum for a sample of a superconducting yttrium barium copper oxide ceramic after an anneal in oxygen. The lower figure shows deconvolution into several peaks which correspond to oxygen in different chemical bonding states. The left most of these peaks is that of barium copper oxide. If the same sample is annealed in carbon dioxide, the spectrum in the upper portion of Figure 9 is obtained. Deconvolution indicates that barium carbonate has formed. Probably this compound should form over a period of time in 123 superconductors exposed to air with a trace of Co_2 present. Studies using angle resolved XPS spectroscopy suggest that the upper 10 to 20 angstrom of the 123 compound after exposure to air do have a different composition from the overall compound. Evidently, this could be critically important to the use of these ceramic superconductors in microelectronics. The power of characterization techniques which reveal such small-scale differences in chemical state is impressive and important.

5. Low Energy Ion Beam Techniques

Secondary ion mass spectroscopy is a technique in which ions with an energy of a few kilovolts are incident and which expel ions from the sample itself. These are resolved and identified in a mass spectrometer. This technique has the advantage of extreme sensitivity and of detection of all elements. A SIMS investigation by Hegde and Caracciolo of Rutgers University, of the surface of 123, shows that a non-conducting surface layer is indeed formed by exposure to air containing carbon dioxide. Molecular ions can be detected and give information on the compounds present in the solid which confirms the XPS results.

6. High Energy Ion Beam Techniques

Rutherford Backscattering Spectroscopy is a technique in which light ions are input at energies of several million electron volts. Some of these ions are scattered backwards into a detector close to the input beam. The energy of the backscattered ions indicates which nucleus they have scattered from. An additional small energy loss indicates the depth of the ion and so allows a depth profile to be constructed.

Figure 10 shows an RBS spectrum for 3 million electron volt helium 2+ scattered from a yttrium barium copper oxide superconductor.[10] The plotted line is the theoretical spectrum for the compounds. The plotted points closely fit the theoretical spectrum showing that the bulk composition is very close to the stoichiometric 123 compound. Note that the RBS technique is non-destructive and of great value but that it is not sensitive to small changes on the surface so that the surface compounds formed in this system must be examined with other techniques such as ESCA or SIMS as we have just seen.

In the technique of elastic recoil detection, ERD, heavy ions are incident at a grazing angle. These ions expel ions from the sample. In this respect the technique is similar to SIMS. However, in ERD the energy of the emitted ion rather than its mass is analyzed. As in the case of RBS a depth profile can be obtained. The advantage here is that hydrogen can be detected which cannot be done with RBS.

Figure 11 shows work by Jeter and Wachtman of Rutgers University in collaboration with G. Arnold of Sandia National Laboratories.[11] In this case glass was exposed to water at 90^oC for a series of times of increased duration. The successive curves show that as the time is increased, the amount of hydrogen and the depth penetration increases. This technique allows the determination of very small diffusion constants.

7. Optical and Infrared Spectroscopy

Micro-Raman spectroscopy can provide non-destructive bond group analysis in volumes on the order of a few cubic microns. Similarly, micro-Fourier transform infrared spectroscopy can be done in conjunction with a microscope and can be used to identify particular bonds.

Exciting work is in progress by Yan, Rydzak and Cannon[11] of

Rutgers University on using micro-Fourier transform spectroscopy to map stress around cracks and inclusions.[12] Figure 12 shows a plot of stress as a function of distance from the crack tip. The slope is close to the theoretical value of 1/2 which is expected on the basis of linear elastic fracture mechanics. Figure 13 shows stress plotted against 1 divided by the square root of 2(3.1416)r. The slope of .71 MPa m should be the critical stress intensity factor and corresponds well with the value of .7 measured by ther techniques.

8. Magnetic Resonance Techniques

Nuclear magnetic resonance involves the input of both static and radio frequency magnetic fields. The output is a radio frequency magnetic field. This technique can be used to determine the average concentration of a sample with magnetically active nuclei present.

A beautiful example of using NMR is shown in Figure 14 by Assink and Kay of Sandia National Laboratory.[13] The upper part of the figure gives a proton spectrum and shows that protons connected to OH groups, methyl groups or ethyl groups can clearly be distinguished. The lower spectrum on the left with an expanded scale shows the nearest neighbor distinctions can be made. These spectra were taken on a tetraethyl orthosilicate reacting with water to form a gel. The lower right spectrum was taken at a later time and shows that the concentration of the various molecular species has shifted as the process of gelling proceeded. This technique permitted the authors to follow a series of simultaneous reactions in detail and to fit a theory for gelling.

9. Neutron Scattering

Small angle neutron scattering measures the scattering of a neutron beam with a position sensitivity detector. Analysis of the angular dependence of the scattering permits a measurement of the average pore size in a material in a complete non-destructive manner. The measurement can even be carried out on a specimen in a furnace so that in-process changes can be followed.

Gabriel Long of NIST and Rosario Gerhardt of Rutgers University have measured the change in pore size during the sintering of micro porous silica by small angle neutron scattering.[14] Figure 15 shows that in this system pores actually expand before shrinking.

A parallel piece of work on sintering of alumina by Long of NIST and Page of Southwest Research Institute[15] is shown in Figure 16. The pore size in alumina first shrinks and then grows substantially as theoretical density is approached.

10. Non-Destructive Evaluation

Non-destructive evaluation is desirable for both process control and examination of finished products. Figure 17 from Ellingson et al. of Argonne National Laboratory[16] indicates different stages in processing and lists the techniques most likely to be useful at various stages e.g., at the green forming stage, the techniques of computed tomography imaging, ultrasonic microscopy, project-on X-ray radiography and magnetic resonance imaging are being explored.

As Figure 18 shows, X-ray computerized tomography is a technique in which both the X-ray source and detector move and a series of X-ray projections are generated. Figure 19 from Ellingson et al. at Argonne and Sawicka [17] et al. at Chalk River Nuclear Laboratory shows X-ray projected radiographs on the left and computed tomography on the right. A large crack can be seen in the left figure while density variations are readily apparent in the right figure.

Magnetic resonance imaging of protons can give a mapping of proton density. This is especially useful for unfired ceramics which still contain the binder. Figure 20 from Ellingson et al.[18] shows density variations in a tile measured by computerized tomography based on magnetic resonance. Voids can be imaged by this technique as shown in Figure 21 which presents the results of magnetic resonance imaging on an unfired sample containing glass beads. These beads are clearly imaged in both the axial and the transverse sections.

In X-ray projection radiography a microfocus X-ray source is used to project portions of the specimen onto a film as shown in Figure 22. In this way a degree of magnification can be obtained. The technique is potentially very useful but is limited by the sensitivity of detection. Gray and Inanc of Iowa State University[19] have modeled the detectability. Figure 23 shows that depending on the X-ray energy, flaws down to a few percent of the sample thickness can be detected. Another complication arises for flaws for a high aspect ratio such as cracks. Figure 24 shows that as the aspect ratio of the flaw increases, the flaw can be detected only for a narrow angle around the plane of the flaw. Flaws of high aspect ratio, such as cracks, which lie outside this angle cannot be detected.

Ultrasonic measures of velocity are commonly used to calculate elastic constants. An interesting new ultrasonic technique developed by David Hsu of Iowa State University[20] relates the ultrasonic attenuation to the void content. Figure 25 shows measurements of porosity by ultrasonic attenuation plotted against porosity measured destructively by acid digestion. The results indicate that quite small porosity percentages can be measured accurately and non-destructively by ultrasonic attenuation.

11. Conclusion

This brief selection of modern characterization techniques is certainly not complete. Many other examples by many other workers could be given. Hopefully, this selection conveys some sense of the power and versatility of modern characterization techniques.

Acknowledgment

The author has benefited from conversations with R. A. Asink, R. Caracciolo, H. M. Chan, D. Clarke, W. Ellingson, J. J. Friel, R. A. Gehardt, J. N. Gray, V. A. Greenhut, M. V. Hegde, A. H. Heuer, A. Inam, D. K. Hsu, B. D. Kay, G. Long, D. E. Newbury, J. Oliver, T. Venkatesan, and X. D. Wu but is solely responsible for the opinions expressed here. Thanks are extended to those who supplied figures and tables used here.

Support from the Center for Ceramic Research, a New Jersey Commission on Science and Technology Center, is gratefully acknowledged.

References

1. "Characterization of Materials" National Materials Advisory Board Report NMAB-229-M, National Research Council, 1967.
2. J.J. Friel, Princeton Gamma Technology, private communication with the author.
3. J. Oliver, N. Giszpenc and L. C. Klein, Rutgers University, private communication with the author.
4. D. E. Newbury, R. B. Marinenko, D. S. Bright, and R. L. Myklebust, "Computer-Aided Imaging: Quantitative Compositional

Mapping with the Electron Probe Microanalyser", Scanning, Vol. 10, p 213-225 (1988).
5. A. H. Heuer, F. Ernst, and P. Pirouz, "High Resolution Electron Microscopy of an Alumina/Copper Interface", to be published.
6. D. A. Bonnell and D. R. Clarke, "Scanning Tunneling Microscopy and Spectroscopy of Ceramics: Silicon Carbide and Zinc Oxide, J. Am. Ceram. Soc. Vol. 71, No. 8, 629-37 (1988)
7. H. M. Chan , Irene Piscopo, D. B. Williams, and M. R. Notis, "AEM STudy of the $DaZrO_3/ZrO_{2(ss)}$ Eutectic: A Comparison between EELS and UTW X-Ray EDS", pages 363-367 in "Analytical Electron Microscopy - 1984", San Francisco Press, Inc.
8. R. Caracciolo, Rutgers University, private communication with the author.
9. M. V. Hegde and R. Caracciolo, Rutgers University, private communication with the author.
10. X. D. Wu, T. Venkatesan, A. Inam, M. S. Hegde, E. W. Chase, C. C. Chang, J. M. Tarascon, J. B. Wachtman, and W. L. Mclean, "High and Low Temperature Processing of High T_c Oxide Films", The Metallurgical Society (TMS) in press.
11. D. Jeter, J. W. Wachtman, and G. Arnold. Water penetration in Glass. Submitted to Communications of the American Ceramic Society.
12. H. Yan, J. W. Rydzak and W. R. Cannon, "A Comparison of FTIR Microspectroscopy and Raman Microprobe Spectroscopy for the Determination of Stress in Ceramic Materials", Ceramic Bulletin, Vol. 68, No. 3, p 704, 1989.
13. R. A. Assink and B. D. Kay, "NMR Studies of the Sol-Gel Transition", Mat. Res. Soc. Symp. Proc. Vol. 32, pages 301-306, Materials Research Society, 1984.
14. G. G. Long, S.T. Krueger and R. A. Gerhardt, Small Angle Neutron Scattering Characterization of Microporous Silica, to be published in transactions of The American Ceramic Society.
15. A. Page, G. G. Long, and S. T. Krueger, "Small Angle Scattering Measurements of Pore Evolution During Sintering of Al_2O_3", Ceramic Bulletin, Vol. 68, No. 3, p 713 (1989).
16. W. A. Ellingson and Michael W. Vannier, "X-Ray Computed Tomography for Nondetructive Evaluation of Advanced Ceramics", Argonne National Laboratory Report ANL-87-52, 1987.
17. "Application of Computer Tomography to the Development of Advanced Ceramics", B.D. Sawicka and B.J. F. Palmer, CSNDT Journal, 24-33, March/April 1989.
18. W. A. Ellingson, J. L. Ackerman, L. Garrido, P. S. Wong, and S. Gronemeyer, "Development of Nuclear Magnetic Resonance (NMR) Imaging Technology for Advanced Ceramics", Argonne National Laboratory Report ANL-87-53, 1987.
19. J. N. Gray and F. Inanc, "Three Dimensional Modeling of

Projection Radiography", Review of Progress in Quantitative Nondestructive Evaluation", edited by D. O. Thompson and D. E. Chimenti, Vol 8A, Plenum Publishing Corporation, 1988.

20. D. K. Hsu, "Ultrasonic Measurements of Porosity in Woven Graphite Polyimide Composites", Review of Progress in Quantitative Nondestructive Evaluation, Vol. 7B, ed. by D. O. Thompson and D. E. Chimenti, Plenum Publishing Corporation, 1988.

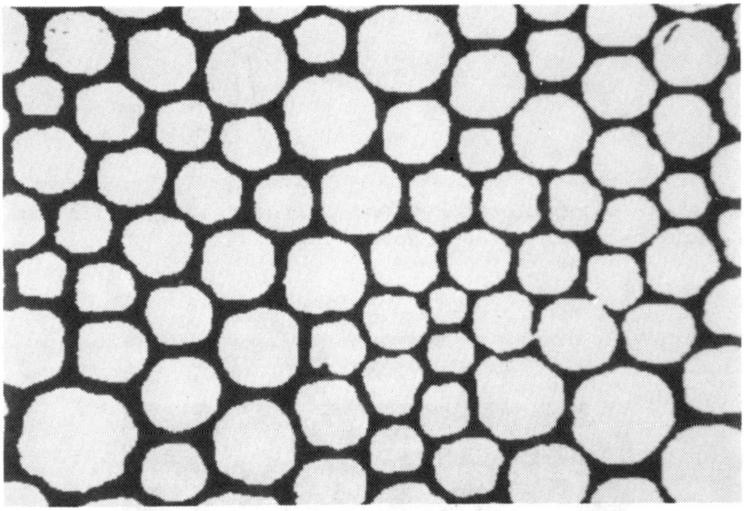

1. Microporous silica gel. From Oliver, Giszpenc and Klein.

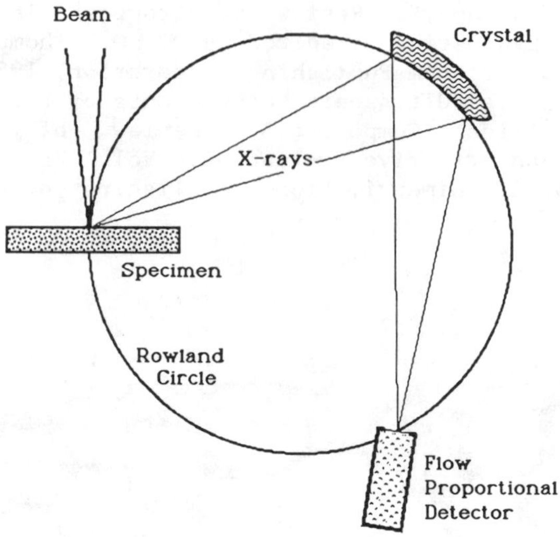

2. Schematic diagram of Rowland circle with specimen, bent crystal, and detector. From Newbury et al.

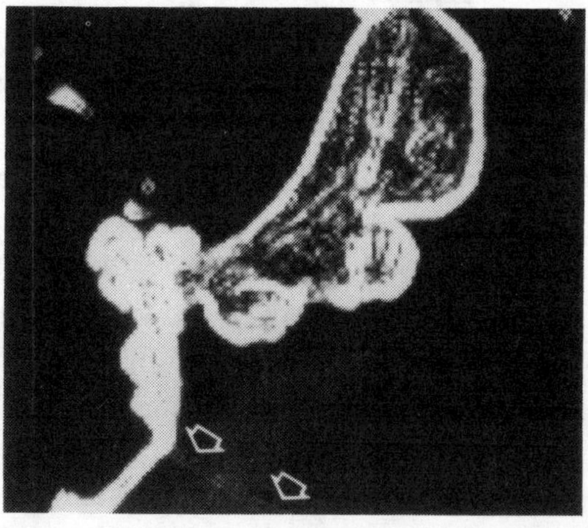

3. Application of thermal color scale to phases and grain boundary of zinc in copper. From Newbury et al.

4. Lattice image showing interface between boundary of aluminum oxide and copper. From Ernst, Pirouz and Heuer.

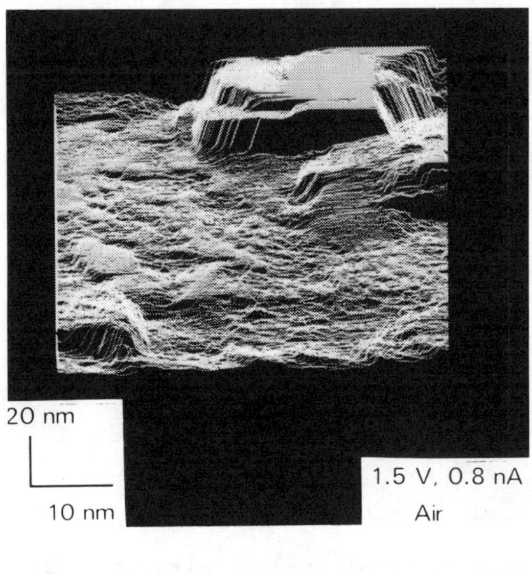

5. Surface of silicon carbide as imaged by scanning tunneling microscopy. From Bonnel and Clarke.

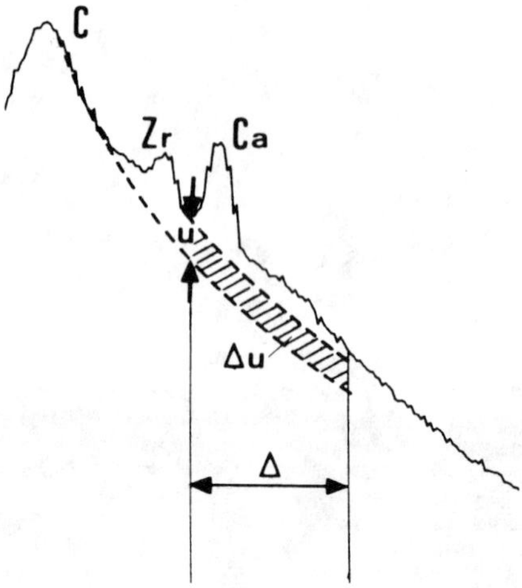

6. Electron energy loss spectrum of calcium zirconate. From Chan et al.

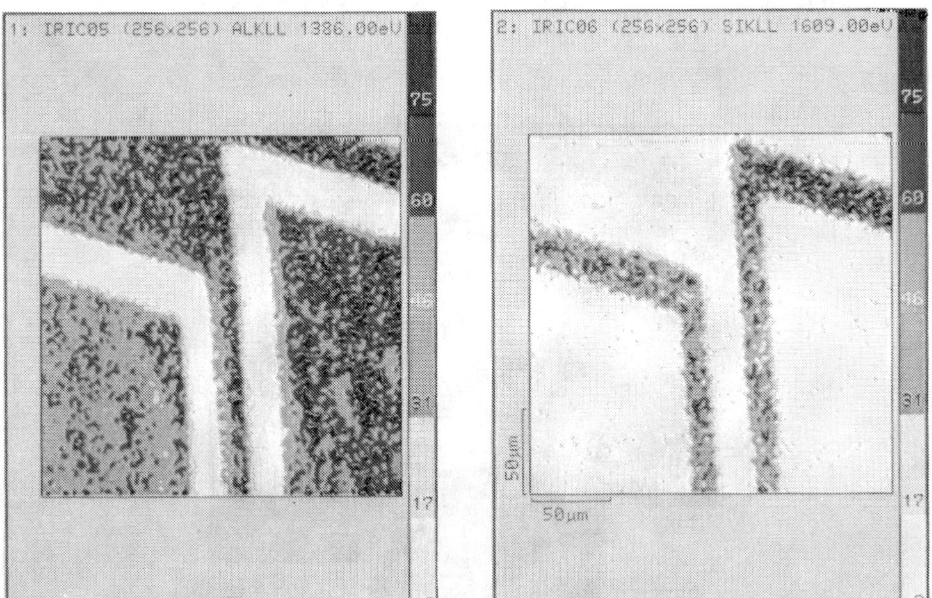

7. Auger image of microelectronic circuit. The left portion is an electron map and the right is a silicon map. From Caracciolo. In color.

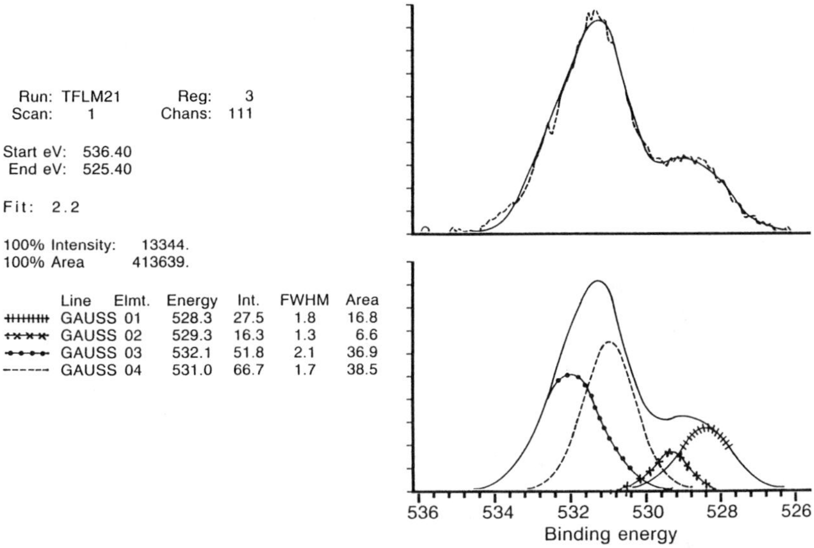

8. X-ray photoelectron spectrum of a "123" superconductor after annealing in oxygen. The lower portion shows peak deconvolution. From Hegde and Caracciolo. In color.

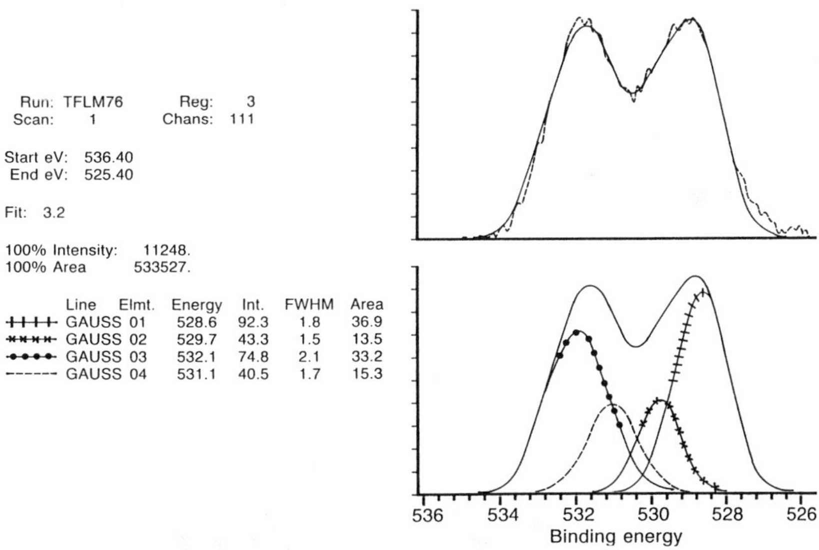

9. X-ray photoelectron spectrum of a "123" superconductor after annealing in carbon dioxide. The lower portion shows peak deconvolution. From Hegde and Caracciolo. In color.

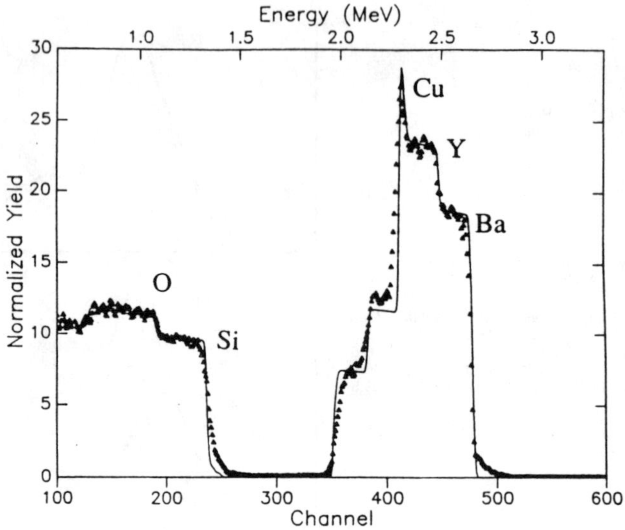

10. A Rutherford Backscattering Spectrum of a "123" superconductor. From Wu et al.

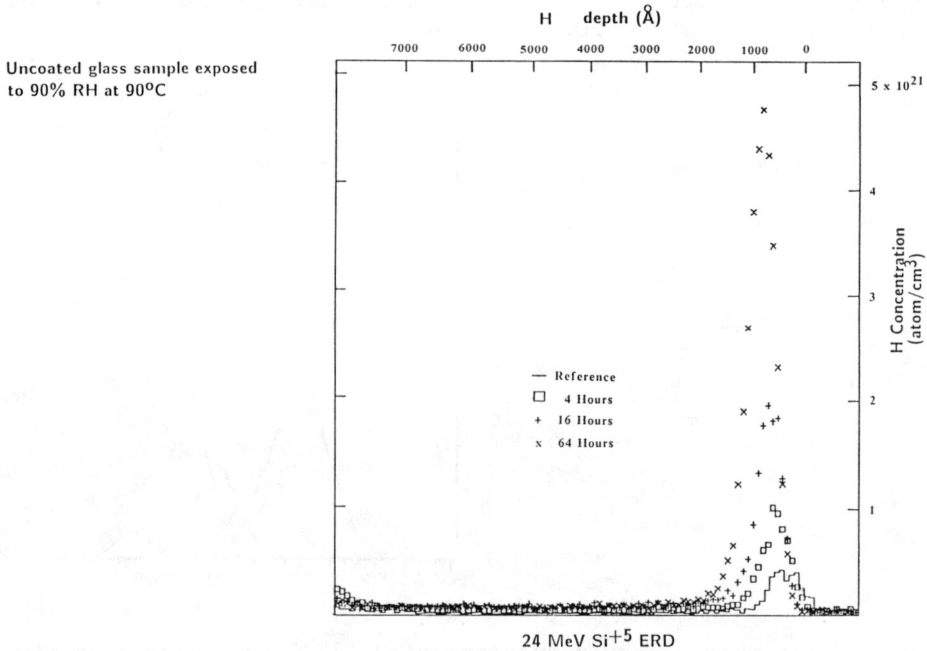

11. An elastic recoil detection spectrum of glass into which water has been diffused for successively longer times. From Jeter, Wachtman, and Arnold.

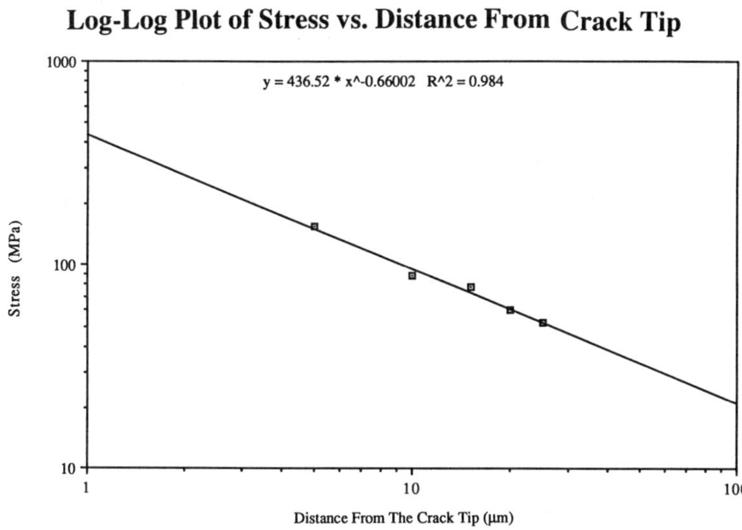

12. A log-log plot of stress as a function of distance from a crack in silicon as measured by FTIR. From Yan, Rydzak and Cannon.

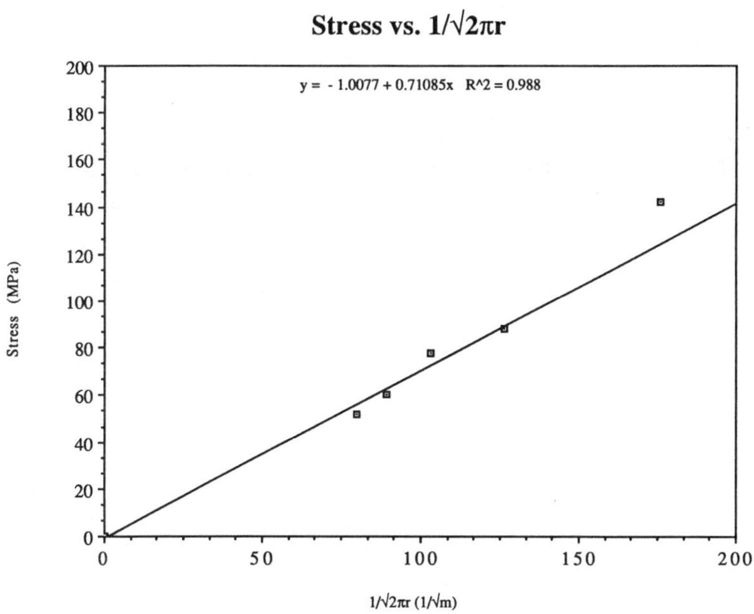

13. A plot of stress measured by FTIR designed to yield the critical stress intensity factor. From Yan, Rydzak and Cannon.

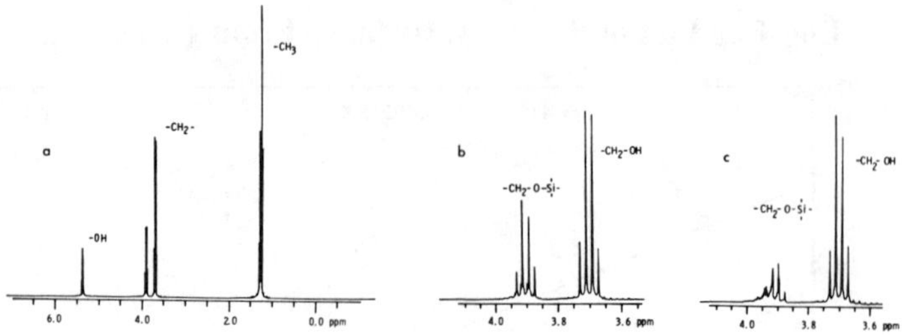

14. NMR spectra on tetraethyl orthosilicate reacting with water to form a gel. From Asink and Kay.

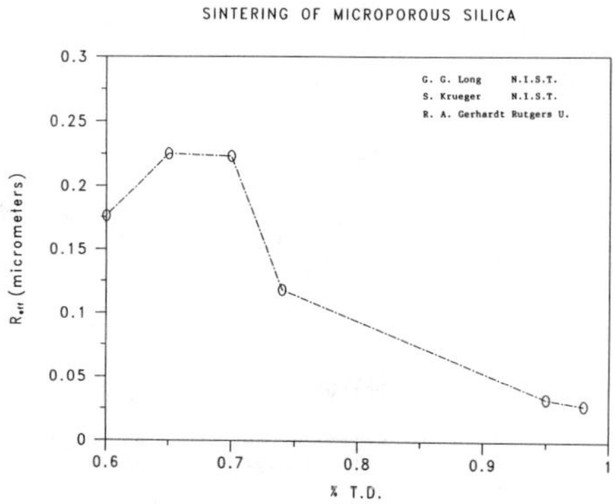

15. Pore size determined by SANS for successive stages of sintering of microporous silica. From Long and Gerhardt.

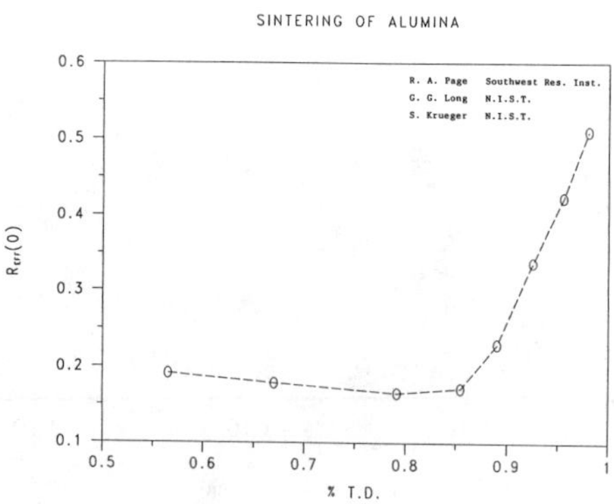

16. Pore size determined by SANS for successive stages of sintering of alumina. From Long and Page.

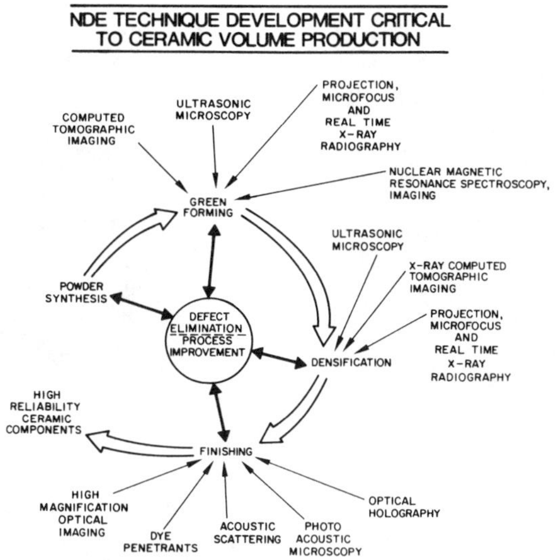

17. Application of NDE methods throughout the cycle of ceramic processing. From Ellingson et al.

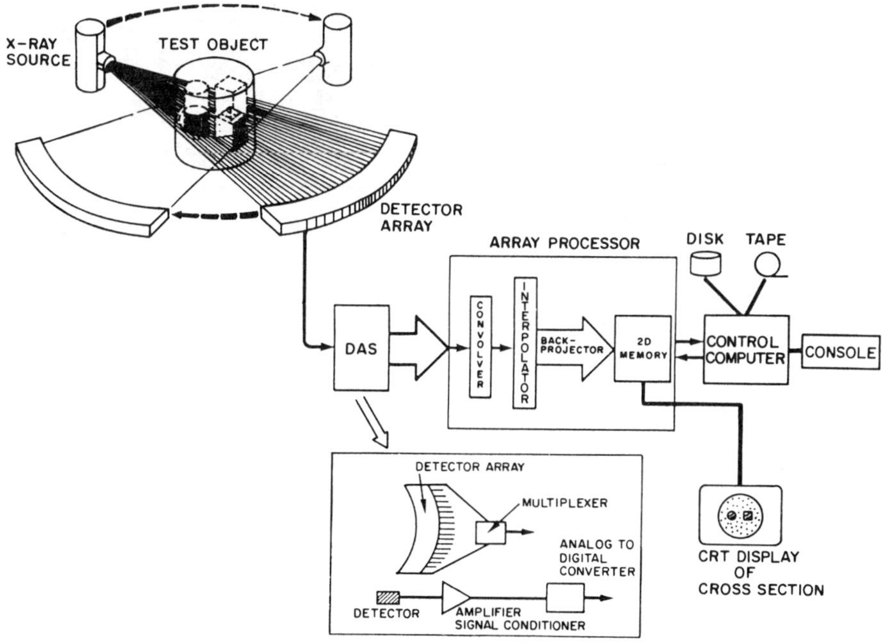

18. Schematic of X-ray Computerized Tomography. From Ellingson et al.

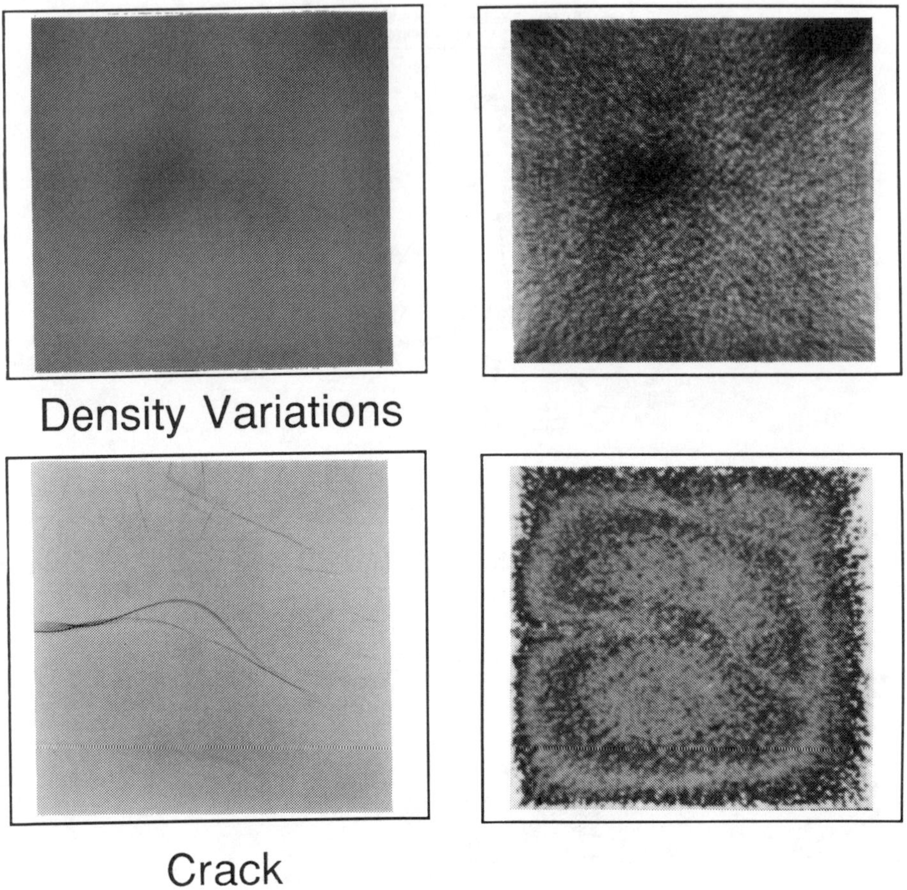

19. Comparison of Radiographs and Computerized X-Ray Tomography showing a crack and density variations. Radiographs from Ellingson et al., tomographs from Sawicka et al. using a Co-60 source.

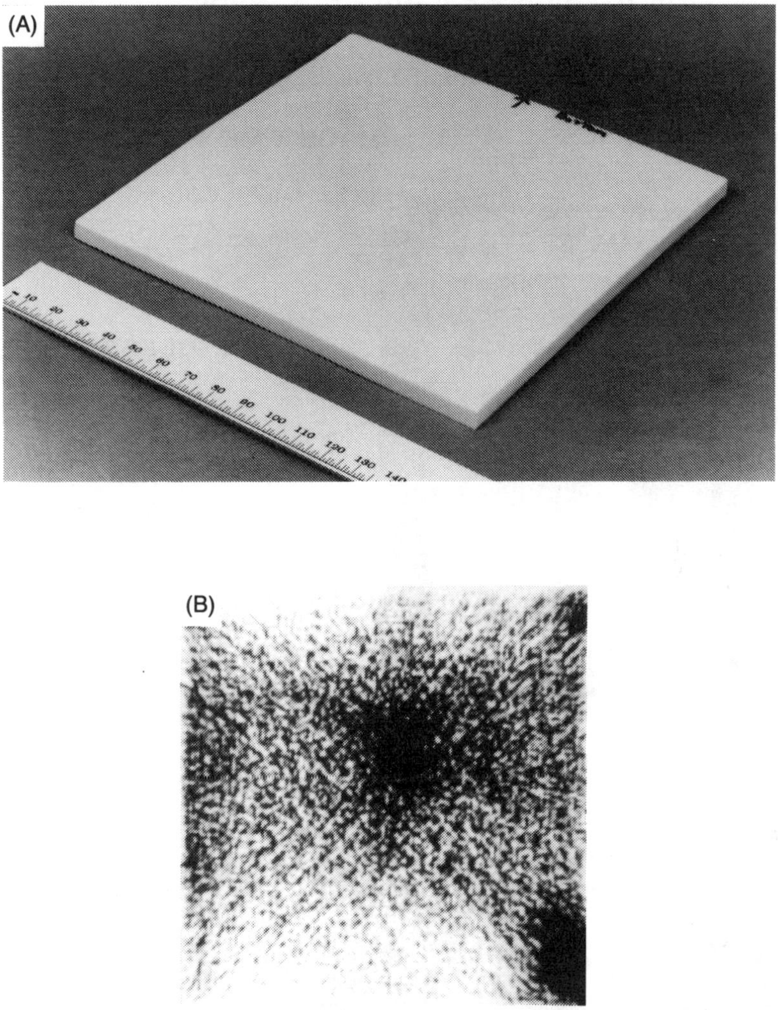

20. Unfired Alumina Ceramic Tile and Computerized Magnetic Resonance Image Showing Density Variations. From Ellingson et al. In color.

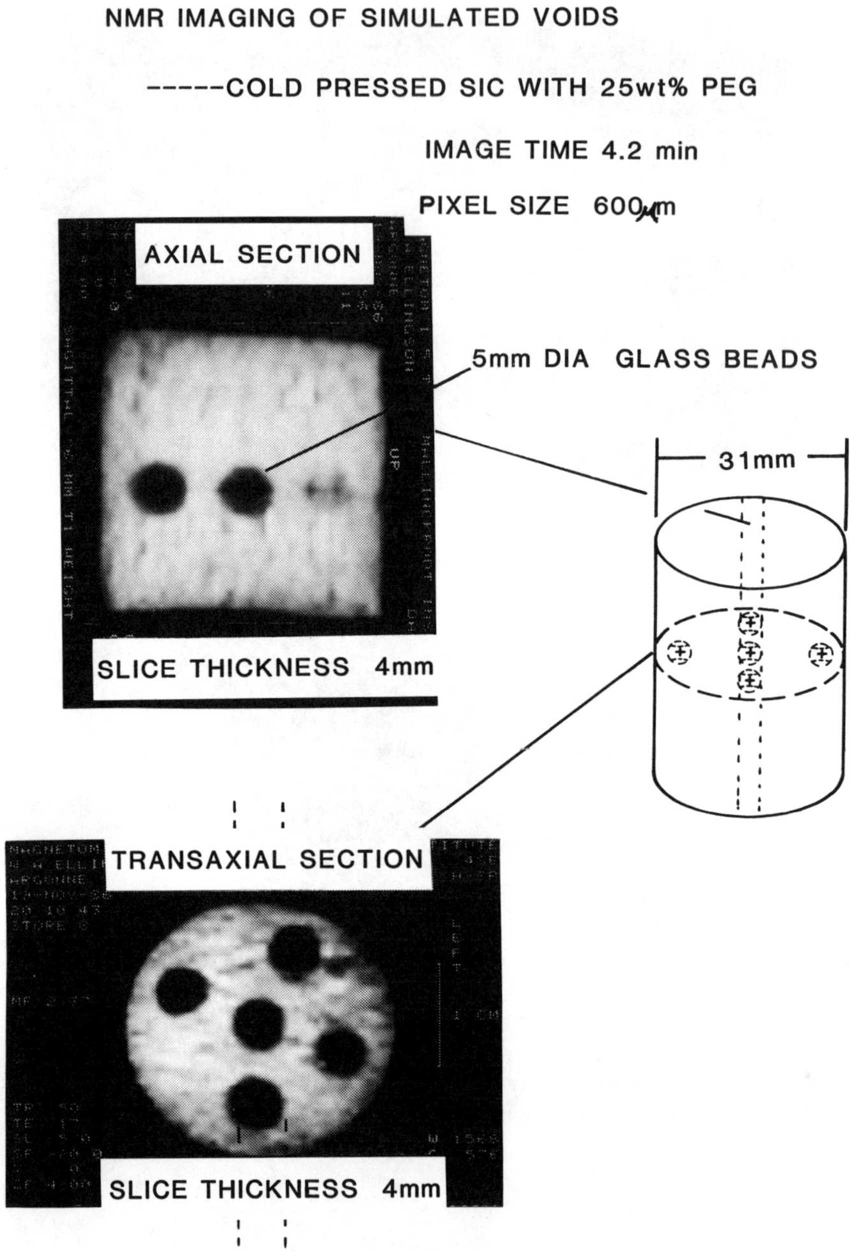

21. Unfired Silicon Carbide Cylinder with Glass Bead Inclusions as Shown by Computerized Tomography Based on Magnetic Resonance Imaging. From Ellingson et al.

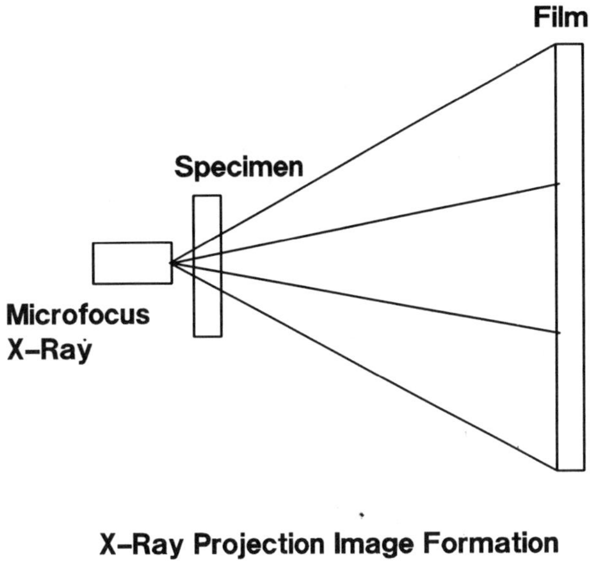

22. Geometry of Microfocus X-Ray imaging.

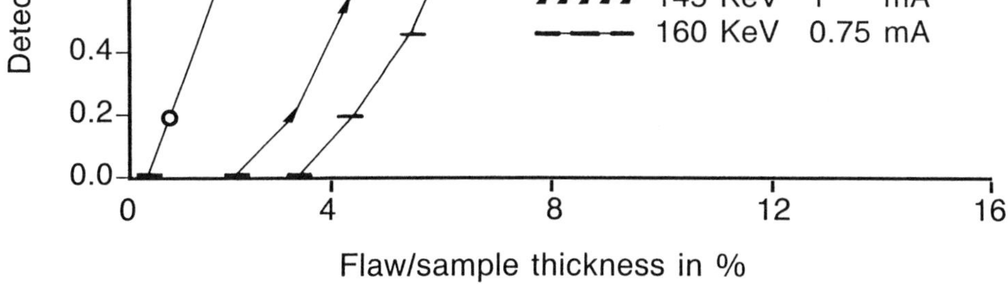

23. Effect of Sample Thickness on Dectability of a Flaw. From Gray and Inanc.

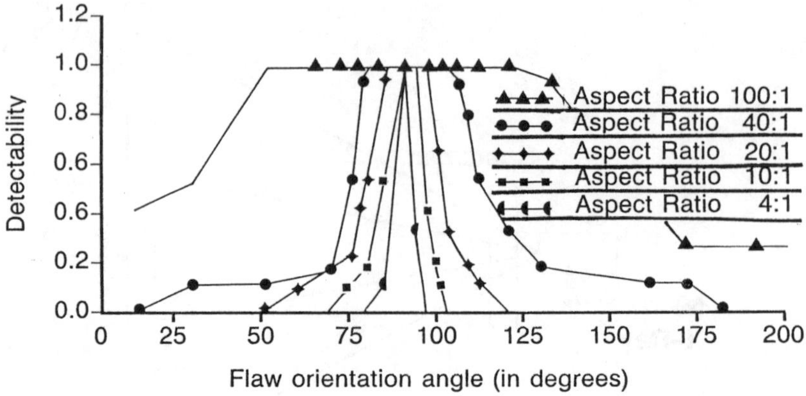

24. The Detectability of Ellipsoidal Flaws as a Function of Flaw Angle for Various Aspect Ratios. From Gray and Inanc.

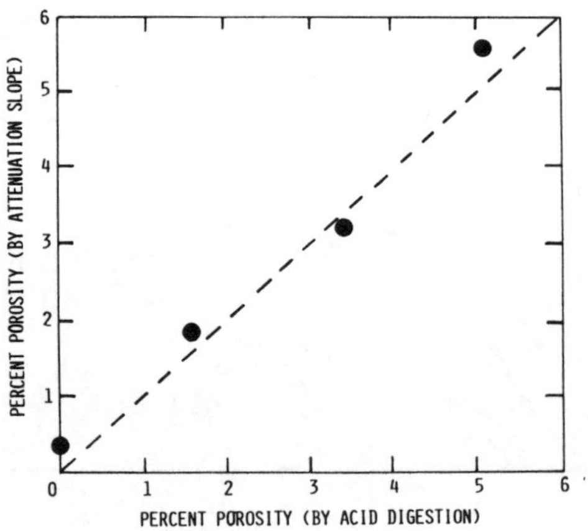

25. Comparison of Ultrasonically Determined Void Contents with Those Measured by Acid Ingestion on Woven Graphite Epoxy Laminate Composites. From Hsu.

TABLE I COMPARISON OF VARIOUS MICROANALYTICAL TECHNIQUES

Name	E in	E out	Resolution Lateral	Resolution Depth	Detect. Limit	Lightest Element	Imaging	Other Info	Advantages
SEM, EDS	e^-(2-30keV)	SE (50eV) BSE (Eo) X-rays (> 1kV)	5nm ~2 μm ~2 μm	5nm ~2 μm ~2 μm	 1000ppm	 Na	yes yes yes		routine, easy sample prep, good imagine & analysis
EPMA, WDS	e^-(5-50keV)	X-rays (> 100eV)	~2 μm	~2 μm	50ppm	Be	yes		quantitative analysis
TEM	e^-(100keV-1meV)	TE (Eo)	0.2nm	<100nm			yes	SAED	high resoltuion, electron diffraction
AEM/ STEM, EDS EELS	e^-(100-400keV)	TE (Eo) SE(50eV) X-rays (>1kV) electrons	0.2nm 3nm 5nm 20nm	<100nm 3nm <100nm <10nm	 1000ppm 1000ppm	 Na Li	yes yes yes no	CBED	analytical resolution small amts. of lt. element
AES	e^-(1-3keV)	e^-(>200eV)	0.2	<3nm	1000ppm	Li	yes		quant. surface analysis
XPS/ ESCA	Xrays(1-1.5keV)	e^-(>10eV)	0.5mm	<3nm	1000ppm	He	no	binding energy	chemical state infomation
SIMS	ions(4-15keV)	secondary ions	5	0.1nm	1ppb	H	some	depth profile, mass	analytical sensitivity for surface & light elements
PIXE	H^+,He^{++}(~3meV)	X-rays	2 μm		1ppm	Na	some		analytical sensitivity
RBS	H^+,He^{++}(2-15meV)	H^+,He^{++}(<Eo)	2mm	10nm	1-1000ppm	Li	no	depth profile mass	non-destructive depth profile
LIMA/ LAMMA	UV(250nm)	ions	~2 μm	2 μm	1ppm	H	yes	mass	analytical sensitivity, Mass spectrum
Micro Raman	vis. light (~500nm)	vis. light	~2 μm	---	---	---	some	molecular spectrum	phase ID
Micro FTIR	IR light	refl. or trans.IR	10 μm	---	---	---	no	molecular spectrum	phase & functional group ID

GLOSSARY OF ACRONYMS

AEM	analytical electron microscopy (STEM)	PIXE	particle (proton) induced X-ray emission
AES	Auger electron spectroscopy	RBS	Rutherford backscattering spectroscopy
BSE	backscattered electrons	SAED	selected area electron diffraction
CBED	convergent beam electron diffraction	SE	secondary electrons
EDS	energy dispersive spectroscopy	SEM	scanning electron microscopy
EELS	electron energy loss spectroscopy	SIMS	secondary ion mass spectrometry
EPMA	electron probe microanalyzer	STEM	scanning transmission electron microscopy (AEM)
ESCA	electron spectroscopy for chemical analysis (XPS)	TE	transmitted electrons
FTIR	Fourier transform infrared spectroscopy	TEM	transmission electron microscopy
LAMMA	laser microprobe mass analyzer (LIMA)	WDS	wavelength dispersive spectroscopy
LIMA	laser ionization mass spectrometry (LAMMA)	XPS	X-ray photoelectron spectroscopy (ESCA)

J. J. Fiiel/PGT

Table 1. Comparison of Various Microanalytical Techniques. By permission of J. J. Friel.

TABLE II

Technique	Ca/O (wt)	Zr/O (wt)
X-ray EDS		
i) Calculated from $(Ca/Zr)_{ss}$	.316 ± .007	2.491 ± .058
ii) UTW ratio method	.346 ± .06	2.797 ± .288
EELS		
Δ = 20 eV	.451 ± .04	-
Δ = 50 eV	.368 ± .06	-

Table 2. Comparison of EDA and EELS Analysis Showing Effect of Choice of Background Correction. From H. Chan.

CHARACTERIZING THE TOP 10,000 Å OF CERAMICS

Richard J. Blattner, Charles Evans & Associates, 301 Chesapeake Dr., Redwood City, CA 94063 USA.

INTRODUCTION

The primary goal of engineered materials is the achievement of some prescribed set of physical properties which may be electrical, optical, magnetic, etc. However, physical properties are nothing more than a colligative manifestation of the elemental, chemical and structural make-up of the material. Thus, it is not surprising that many different analysis methods have been developed over the years to provide information on these various fundamentally different levels. Most importantly, materials analysis has consistently played an important role in research and development, process monitoring, production quality control and failure analysis.

The foregoing is true regardless of the physical dimensions or nature of the materials in question. However, during the last two decades there has been a growing scientific and technological interest in the outermost 10,000 Å of solids, including ceramic materials, spawning the development of analysis methods appropriate to this specific regime. They are collectively referred to as the "surface analysis" techniques but as we shall see there is some ambiguity in that terminology which we hope to clarify. Because elemental analysis is a common requirement for all types of materials and because elemental information is the type provided by the more common techniques available, we will focus our attention in this area. However where chemical or structural information can also be obtained, these capabilities will be mentioned as well.

What Do We Mean by "Surface"?

Strictly speaking the term "surface" applies to the boundary between two dissimilar media and thus could be vanishingly small. We could call this imaginary line the "surface", but it would really have little physical significance. A more meaningful definition might be the last atomic layer between the solid and ambient, but its practical value would be limited as well. Another type of "surface" might be a wear resistant coating such as TiN on tool steel that could be thousands of Angstroms thick. To the production engineer, the entire coating is the "surface." The conclusion from these two rather extreme cases is that the meaning of the term surface is largely a matter of context which leads to a certain amount of ambiguity.

Within the top 10,000 Å of a solid, we conveniently have five decades in depth. For simplicity, we can assign a name to each of these five regimes as follows:

1 Å	=	physical surface
>1-10 Å	=	surface
>10-100 Å	=	surface layer
>100-1000 Å	=	thin film
>1000 Å-10000 Å	=	coating

Although this may appear to be a trite exercise, hopefully the motivation will become obvious when we discuss real surface analysis measurements.

But First the Alphabet Soup.

There are numerous analytical methods capable of providing elemental, chemical, structural or physical information about the top 10,000 Å of ceramic materials. Space does not permit inclusion of a completely exhaustive list, but there is a subset of some 10 techniques ranging from pervasive to emerging that we will consider herein. Each of these techniques is commonly referred to by its acronym as follows:

AES	Auger Electron Spectrometry
ESCA	Electron Spectroscopy for Chemical Analysis
EPMA	Electron Probe Microanalysis
LIMS	Laser Ionization Mass Spectrometry
RBS	Rutherford Backscattering Spectrometry
SEM/	Scanning Electron Microscopy/
EDX	Energy Dispersive X-Ray Spectrometry
SIMS	Secondary Ion Mass Spectrometry
SNMS	Sputtered Neutrals Mass Spectrometry
TEM	Transmission Electron Microscopy
TXRF	Total Reflection X-Ray Fluorescence Spectroscopy

Because of their commonality SEM/EDS, EPMA and TEM will not be discussed, except insofar as they relate to the other techniques. In the paragraphs that follow each of the remaining methods will be described. The more complicated and/or uncommon the technique the more detailed will be the description. But before we proceed, we must digress for a moment to mention the concept of the analytical volume.

Each of the techniques listed above has associated with it an analytical volume. The analytical volume is that region of the specimen from which the analysis is derived. The composition obtained for a given analytical volume represents some sort of average over all the atoms contained therein. If the material within the analytical volume is inhomogeneous, these inhomogeneities will: a) not be perceived, and b) may give rise to serious inaccuracies in the analytical results.

The Tool Kit.

Figure 1 shows a simple schematic representation of each of the ten techniques listed above. The most important features to notice are the incident excitation media and the resultant responses that are used for measurement. In every case, photons, electrons or ions are caused to strike the sample and photons, electrons or ions subsequently emanating from the sample are detected constituting the analysis.

AES - Auger electron spectrometry is one of the most common and widely used surface analysis techniques. A beam of energetic electrons (typically 1-10 keV) bombards the sample producing various electron emissions including Auger electrons. The energy of the Auger electron, assuming it has not undergone an energy loss collision along its outward path in the solid, will be characteristic of the emitting atom. The limitation on energy loss collisions defines the escape depth of the analytically useful Auger electrons which ranges from about 3-50 Å depending on the electron energy and the material. AES is primarily an elemental analysis technique although there are instances where the Auger electron energy or energy distribution may be influenced by the chemical environment of the emitting atom thereby providing some chemical information in selected cases. The Auger transition is a three electron process thereby excluding hydrogen and helium from the list of detectable elements. Detection limits vary from element to element but in general are in the 0.1-1% range for primary beam currents on the order of 1 μamp. Spatial resolution is limited by the beam diameter which can range from a few hundred Angstroms or less at the highest resolution (with degraded detection limits) up to several tens of micrometers. The technique is relatively free of spectral interferences although they can occur. A typical AES vacuum chamber operates in the 10^{-9}-10^{-10} Torr range allowing routine analysis of the light elements C, N and O. In order to analyze depths beyond the Auger escape depth one must somehow expose those underlying regions. A number of methods have been employed including fracture or sectioning, but by far the most common is auxiliary ion beam sputtering. By combining ion beam sputtering with AES analysis the technique can provide in-depth profiling. Detection limits are not influenced by ion sputtering because the Auger process and the sputtering process are uncoupled, but depth resolution depends on crater flatness within the analyzed area and can be degraded by ion beam mixing and morphological texturing of the physical surface.

ESCA - In many ways ESCA or XPS is similar to AES although as the name implies there is an emphasis on chemical rather than elemental analysis. In the ESCA technique the sample is irradiated with a monochromatic x-ray beam producing photoelectrons by the photoelectric effect. Like Auger electrons these photoelectrons can possess chemical and/or elemental information provided they have not undergone an inelastic collision prior to reaching the detector. Escape depths in ESCA are comparable to those found for Auger electrons of the same energies. Energy analysis of the photoelectrons, suitably calibrated, provides a direct measure of the electron

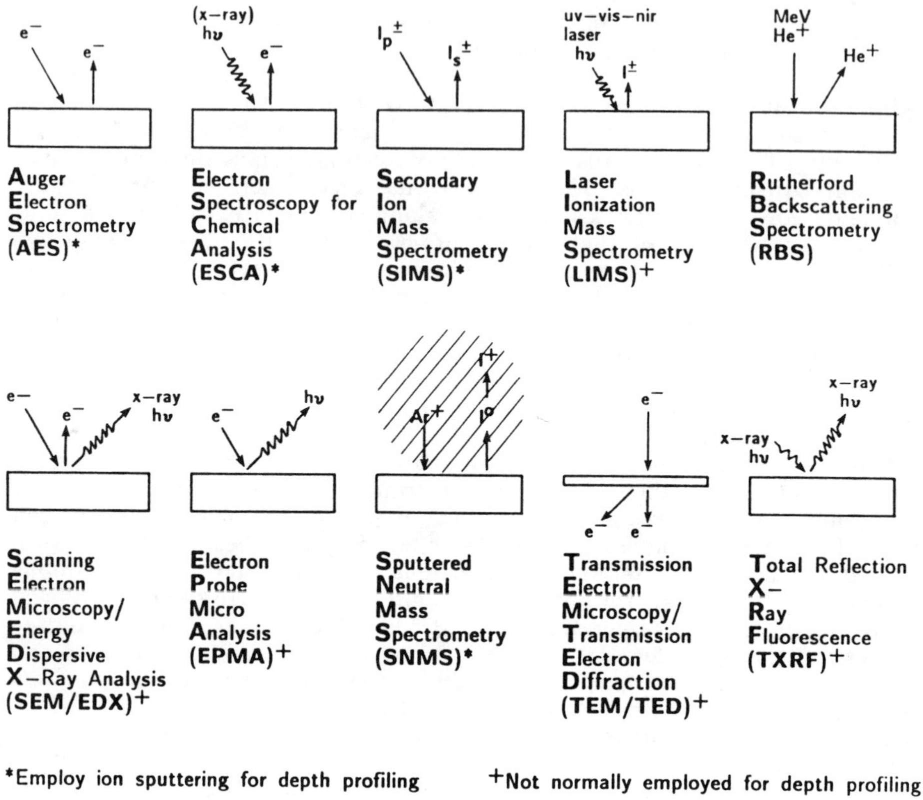

Figure 1. Simple schematic representation of the primary and secondary radiations employed in ten different methods of "surface" analysis ranging from very common (SEM/EDX) to quite recently developed (TXRF).

binding energy in the atom which can be quite sensitive to chemical state or bonding environment. Because the technique employs photon rather than electron excitation there is a reduced susceptibility to charging problems with insulating samples compared to AES. X-radiation can also be less damaging to the sample than an intense electron beam particularly for materials like polymers or organics. Spatial resolution is limited by the primary beam spot size, but since x-rays cannot be well focused it is achieved mainly through beam collimation. Typical analyzed areas range from about 40 μm up to mm dimensions. Detection limits will vary with analyzed area but are generally also in the 0.1-1% range. Like AES, ion beam sputtering is used in ESCA for depth profiling although the data acquisition times tend to be longer for comparable sensitivity and chemical bonding information can be destroyed or distorted by ion beam induced surface damage.

LIMS - Laser ionization mass spectrometry is one of the relatively new methods for surface analysis. In this technique the sample is irradiated with a single pulse (~10 nsec) of high intensity laser light (typically Nd:YAG at 266 or 532 nm). The light is absorbed by the solid and if the energy density exceeds the ablation threshold a small amount of sample material will be ablated as ions and neutrals. By suitably biasing the sample the ion component (positive or negative) is accelerated away from the sample into a time-of-flight mass spectrometer where the ions are separated and collected according to their mass-to-charge ratio by a transient recorder-based detection system. The detectable mass range can be between 500 and 1000 amu with nominally unit mass resolution at mass 200 amu. Sampling depth is a function of the depth of the ablated crater which varies with laser power and certain physical properties of the material such as optical absorptivity, but typical values range from a few hundred to several thousand Å per pulse. Spatial resolution is basically a function of the laser beam spot size which ranges from about 1-5 μm. The technique is used primarily for elemental analysis (H-U) but copious molecular ions appear in the mass spectra which may be related to sample chemistry. Detection limits vary with laser power, sample absorptivity, spectroscopy (+ or -) and the ionization potentials and electron affinities of the elements. However for some of the more favorable cases such as Na^+ or Cl^-, detection limits in the 1-10 ppm range are obtained. The major strengths of the technique are the wide range of sample types that can be analyzed, analysis speed (<1 msec spectral acquisition time), relative freedom from sample charging problems with nonconducting samples, simultaneous multi-element detection and relatively small sampling volume. The major weaknesses involve poor depth resolution for depth profiling, the complexities of spectral interpretation for complex samples, and difficulties associated with quantitation. When operated below the laser ablation threshold (laser desorption), LIMS is one of the few techniques capable of analyzing adsorbates on the physical surface.

A variant on conventional LIMS is multiphoton laser post ionization(1). With this technique a second laser beam is directed parallel to the sample surface so as to intersect the expanding vapor plume produced by a single laser pulse from the normal incidence laser. The second laser is pulsed after a brief time delay with respect to the first laser pulse in order to maximize the interaction with the dense vapor cloud. The power density of the first laser beam can be reduced in this mode because ionization is provided by the second laser. Two benefits of the two laser approach are that smaller sampling volumes can be achieved because of the reduced ablation laser power (smaller spot sizes and shallower ablation depths) and more uniform and efficient ionization can be achieved by decoupling the ablation and ionization mechanisms.

RBS - Rutherford backscattering spectrometry as an analytical tool dates back to the early days of MeV ion accelerators in nuclear physics. While it may be one of the oldest techniques, it is not particularly common especially for industrial applications because of the requirement for a high energy accelerator. In the last ten years, however, commercial instruments have been produced that are sufficiently small, inexpensive and reliable to make the technique practical for any organization wishing to acquire the capability. The technique begins with the production of an energetic ion beam (typically He^+ or He^{++} with projectile energies on the order of 1-3 MeV). This beam is directed onto the sample where incident ions collide with target atoms and are backscattered to a surface barrier detector where they are recorded by energy. Two energy loss mechanisms in the RBS process give rise to a reduction in the incident particle energy: 1) kinematic loss that occurs during the scattering collision and 2) electronic energy loss that occurs continuously while the projectile is traversing the solid along the inward and outward paths. The kinematic energy loss for 180° backscattering is given by the equation:

$$E = E_o \left(\frac{M-m}{M+m} \right)^2$$

where E is the energy of the particle just after scattering, E_o is the energy just prior to scattering (the beam energy for scattering from near surface atoms), M is the target mass and m is the mass of the projectile. The kinematic factor dominates the total energy loss for near surface scattering events, and for heavy target atoms the kinematic loss is small. Electronic energy loss gives rise to a convoluted mass and depth relationship contained within the scattered particle energy. Scattering by a light atom at the surface can produce the same particle energy as that from a heavy atom buried deeper in the target because of the additional electronic energy loss terms. The intensity axis is a simple function of concentration and the Rutherford scattering cross section. Because of this relationship the technique is quite sensitive to heavy elements (~100ppm) and relatively insensitive to light elements (1-20%).

There are two major advantages to the RBS technique compared to other surface analysis methods. The process can be represented by relatively simple mathematical relationships in closed form permitting standardless quantitation, and depth profiling is accomplished without sputtering (non-destructive). Depth resolution is limited by detector energy resolution, typically 100-200 Å. A grazing exit detector may be employed which gives better depth resolution (~50-100 Å) within the top 500-1000 Å. For near normal exit RBS the sampling depth is about 10,000 Å depending on the incident beam energy and the stopping power of the matrix. Spatial resolution is limited by beam diameter which is on the order of 1-3 mm for most instrumental configurations. Given the appropriate focusing hardware smaller beam diameters are possible at increased cost and operational complexity. Data acquisition time for a routine analysis is 30 minutes which generally provides 2-5% counting statistics for major components. Film thickness can be determined if the film density is known. Conversely, if film thickness is known RBS can be used to determine film density. Insulating samples are generally not a problem by this method.

Two ancillary techniques that can be performed with an RBS analytical system are channelling and hydrogen forward scattering. Channelling is a procedure used to measure crystalline damage in single crystal targets like Si or GaAs wafers. A collimated parallel beam of helium ions is directed onto the sample so that the beam axis and a major crystallographic axis are parallel. When this condition is satisfied and if the crystal is reasonably perfect the yield of backscattered particles will decrease because the probability of a Rutherford collision is reduced by the high density of open "channels" seen by the beam in this orientation. Non-ordered atoms such as an interstitial impurity will still backscatter with a high probability making the technique very sensitive to small numbers of off-lattice site dopants or matrix atoms (self-interstitials). Channelling can also be used to reduce substrate yield for analyzing surface layers or heavy impurities (relative to substrate material) at or near the surface.

Hydrogen forward scattering is a relatively new technique specific for the analysis of hydrogen in surface layers or thin films(2). If the incident beam is allowed to strike the surface at a glancing angle, hydrogen atoms can be forward scattered upon collision with a helium projectile. By placing a thin foil in front of a detector positioned so as to accept the recoiled hydrogen, the foil thickness can be chosen so as to transmit hydrogen but not helium. This technique is currently a research topic but successful hydrogen depth profiling experiments have been performed with detection limits in the 0.05 at-% range and depth resolutions of a few hundred Angstroms.

<u>SIMS</u> - Of all the surface analysis techniques, SIMS is undoubtedly the most complicated, in part because it is the only technique capable of measuring the entire periodic table from major constituent to ultra trace impurity concentrations in a single analysis as well as organic or inorganic molecules. It is further complicated because exact capabilities are a strong function of

the type of instrument employed. In a typical SIMS analysis, an energetic primary ion beam (~1-20 keV) bombards the sample inducing sputtering. Some fraction of the sputtered material leaves the sample as positive or negative secondary ions. Depending on the polarity of the sample potential, these positive or negative ions are accelerated away from the sample into a mass spectrometer where they are separated according to their mass-to-charge ratio. Both atomic ions and molecular ions are formed in the sputtering process giving rise to the possibility for molecular ion interferences with elemental ions of the same nominal mass. On the other hand molecular ions can be used for characterizing sample chemistry, particularly when analyzing organic or inorganic compounds. Another complicating factor arises from the fact that secondary ions are emitted from the sample with a range of emission angles and energies, not all of which are simultaneously accepted and transmitted by the mass spectrometer.

The chemical nature of the primary beam also plays an important role in SIMS. If a chemically inert species such as argon (Ar^+) or gallium (Ga^+) is employed, the ratio of sputtered ions to neutrals of the same species (ion yield) will be relatively low (10^{-3}-10^{-6} or lower). The dominant factors are the ionization potential for positive ion emission and electron affinity for negative ion emission. Moreover, the ion yield under these conditions can be strongly influenced by the instantaneous chemical environment of the surface which can change during the course of an analysis. Matrix effects can be reduced and ion yields enhanced by the use of reactive primary ion bombardment. The most common are oxygen (0_2^+ or O^-) for positive ion emission (electropositive element analysis) and cesium (Cs^+) for negative ion emission (electronegative element analysis). These issues about ion yield, ion yield enhancement and ion acceptance and transmission efficiency through the ion optics and mass spectrometer are critically important because the goal of a SIMS analysis is very often ultra-trace element detection (<<10 ppm) requiring the maximum possible analytical signal. However, if detection limits at or above 10 ppm are sufficient, they all become much less important.

Because SIMS is a mass spectrometric technique it is capable of measuring all elements (H-U) and isotopes, and with the appropriate type of mass spectrometer can make exact mass measurements with accuracies on the order of millimass units (0.001 amu). This capability can be useful in deducing the atomic composition of molecular cluster ions and can also aid in separating atomic and molecular ions of the same nominal mass. However, in order to make these types of determinations the spectrometer must be capable of high mass resolution ($M/\Delta M$ in the 1000 to 10,000 range). Three types of mass spectrometers are currently used for SIMS: quadrupole, magnetic/electrostatic and time-of-flight (TOF). The main points for comparison are mass resolution, ion extraction and transmission efficiency and mass range. As a practical matter the mass range of magnetic SIMS instruments designed mainly for elemental analysis is constrained by the strength and physical dimensions of the magnetic and electrostatic analyzers, usually 1-250 amu. Quadrupole SIMS mass spectrometers operate in the 1-500 or 2-1000 amu range and are thus better suited for organic analysis. For the largest possible mass range, TOF spectrometers are superior. In fact, the mass range is not limited by the spectrometer but by the ion detection system. Both magnetic and TOF spectrometers can be operated effectively with relatively high ion acceleration fields at the sample (~10,000 V/cm). This increases ion extraction efficiency hence increasing sensitivity, but also makes the analysis more susceptible to charging problems with insulating samples. Charge compensation techniques are therefore important adjuncts with these types of systems. Quadrupole spectrometers on the other hand generally operate near ground potential and thus are limited to ions in the 0-100 eV range. Since the sample is not a significant ion optical component and is electrically maintained near ground potential, quadrupole SIMS is usually less susceptible to charging effects with insulating samples. Mass resolution is typically only a few hundred for quadrupole SIMS systems whereas it can be as high as 10,000 or more in a magnetic instrument. Consequently, quadrupole systems are more susceptible to molecular ion interferences than either magnetic or TOF spectrometers. One other disadvantage of the quadrupole compared to magnetic or TOF spectrometers for SIMS applications is the mass dependence of the ion transmission. Heavier ions are not transmitted as efficiently as lighter ions systematically biasing sensitivity as a function of ion mass.

One of the most important considerations relating to sensitivity (detection limits) for trace element analysis is the primary beam current delivered to the sample within the acceptance area of the secondary ion extraction optics. Unlike AES or ESCA, the sputtering process and analytical signal production are coupled in SIMS. The higher the sputtering rate the larger will be the instantaneous ion signal. A typical depth profiling primary beam spot size and current would be 1-5 μamp into 5 to 50 μm (FWHM) with a nominal Gaussian beam profile. The sputtering rate will depend on such factors as primary beam species and energy, incidence angle, primary beam raster area, whether the beam is pulsed or continuous (duty cycle) and the matrix material itself. Sputtering rates can vary widely from much less than a monolayer per second to hundreds of Angstroms per second. In polycrystalline, multiphase or disordered materials different regions can sputter at different rates within the same field of view complicating quantitation and limiting depth resolution when depth profiling. All of these factors combine to necessitate the careful use of standards for quantitation. Nonetheless, accuracies of a factor of two or three (multiply or divide) are readily achievable for impurity levels below 0.1% down to sub ppb concentrations based on a single calibration standard or an archival relative sensitivity factor (RSF). Highly accurate measurements ($\leq$2% accuracy and precision) can be achieved in homogeneous materials using rigorously controlled analytical protocols when reliable standards are available.

A developing area in SIMS is ion imaging or microanalysis at high spatial resolution. There are two fundamentally different approaches: stigmatic and microbeam. In stigmatic imaging, spatial resolution is achieved by the ion optics of the instrument. Except for the mass spectrometry part of the process stigmatic ion imaging is the ion analogue of light microscopy.

Resolution is limited by the ion optics to 0.5-1 μm under favorable conditions (conducting sample and smooth planer surface). Image contrast is a function of relative ion count rate picture element by picture element (pixels). As we have discussed before, however, relative ion intensity can be influenced by many factors so that the interpretation of image contrast is subject to the same limitations as quantitation. Two types of stigmatic imaging detectors are currently in use: 1) microchannelplate/fluorescent screen image intensifiers used in conjunction with a conventional video camera and CRT display, and 2) resistive anode encoder (RAE) consisting of a set of microchannelplate transducer/amplifiers followed by a resistive film from which an ion arrival event position can be encoded by a set of four electrical output pulses. The RAE output signal can be easily digitized and transferred to a computer via fast pulse processing electronics for image acquisition, storage and manipulation. The RAE dynamic range is constrained by the pulse pair resolution of the signal processing electronics which saturate at a total count rate of around 10^5 cps with single ion detection sensitivity on the low end. The microchannelplate/camera combination on the other hand has a narrower linear dynamic range of 1-2 orders of magnitude (camera limitation) but these 1-2 decades can be scaled over a very large range ($\sim 10^6$ or more) by adjusting the accelerating voltage on the microchannelplates in a feedback loop with the camera output. In general, the RAE is preferable for low level imaging such as for trace impurities, while the camera based system has greater utility for high count rate imaging such as major constituents at high sensitivity ion emission conditions.

The alternative approach to SIMS ion imaging employs a microfocused ion beam analogous to an SEM. Like any microbeam imaging technique spatial resolution is limited by beam spot size. However, because the image signal is the detected secondary ion intensity, primary beam current is a determining factor in sensitivity with all the concomitant caveats about ion yield, spectrometer transmission , etc. Owing to the optical performance of ion sources and ion optical columns, current density rather than total current is the most important factor regarding microbeam SIMS image resolution/sensitivity. In this regard, source brightness is very important. To date, the highest resolution results have been achieved with liquid metal ion guns utilizing Cs or Ga ($\leq$2000 Å resolution).

<u>SNMS</u> - Sputtered neutral mass spectrometry (SNMS) utilizes a low pressure (10^{-4} Torr) rf argon plasma to sputter the sample surface and ionize neutral species produced in the sputtering process(3). The ionized sputtered neutrals are then extracted from the plasma and analyzed by a quadrupole mass spectrometer similar to those employed in quadrupole SIMS. Ion yields are relatively uniform and matrix independent, but the ionization mechanism in SNMS is believed to be mainly electron impact due to the low concentration of excited state atoms in the plasma.

There are two modes of operation for SNMS: the direct bombardment mode (DBM), which is the normal method of operation for analyzing electrically conducting or semi-conducting samples, and the separate bombardment mode

(SBM), which is used to analyze insulating samples. In bulk and thin-film analyses of conducting materials, the same rf plasma is used both to sputter the sample and ionize the sputtered neutral atoms. The direct bombardment mode requires that the sample be at a more negative potential with respect to the plasma in order to accelerate inert gas ions to the sample surface. Most of the positive secondary ions emitted from the surface are attracted back onto the sample, and so the contribution of these ions, and consequently the SIMS matrix effects, are minimal. Certain combinations of sample-plasma spacings and voltages yield a current density of $1mA/cm^2$ over a 5 mm ϕ spot at an energy of 300eV. This use of low impact energy sputtering minimizes ion beam induced mixing and possibly texturing effects. Additionally, very little redeposition of sputtered particles onto the sample surface takes place. Hence, the depth resolution attainable by the direct bombardment mode could approach 10 Å.

When the sample potential is biased slightly positive relative to the plasma potential, Ar^+ from the plasma will not be attracted to the sample, and a separate collimated ion beam of several keV energy must be employed to sputter the sample. While the sputtered neutrals that are ejected are ionized in the adjacent rf plasma region, any sample charging created during ion beam bombardment is neutralized by plasma electrons naturally attracted to the sample surface due to the positive bias. Therefore, the separate bombardment operating mode is suitable for the analysis of insulating materials. Direct bombardment analysis of insulators can be accomplished by placing a conducting grid over the specimen, but the grid will contribute to the analysis and degrade depth resolution.

TXRF - Total reflection X-Ray Fluorescence (TXRF) is a very new technique for measuring trace levels of metallic surface contamination on silicon wafers or other highly polished surfaces (e.g., sapphire or quartz windows)(4). In the TXRF technique, an X-ray beam from a conventional X-ray tube, typically a molybdenum anode, illuminates the wafer surface at grazing incidence below a critical angle so that the condition for total external reflection of the X-rays is met, thereby limiting the analysis to the top 30 Å of the wafer. The fluorescence radiation from surface atoms is detected using a conventional Si(Li) X-ray detector which simultaneously detects all transition and heavy metal impurities, similar to EDX analysis. The analytical area is determined by a mechanical aperture between the sample and the detector, normally allowing an area 8 mm in diameter to be measured. The improved detection limits as compared to conventional X-ray fluorescence (XRF) results from a major reduction in the Si background signal from the substrate. Quantitation is performed using standards prepared by intentionally contaminating control wafers with a known number of impurity atoms. The achievable precision of the measurement is a function of the signal-to-background ratio. For example, the precision is about 2% (one relative standard deviation) for Zn at $1.5x10^{12}$ atoms/cm^2 and 17% for Fe at $0.35x10^{12}$ atoms/cm^2.

CONCLUSION

In this paper we have endeavored to better define the notion of surface analysis and how the concept relates to the top 10,000 Å of ceramic or ceramic-like materials. A number of surface analysis techniques varying from quite common to very new have been described. Because many of these techniques involve the use of charged particle beams (electrons or ions) in some way, sample charging during the analysis is a serious consideration when analyzing non-conducting materials. Sample charging along with the inherent capabilities/limitations of the techniques (including the analytical volume) available must be taken into account when designing any surface analysis measurements or choosing an instrument/technique for purchase.

REFERENCES

[1] B. Schueler and R. W. Odom, "Non-resonant Multiphoton Ionization (NRMPI) of the Neutrals Ablated in Laser Microprobe Mass Spectrometry Analysis of GaAs and $Hg_{0.78}Cd_{0.22}Te$," *Journal of Applied Physics*, 61, 4652 (1987).

[2] T. T. Bardin, J. G. Pronko and A. Joshi,"Quantification of Hydrogen in Surfaces and Thin Films Using a Non-destructive Forward Scattering Technique," *Thin Solid Films*, 119, (1984).

[3] P. K. Chu, J. C. Huneke and R. J. Blattner, "Application of Glow Discharge Mass Spectrometry and Sputtered Neutral Mass Spectrometry to Materials Characterization," *Journal of Vacuum Science and Technology* A5, 295 (1987).

[4] R. S. Hockett, S. M. Baumann and E. Schemmel, "Diagnostic Techniques for Semiconductor Materials and Devices," *The Electrochemical Society Proceedings*, 88-20 (1988).

X-RAY PHOTOELECTRON SPECTROSCOPY OF ELECTRONIC CERAMIC MATERIALS: COMPLEX CHROMITES AND MANGANITES

Larry R. Pederson and Linda S. Dake
Pacific Northwest Laboratory, Richland, WA 99352

ABSTRACT

X-ray photoelectron spectroscopy (XPS) has been used to probe valency in substituted perovskite oxides of the general formula ABO_3, including $(La,Sr)CrO_3$, $(Y,Ca)CrO_3$, and $(Y,Ca)MnO_3$. When Sr was substituted for La or when Ca was substituted for Y in the chromite series, a second Cr $2p_{3/2}$ peak appeared 3-4 eV higher in binding energy than the principal peak, consistent with oxidation of Cr^{3+} to Cr^{4+} and with previously proposed conduction mechanisms involving a highly localized electron hole (small polaron). Similar substitutions in the manganite series gave no observable energy shifts in the Mn 2p peak and negligible satellite intensity. For the manganites, a highly localized electron hole on an Mn site does not appear to be formed. Rather, results indicate a higher degree of covalency than found in the chromites.

INTRODUCTION

Perovskite-type mixed oxides of the general form ABO_3 have received considerable attention because of their unusual electrical, magnetic, and/or catalytic properties and their often refractory nature [1-4]. Rare earth, alkaline earth, and/or Pb ions usually occupy the A-site, while the B-site has contained Cr, Mn, Fe, Co, Ni, Ti, and Cu, among others. Properties of these materials can be tailored by substitution onto either the A- or B-sites with acceptor or donor cations. For example, in the $(La,Sr)CrO_3$ system, substitution of Sr^{2+} for La^{3+} is generally believed to result in a charge-compensating transition from Cr^{3+} to Cr^{4+} [5,6]. This gives rise to the formation of small polarons localized at the Cr^{4+} sites, which act as charge carriers; the transport mechanism is thermally activated hopping. In principle, charge compensation can also be provided by formation of oxygen vacancies, but such sites would not contribute to the electrical conductivity [7]. Charge carrier concentrations in complex manganites, however, are largely independent of the extent of divalent for trivalent ion substitution on the A-site; in the absence of substitution, manganites have a much greater intrinsic electrical conductivity than have chromites [4]. Thus, even if localized Mn^{4+} sites are formed in complex manganites, it is less clear that such sites are important charge carriers in electrical conduction.

In the present study, X-ray photoelectron spectroscopy (XPS) was used to probe valency changes at the B-site in substituted chromites and manganites, with the purpose of providing insight into conduction mechanisms. XPS has proved very useful to characterize the electronic structure of many solid-state systems, including complex perovskites [8-12]. During XPS analysis, samples are irradiated in vacuum with monoenergetic soft x-rays, and energies of ejected valence-band and inner shell photoelectrons determined [13]. The binding energy for a specific electron in a given element is modified by the chemical environment seen by that element. Information can be obtained relating to valency, ionicity of chemical bonds, and polarization energy, all of which are related to dielectric properties of the material [14, 15].

EXPERIMENTAL

Chromite and manganite powders were prepared using the glycine-metal nitrate combustion synthesis route [16]. An aqueous solution of metal nitrates and glycine (1:3 molar ratio) was concentrated by evaporation and then ignited. The resulting ash consisted of submicron primary ceramic particles of a single phase, with surface areas in excess of 20 m^2/gram. The powders were calcined for several hours at 800°C in air to remove any residual organic materials, which usually constituted less than 1 percent by weight of the ash. In contrast, the usual method of preparing complex chromites by solid state reaction of oxides yields several solid phases, even when heated to 1400°C in air [17].

XPS spectra were obtained using a Perkin-Elmer Model 560 Multiprobe system, equipped with a double-pass cylindrical mirror electron energy analyzer and a magnesium anode (Mg Kα = 1253.6 eV). Wide energy scans were obtained at 100 eV pass energy, while 25 eV pass energy was used when focusing on individual peaks of interest. The energy scale of the instrument was calibrated to Cu $2p_{3/2}$ (metal) at 932.6 eV and Au $4f_{7/2}$ at 84.0 eV [10]. XPS peak energies were referenced to the adventitious carbon 1s peak at 284.8 eV.

RESULTS AND DISCUSSION

Chromites

XPS spectra were obtained for two series of complex chromites: $La_{1-x}Sr_xCrO_3$ and $Y_{1-x}Ca_xCrO_3$. For purposes of comparison, data was also obtained for Cr_2O_3. X-ray diffraction analysis showed the products to consist of a single phase, when substituted from 0 to 50 percent of divalent cations onto the A-site. The crystal structure of $(Y,Ca)CrO_3$ is orthorhombic perovskite; substitution of Sr for La in $(La,Sr)CrO_3$ is known to stabilize the rhombohedral symmetry at room temperature [18].

The Cr 2p spectra proved quite sensitive to substitution of divalent for trivalent cations on the A-site for both $(La,Sr)CrO_3$ and $(Y,Ca)CrO_3$. Representative spectra, given in Figures 1 and 2 for $(La,Sr)CrO_3$ and $(Y,Ca)CrO_3$, respectively, show two main peaks in the Cr $2p_{3/2}$ spectra, separated in energy by approximately 3-4 eV. The position of the feature at the lower binding energy, attributed to Cr^{3+}, was unaffected by the extent of A-site substitution, within an estimated measurement error of ±0.3 eV. Binding energies for the Cr $2p_{3/2}$ and other relevant photoelectron peaks are listed in Tables 1 and 2. For both series, the feature at higher binding energy grew in intensity and shifted to somewhat higher energies as the extent of substitution increased. This feature contains contributions both from a Cr^{3+} shake-up satellite (a result of an electronically excited final state [19]), and photoelectron emission from Cr^{4+} sites, the relative amounts of which depend on the extent of substitution. Shake-up satellites are quite common for paramagnetic compounds, but are not always as prominent as those obtained for the chromites. (See, for example, the results for Cr_2O_3 in Figure 1.) The Cr^{4+} contribution occurs at higher binding energy than the Cr^{3+} satellite; hence, the shift towards higher binding energy as the extent of substitution is increased.

The relative intensities of the peaks comprising the Cr $2p_{3/2}$ spectra for both series varied in a manner consistent with the formation of localized Cr^{4+} sites as a result of A-site substitution. In Figure 3, the integrated areas of the feature at the higher binding energy (Cr^{4+} plus Cr^{3+} satellite) divided by the total Cr $2p_{3/2}$ peak areas are plotted against the expected fraction of Cr^{4+} in the sample, yielding essentially a linear relation for each series. Non-zero intercepts for both series are, of course, due to the presence of shake-up satellites, which were more prominent for the $(La,Sr)CrO_3$ than for the $(Y,Ca)CrO_3$ series.

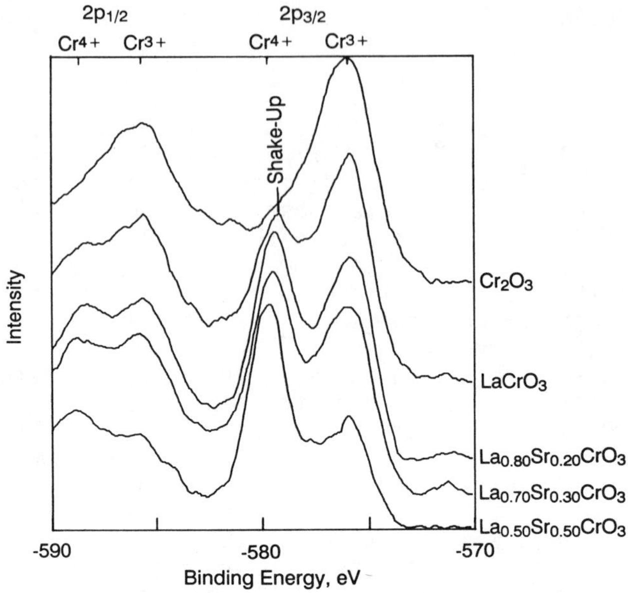

Fig. 1. Cr 2p spectra for selected $(La,Sr)CrO_3$ compounds and Cr_2O_3. The $2p_{3/2}$ peak at approximately 576 eV is attributed to Cr^{3+} while that at 579-580 eV is attributed to a shake-up satellite plus Cr^{4+}.

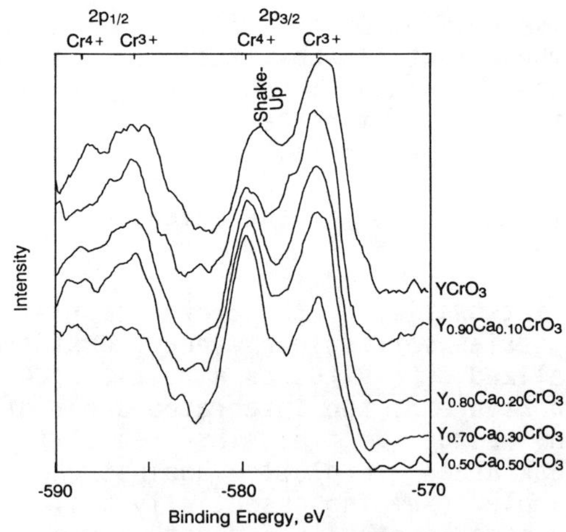

Fig. 2. Cr 2p spectra for selected $(Y,Ca)CrO_3$ compounds. Results are similar to those for $(La,Sr)CrO_3$, except that shake-up satellite intensities were greater for the same extent of substitution.

Table 1. Binding energies of core orbitals for (La,Sr)CrO_3[a]

Compound	Cr $2p_{3/2}$[b]	Cr $2p_{3/2}$[c]	O 1s	La $3d_{5/2}$	Sr $2p_{3/2}$
Cr_2O_3	576.0	-	529.8	-	-
$LaCrO_3$	576.0	579.4	529.3	834.2	-
$La_{0.05}Sr_{0.95}CrO_3$	575.9	579.4	529.2	834.1	133.6
$La_{0.90}Sr_{0.10}CrO_3$	575.8	579.5	529.3	834.4	133.6
$La_{0.80}Sr_{0.20}CrO_3$	575.9	579.6	529.4	834.2	133.7
$La_{0.70}Sr_{0.30}CrO_3$	575.9	579.7	529.5	834.1	133.8
$La_{0.50}Sr_{0.50}CrO_3$	576.0	579.8	530.2	834.3	133.5

Table 2. Binding energies of core orbitals for (La,Sr)CrO_3[a]

Compound	Cr $2p_{3/2}$[b]	Cr $2p_{3/2}$[c]	O 1s	La $3d_{5/2}$	Sr $2p_{3/2}$
Cr_2O_3	576.0	-	529.8	-	-
$LaCrO_3$	576.0	579.4	529.3	834.2	-
$La_{0.05}Sr_{0.95}CrO_3$	575.9	579.4	529.2	834.1	133.6
$La_{0.90}Sr_{0.10}CrO_3$	575.8	579.5	529.3	834.4	133.6
$La_{0.80}Sr_{0.20}CrO_3$	575.9	579.6	529.4	834.2	133.7
$La_{0.70}Sr_{0.30}CrO_3$	575.9	579.7	529.5	834.1	133.8
$La_{0.50}Sr_{0.50}CrO_3$	576.0	579.8	530.2	834.3	133.5

a C 1s adjusted to 284.8 eV
b Cr^{3+}
c Cr^{4+} plus Cr^{3+} shake-up satellite

By subtracting shake-up satellite contributions to the $2p_{3/2}$ peaks, intensity ratios were generally in good agreement with expected Cr^{4+} fractions, as shown in Figure 4.

Observed chemical shifts in core levels of 3-4 eV (binding energies of 579-580 eV) accompanying Cr^{3+} to Cr^{4+} conversion in complex chromites are larger than might be expected, considering Cr $2p_{3/2}$ binding energies and calculated formal charges for selected chromium compounds listed in Table 3 [20]. Binding energies obtained for the chromites more closely resemble values listed for hexavalent Cr salts than for the formally tetravalent Cr in CrO_2; results for CrO_2 are essentially indistinguishable from Cr_2O_3.

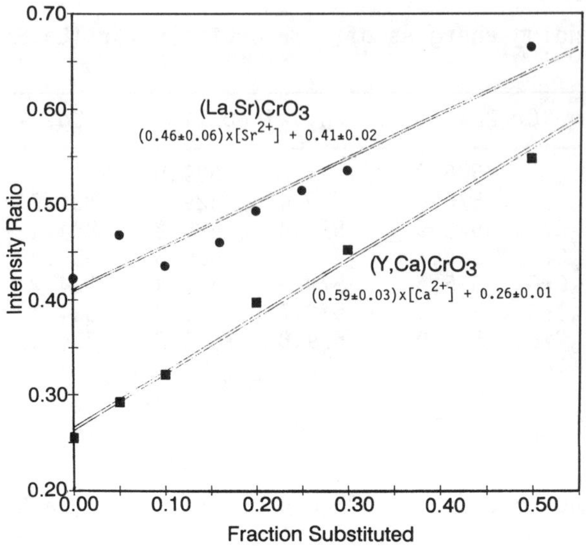

Fig. 3. Ratio of Cr^{4+} plus shake-up satellite intensity to total $2p_{3/2}$ intensity versus divalent cation substitution fraction for $(La,Sr)CrO_3$ and $(Y,Ca)CrO_3$. Lines given were calculated by least-squares.

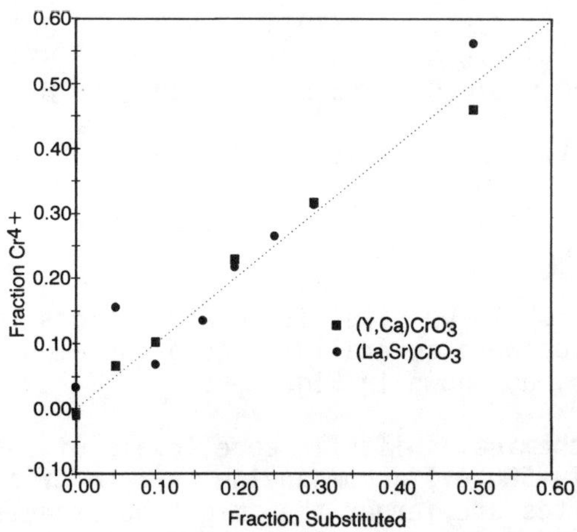

Fig. 4. Fraction Cr^{4+} determined from peak intensity ratios, with shake-up satellite contributions eliminated, versus divalent cation substitution fraction. The line represents the expected Cr^{4+} fraction.

Table 3. Cr $2p_{3/2}$ Binding Energies for Chromium Compounds [20]

Compound	Binding Energy, eV	Formal Valence
Cr_2O_3	576.8	+3
$Cr(OH)_3$	577.4	+3
CrF_3	580.3	+3
$CrCl_3$	577.8	+3
$CrBr_3$	576.2	+3
CrI_3	576.7	+3
CrO_2	576.3	+4
CrO_3	579.8	+6
K_2CrO_4	579.4	+6
$K_2Cr_2O_7$	579.9	+6

However, the size of the chemical shift often is not well-correlated with the formal valence that is calculated from the molecular formula. Rather, the dielectric properties of the solid will affect the magnitude of the chemical shift, as described below.

Chemical shifts in core-level binding energies are an indicator of the effective valence on a given atom, which may differ considerably from the formal valence. Initial studies by Siegbahn [21] related chemical shifts on core-level binding energies with effective valency by

$$\Delta E_b = -aq_i + \Delta V \tag{1}$$

where ΔE_b is the binding energy shift from some reference value, "a" is a constant, q_i is the effective charge, and ΔV is a Madelung-type potential term. This accounts for effects of localized charge densities on neighboring atoms. Shifts in binding energy can also be expressed as [14,15]

$$\Delta E_b = \delta\epsilon_b - R_{ext} \tag{2}$$

where $\delta\epsilon_b$ is the shift in the orbital energy of a free atom, and R_{ext} is the extra-atomic relaxation energy, given by

$$R_{ext} = q^2(1 - 1/k)/2r \tag{3}$$

where k is the dielectric constant and r is the radius of minumum electron screening distance. For the series of Cr(III) halides listed in Table 3, Cr $2p_{3/2}$ binding energies vary by more than

4 eV, despite all being formally trivalent. The highest binding energy corresponds to the fluoride, which has the largest electronegativity difference and smallest polarizability, and thus is least able to offset the electrical charge separation. Binding energies for the other Cr(III) halides generally decrease with increasing atomic weight of the halide, which corresponds to an increasing polarizability and decreasing electronegativity difference. Thus, shifts in core level binding energies relative to some standard reflect the actual share of bonding electrons and dielectric properties of that material. The appearance of two well-separated peaks for the chromites accompanying the conversion of Cr^{3+} to Cr^{4+} is consistent with a highly-localized electron vacancy being located on a Cr atom in the perovskite structure. This interpretation is consistent with the generally accepted conduction mechanism and the relation of A-site substitution to carrier density [4-6].

A simplified electron energy level diagram, given in Figure 5 for Cr^{3+} and Cr^{4+}, further illustrates how a localized electron vacancy (small polaron) can enhance conductivity. The Cr 3d orbitals are split by the octahedral oxygen field of the chromites into triply degenerate t_{2g} and doubly degenerate e_g levels [13,14]. For Cr^{3+}, one electron populates each of the three t_{2g} levels and the e_g levels are vacant, whereas for Cr^{4+}, only two of the three t_{2g} levels are filled. For chromites having only Cr^{3+} present, electrical conductivities are very small [4-6], because electron transfer from one Cr^{3+} site to another would require either promotion to vacant e_g levels or spin pairing in the occupied t_{2g} levels, both of which are energetically unfavorable. Substitution to create Cr^{4+}, however, provides a localized vacancy in the t_{2g} level.

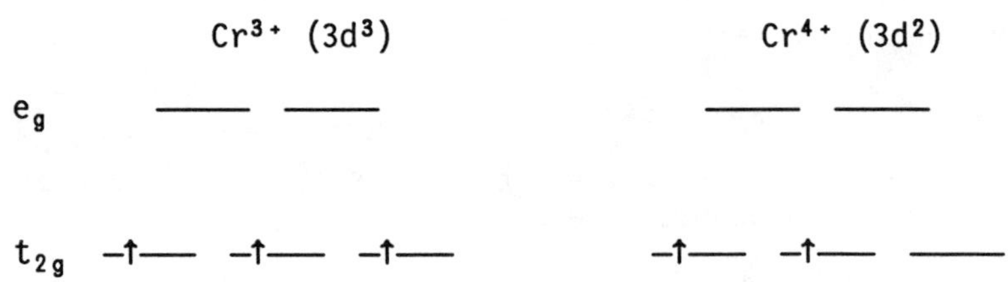

Fig. 5. Electron Energy Levels for Cr in an Octahedral Field

Manganites

XPS spectra were obtained for the complex manganite series $Y_{1-x}Ca_xMnO_3$ and for MnO_2. XRD analysis showed the manganites to consist of a single phase, when from 0 to 50 percent of Y was substituted by Ca on the A site. The crystal structure of $(Y,Ca)MnO_3$ is hexagonal perovskite, while $CaMnO_3$ is orthorhombic [22].

In contrast to chemical shifts and satellite structure associated with oxidation of Cr in complex chromites, no significant changes in the XPS spectra of Mn were observed as Ca was substituted for Y in $(Y,Ca)MnO_3$. The Mn $2p_{3/2}$ spectra for selected members of this series and for MnO_2, given in Figure 6, show the Mn $2p_{3/2}$ peak to be broadened by multiplet splitting and to exhibit negligible satellite intensity, as has been found previously for $LaMnO_3$ [11]. Principal peak energies for these are given in Table 4 and show no obvious sensitivity to substitution.

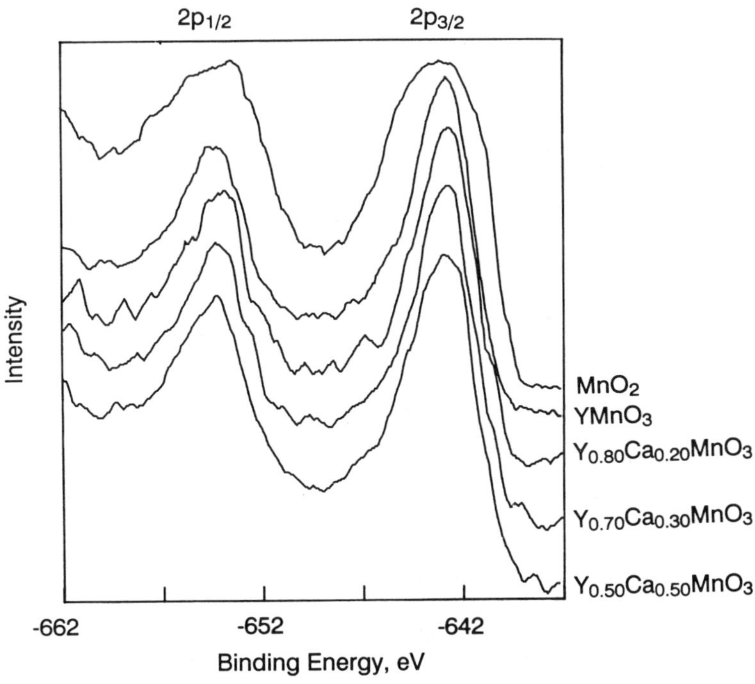

Fig. 6. Mn 2p spectra forselected $(Y,Ca)MnO_3$ compounds. Unlike the chromites, no second $2p_{3/2}$ peak was observed as a result of Ca substitution for Y.

Table 4. Binding energies of core orbitals for (Y,Ca)MnO_3 [a]

Compound	Mn $2p_{3/2}$	O 1s	Y $3d_{5/2}$	Ca $2p_{3/2}$
MnO_2	642.2	529.8	-	-
$YMnO_3$	642.1	529.5	156.6	-
$Y_{0.95}Ca_{0.05}MnO_3$	642.0	529.6	156.4	347.1
$Y_{0.90}Ca_{0.10}MnO_3$	641.8	529.5	156.6	347.3
$Y_{0.80}Ca_{0.20}MnO_3$	642.0	529.8	156.6	347.2
$Y_{0.70}Ca_{0.30}MnO_3$	642.1	529.6	156.8	347.3
$Y_{0.50}Ca_{0.50}MnO_3$	642.0	529.8	156.5	347.0

a C 1s adjusted to 284.8 eV.

A probable explanation for the lack of any significant changes in the Mn 2p spectra is that substitution of Ca for Y does not lead to highly localized Mn^{4+} sites in complex manganites. Rather, Mn-O bonds have a strongly covalent character, similar to Cu in the high T_c superconductor $YBa_2Cu_3O_{7-x}$ [12,23]. Satellite intensity as well as energy separation between the main peak and the satellite has been shown to decrease with increasing covalency of chemical bonds in a large number of solids [24], consistent with the present data and explanation. Band structure calculations for $LaMnO_3$ and related compounds have shown considerable overlap between the O 2p and Mn 3d orbitals [11], which could destabilize a localized electron vacancy.

Creation of an additional electron vacancy in substituted manganites has a smaller effect on electrical conductivity than in the chromites [4]. Manganites already have an available vacancy in the e_g level without substitution of a divalent for trivalent cation on the A-site. The electron energy level diagram for Mn^{3+} of Figure 7 shows one electron occupying each of the three t_{2g} levels and one e_g level, whereas for Mn^{4+}, both of the e_g levels are vacant. Following this model, creation of an additional vacancy in the e_g level by substitution is unlikely to affect electrical conductivity as strongly as creation of a vacancy in the chromites, consistent with experimental findings that charge carrier concentrations are essentially independent of the extent of A-site substitution [4].

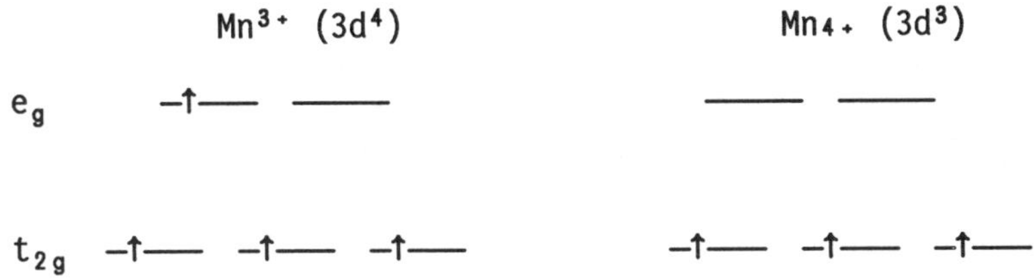

Fig. 7. Electron Energy Levels for Mn in an Octahedral Field

SUMMARY

XPS analysis revealed significant differences in the effects of A-site substitution on B-site valency in complex chromites and manganites. For (La,Sr)CrO_3 and (Y,Ca)CrO_3, spectra were consistent with the formation of a highly localized electron vacancy on a Cr atom (discrete Cr^{4+} created) as Sr or Ca was substituted for La or Y, respectively. These sites enhance conductivity via small polaron hopping. For (Y,Ca)MnO_3, formation of discrete tetravalent Mn sites was not apparent. Rather, the bonds formed were highly covalent, as evidenced by a lack of intense satellites and no apparent chemical shifts in the core spectra. Results for the manganites are similar to those being reported for the cuprate perovskites, where only divalent Cu has been found by XPS despite an expectation of trivalent Cu.

ACKNOWLEDGMENTS

The authors wish to thank J.P. Gardner, Jr. for preparation of chromite and manganite samples. This work was supported by the Division of Basic Energy Sciences of the U.S. Department of Energy.

REFERENCES

1. E. O. Wollan and W.C. Koehler, Phys. Rev. 100, 545 (1955).
2. D. B. Meadowcraft, Brit. J. Appl. Phys. 2, 1225 (1969).
3. M. Kertesz, I. Riess, D. S. Tannhauser, R. Langpape, and F. J. Rohr, J. Solid State Chem. 42, 125 (1982).
4. W. J. Weber, C. W. Griffin, and J. L. Bates, J. American Ceramic Society 70(4), 265 (1987).
5. J. B. Webb, M. Sayer, and A. Masingh, Can. J. Phys. 55, 1725 (1977).

6. D. P. Karim and A. T. Alred, Phys. Rev. B 20(6), 2255 (1979).
7. H. U. Anderson, R. Murphy, K. Humphrey, A. Alred, W. L. Procarine, R. J. Ackerman, and J. L. Bates, in Rare Earths in Modern Science and Technology, eds. G. J. McCarthy and J. J. Rhyne, Plenum Press, New York, p. 55 (1978).
8. W. Y. Howng and R. J. Thorn, J. Phys. Chem. Solids 41(1), 75 (1980).
9. J. L. Fierro, G. Tejuca, and L. Gonzalez, Appl. Surf. Sci. 27(4), 453 (1987).
10. K. Tabata and S. Kohiki, J. Mater. Sci. 23(1), 343 (1988).
11. D. J. Lam, B. W. Veal, and D. E. Ellis, Phys. Rev. B 22(12), 5730 (1980).
12. D. E. Ramaker, N. H. Turner, J. S. Murday, L. E. Toth, M. Osofsky, and F. L. Hutson, Phys. Rev. B 36(10), 5672 (1987).
13. C. D. Wagner, W. M. Riggs, L. E. Davis, J. F. Moulder, and G. E. Muilenberg, Handbook of X-Ray Photoelectron Spectroscopy, Perkin-Elmer Corp., Physical Electronics Division, Eden Prairie, MN (1979).
14. C. D. Wagner and A. Joshi, J. Elec. Spectr. 47, 283 (1988).
15. D. A. Shirley, Phys. Rev. A, 7(5), 1520 (1973).
16. L. R. Pederson, L. A. Chick, and G. J. Exarhos, U. S. Patent pending (1988).
17. M. Yoshimura, S. Song, and S. Somiya, Ferrites: Proc. Int. Conf. Japan 429 (1980).
18. E. M. Levin and H. F. McMurdie, Phase Diagrams for Ceramists, 1975 Supplement, ed. M. K. Reser (American Ceramic Society, Columbus, OH, p. 144 (1975).
19. G. A. Vernon, G. Stucky, and J. A. Carlson, Inorg. Chem. 15, 278 (1976).
20. C. D. Wagner, Proc. 10th Intl. CODATA Conf. 1986, (Computer Handling Dissemination Data), p. 294, Elsevier, Amsterdam (1987), to be distributed on computer disk by U. S. National Bureau of Standards, Gaithersburg, MD 20899.
21. K. Siegbahn, C. Nordling, A. Fahlman, R. Nordberg, K. Hamrin, J. Hedman, G. Johansson, T. Bergmark, S. E. Karlsson, I. Lindgren, and I. Lindberg, ESCA, Atomic, Molecular, and Solid State Structure Studied by Means of Electron Spectroscopy, Almquist and Wiksell, Uppsala, Sweden (1967).
22. E. Poliert, S. Krupicka, and F. Kuzmicova, J. Phys. Chem. Solids 43, 1137 (1982).
23. D. E. Ramaker, Proc. AIP Conf. Thin Film Process. Charact. High-Temp. Supercond. 165, 284 (1988).
24. S. Larsson, Chem. Phys. Lett. 40, 362 (1976).

CHEMICAL CHARACTERIZATION OF CERAMIC SUBSTRATES FOR MICROELECTRONIC PACKAGING

Guna Selvaduray and Mary Beth Lange, Department of Materials Engineering, San Jose State University, San Jose, CA 95192.
Dan Sullivan, Surface Science Laboratories, Inc., Mountain View, CA 94043.

INTRODUCTION

As part of a larger study aimed at identifying materials selection parameters for ceramic microelectronic packaging substrates, chemical analysis of these substrates was undertaken. AlN, Al_2O_3, and BeO substrates from commercial manufacturers were analyzed for their surface and shallow subsurface chemical composition, utilizing Electron Spectroscopy for Chemical Analysis (ESCA) and Auger Electron Spectroscopy (AES). The samples were handled in a manner that can be expected in a manufacturing environment, with no special precautions taken. The results are presented and discussed from the perspective of the user, with the intent of educating the user on how to interpret the results obtained.

The effect of chemical composition on the desired properties of ceramic microchip packages is first discussed. The results of the analysis are presented next. The results are then discussed and interpreted.

CERAMIC SUBSTRATES FOR MICROELECTRONIC PACKAGING

The most common material of construction for ceramic substrates today is Al_2O_3. Other frequently used materials include BeO, AlN, and SiC. The package protects and isolates the microchip from the environment. As such, the package must not only be able to endure the effects of the environment, but it must also not be the source of contaminants that could lead to failure. Some of the functions the package is to perform include (1) Dissipation of generated heat, (2) Resistance to thermal cycling, (3) Resistance to corrosive environments, and (4) Provide hermetic sealing.

The physical properties of the material that affect its ability to perform the above functions include: (1) Thermal conductivity, (2) Dielectric properties (dielectric constant, dielectric loss, dielectric strength),

(3) Resistivity, (4) Coefficient of thermal expansion, (5) Flexural strength, and (6) Thermal shock resistance.

CHEMICAL COMPOSITION DEPENDENCE OF PHYSICAL PROPERTIES

The above properties are very strongly dependent not only on the chemical composition of the specific material, but also on the nature and extent to which impurities are present. The thermal conductivity of BeO and Al_2O_3 as a function of purity is shown in Table 1.[1]

Table 1: Thermal conductivity, k (W/°C cm), of BeO and Al_2O_3 as a function of purity.

BeO		Al_2O_3	
Purity	k	Purity	k
99.9	2.8	99.5	0.37
99.5	2.6	94	0.28
99	2.4	91	0.17
97	2.1		

The dielectric properties, including the breakdown voltage can also be affected by impurities, sometimes even when they are present in small amounts. Impurity compounds dissolve in the matrix, leading to the formation of a variety of point defects. In the case of single crystals, dielectric losses have been found to be determined mainly by minor impurities. [2] The presence of Cl^- ions have also been known to lead to the corrosion of solder joints and other circuit paths after hermetic sealing.

A critical necessity for reliable performance of microchips is their adherence to the substrate, with the two performing as one integral unit. The bonding mechanisms for the various classes of adhesives used has been discussed by Selvaduray. [3] The surface chemical composition of the chip and the substrate are critical in determing the reliability of the adhesion joint.

REQUIREMENTS OF ANALYTICAL TECHNIQUES

The above examples demonstrate the need for accurate chemical analyses of ceramic substrates that are used in microelectronic packaging. It is also important that these techniques require a minimum of sample preparation while yielding accurate information.

In addition to information on concentrations of chemical constituents, some of the other parameters of importance include: (1) Capability to detect light elements, such as carbon and nitrogen, (2) Limits of detection, (3) Ability to obtain quantitative information, as opposed to qualitative or semi-quantitative information, and (4) Ability to obtain surface and subsurface information.

ESCA AND AES INSTRUMENTS USED IN THIS ANALYSIS

The ESCA instrument used in this analysis was a Surface Science Instruments SSX-100 small spot spectrometer using a monochromatized Al K-alpha source. The beam has a major axis diameter of 250 μm and a minor axis diameter of 150 μm. The limits of detection are between 0.5 to 1.0 atomic percent for most elements, with a 3 to 5 atomic percent limit of detection for Be and B.

The AES employed in this analysis was a Perkin Elmer S90 Scanning Auger Spectrometer. The beam diameter is between 2 to 3 μm, and the limit of detection is 1 atomic percent.

From the beam diameters, it can be seen that AES provides chemical information from a much smaller area than ESCA. While the AES provides primarily elemental information, ESCA is capable of providing chemical compound information, which is extremely useful in determing the source(s) of contamination. In this investigation ESCA analysis was performed at the surface only, while AES analysis was done at the surface and at etch depths of 30Å and 60Å. An Argon ion etching gun was used.

ANALYTICAL RESULTS AND THEIR INTERPRETATION

The chemical analyses from ESCA and AES are shown in Tables 2 and 3 respectively. The tabulated results are for an AlN, a 90% Al_2O_3, and a BeO substrate. As can be seen from these Tables, the major surface component, regardless of bulk constitution, was found to be carbon. Major sources of carbon contamination are dust, fingerprints, and contamination from plastic bags. Residual carbon from improper burn-off of organic binders during sintering is another possible source of carbon contamination.

Table 2: Chemical Composition of Substrates by ESCA (in atomic percent units for elements detected).

Substrate	Al	Si	Cl	C	Ca	K	N	O	Na	Be
AlN	5.6	2.4	0.7	63	0.6	0.8	4.8	21	0.7	-
Al_2O_3 (90%)	1.6	6.4	-	72	-	-	0.6	19	-	-
BeO	-	1.2	-	66	-	-	1.4	20	0.3	8.4

Table 4 reports the elemental composition reported in Table 2, but after removing carbon from consideration, and re-normalizing the other elements. Three possible contaminants now dominate the spectrum, namely Si, N, and O. Silicon as a contaminant could have originated at one of the following sources: (1) The aluminum or alumina, (2) The binders used in fabrication, or (3) Packaging materials, especially plastic bags. Both nitrogen and oxygen, being the two most predominant elements in the atmosphere, are generally adsorbed onto the surface. The extent to which these elements are adsorbed depends a great deal on the reactivity of the surface and the age of the surface.

The results from AES are even more dramatic, than those obtained from ESCA, in some respects. The surface composition of the AlN substrate, in the as-received condition, is found to be comprised of C, O, and N, with no Al signals detected. Freon wash of the surface changes the results, with the carbon content dropping considerably and the detection of a rather weak Al signal. Etching the surface to a depth of 30Å also "changes" the chemical composition significantly. A combined wash and 30Å etch, however, results in the carbon concentration dropping significantly.

The effect of surface porosity of the samples being analyzed also has an effect on the results. In the case of metallic samples, surface-adsorbed contaminants are generally between 10 to 40 Å thick, and an etch depth of 60Å will begin to yield information on bulk chemical composition. In the case of ceramic substrates pressed and sintered from powders,however, this etch depth has to be significantly greater. As can be seen from Table 3, all the substrates show significant carbon contamination on the surface. Though this concentration drops with the etch depth, there are still appreciable amounts of carbon at a depth of 60Å.

Table 3: Chemical Composition of Substrates by AES (in atomic percent units for elements detected).

Substrate	Etch Depth	O	N	Ca	C	K	Cl	S	Si	Be†	Si†	Al	Al†
AlN	As Rcvd	3.8	3*		93								
	30Å	12	15	2.4	50	2*	2.5	1.1				14	
	Wash	14	12		66			1*	2*			5*	
	Wash & 30Å	12	32		19			0.8				37	
90% Al_2O_3	As Rcvd	10			83						7*		
	30Å	18			77						5*		
	60Å	22			56						5*		18
BeO	As Rcvd	7.5			78					15			
	60Å	19			30					52			

* Indicates a weak signal

† The elements detected are in the oxide form

Aluminum Nitride

The detection of chlorine is significant because if it is not removed, it can lead to corrosion of circuit traces and joints during service. The AES analysis after a Freon wash and a 30Å etch, however, does not detect chlorine. This indicates that the chlorine is not a part of the bulk composition, but is a surface contaminant. Therefore, the substrates should be cleaned before die-attachment and hermetic sealing.

Table 4: ESCA Chemical Composition renormalized after Carbon

Substrate	Al	Si	Cd	Ca	K	N	O	Na	Be
AlN	15.3	6.6	1.9	1.6	2.2	13.1	57.3	1.9	-
90% Al_2O_3	5.8	23.2	-	-	-	2.2	68.8	-	-
BeO	-	3.8	-	-	-	4.5	63.9	0.9	26.8

Note: If Si is present as an organic siloxane, some carbon content will be associated stoichiometrically with Si, e.g., $(R_2SiO_2)_m$.

A comparison of the atomic percentages of Al and N indicates that there is insufficient N content for all of the Al to be in the form of AlN. Based on thermodynamic reasoning, some oxidation of AlN can be expected. Significant oxidation can result in a drop in the thermal conductivity of the substrate, thus affecting its heat dissipation capability. When the results from AES are taken into consideration, it is seen that the aluminum content is not bound in the form of the oxide. The discrepancy in the aluminum and nitrogen content can now be accounted for in terms of inherent errors in the readings. It can now be safely assumed that any oxidation is minimal, i.e., below detection limits.

90% Aluminum Oxide

ESCA data for this sample indicates the presence of surface adsorbed oxygen and nitrogen in addition to carbon. The silicon content, when reviewed together with AES results, is shown to be in the oxide form. The reduction in silicon content with increasing etch depth is interpreted as an indication that surface contamination from organic siloxanes is present, but that subsurface silicon exists in the form of SiO_2.

Beryllium Oxide

ESCA results for this sample were interpreted with the aid of high resolution scans and AES data. A high resolution ESCA spectrum for the binding energy range from 388.0 to 408.0 eV which covers the N 1s peak in the survey scan shows that the source of nitrogen is organics. Similarly, a high resolution scan over the100.0 to 120.0 eV binding energy range, which covers the Be 1s peak shows that the Be is present in the form of BeO. These conclusions are corroborated by the results from AES which show no nitrogen or silicon in depth etching, and also shows the Be to be in the form of an oxide.

CONCLUSION

Both the bulk and surface chemical composition of ceramic substrates used in microelectronic packaging have a significant impact on the reliability of the microchip during its service life. Chemical analysis of an AlN, and Al_2O_3, and a BeO substrate, by ESCA and AES, show significant levels of surface contamination, notably by carbon from various sources, and adsorbed gases. Due to the porous nature of the substrates, surface contamination was found to extend deeper into the surface than usually expected with metallic samples. Chlorine contamination was found to be a surface feature which can be removed easily. Oxidation of AlN was found to be below detection limits.

ACKNOWLEDGEMENTS

The authors wish to thank Robert Cormia and Jared Smith, both of Surface Science Laboratories, for their kind assistance in the interpretation of the analytical results.

REFERENCES

1 C. A. Neugebauer, A.F. Yerman, R.O. Carlson, J.F. Burgess, H.F. Webster, and H.H. Glascock, The Packaging of Power Semiconductor Devices, Electrocomponent Science Monographs Vol. 7, Gordon and Breach Science Publishers, New York (1986), p.10.

2 W.D. Kingery, H.K. Bowen, D.R. Uhlmann, Introduction to Ceramics, John Wiley & Sons, Inc., New York, (1976), p. 940.

3 G. S. Selvaduray, "Die Bond Materials and Bond Mechanisms in Microelectronic Packaging", Thin Solid Films, 153(1987), pp. 431-445.

BIBLIOGRAPHY

1. D. Briggs and M.P. Seah, Practical Surface Analysis, John Wiley & Sons, New York, (1983).

2. American Society for Metals, Metals Handbook 9th Edition, Volume 10, Materials Characterization, (1986).

SURFACE ANALYSIS BY MASS SPECTROMETRY OF POST-IONIZED SECONDARY NEUTRAL SPECIES

J. B. Pallix and C. H. Becker
Molecular Physics Laboratory
SRI International
Menlo Park, California 94025

ABSTRACT

A comparison of methods will be presented for general surface analysis involving mass spectrometry of ionized sputtered or desorbed neutral species. These methods overcome the often severe problems associated with matrix effects which are commonly observed in secondary ion mass spectrometry (SIMS) analyses. Decoupling of the ionization and desorption processes allows more quantitative information to be obtained and facilitates investigations of the desorption and ionization processes themselves. The capabilities and limitations of the different techniques will be discussed in terms of depth and lateral resolution, quantitative ability, and sensitivity.

I. INTRODUCTION

Increasing demands for combined high sensitivity and quantification in inorganic materials characterization have led to the development of a number of mass spectrometric methods for the detection of secondary neutral atoms and molecules. While secondary ion mass spectrometry (SIMS) has long been recognized as the most sensitive surface analytical tool, problems arise in studies of many chemical systems which exhibit severe matrix effects. Matrix dependent secondary ion yields are quite unpredictable and can vary even as much as seven orders of magnitude. Therefore, standard reference materials are required with composition very similar to the unknown, but are unavailable or impractical in many cases. These difficulties place severe limitations on the ability to obtain quantitative information from a wide range of SIMS analyses.

An obvious solution to these problems is to decouple the ionization and desorption processes and to detect secondary neutral species, which typically comprise more than 90% of the material removed from the surface during ion beam sputtering. This approach was first investigated in the work of Honig where sputtered neutrals were ionized by electron impact [1,2]. Because the neutral species dominate the emission process regardless of the chemical matrix, their detection largely eliminates matrix effects. Thus semiquantitative data can be provided for relative concentrations when nonselective ionization is employed. Decoupling the desorption and ionization processes also permits detailed investigations of the desorption process itself.

The acronym SNMS sometimes refers to "secondary neutral mass spectrometry" but is also commonly used in reference to only one approach, the "sputtered neutrals mass spectrometry" technique which employs a low pressure rf plasma for postionization [3,4]. This acronym will not be used in this review in order to avoid confusion. Actually a variety of techniques are currently in use for mass spectrometric detection of secondary neutrals [5]. Not all of these methods employ ion beam sputtering for removal of surface material. In fact, desorption by laser [6,7], electron beam [8,9], rf or dc plasma [1,3,4,10-12] or thermal desorption [13] are quite common in secondary neutral mass spectrometry. Desorbed neutrals may be subsequently ionized by nonselective processes such as by electron beam [14,15], intense untuned laser beam [6-9], hot electron gas [1,3,4,10] or thermal ionization [16]. The selective resonantly-enhanced multiphoton ionization technique is also useful in many investigations but its applications are more specialized [17,18]. Of these methods, only those which permit the study of near surface regions and interfaces will be discussed here and compared in terms of general applicability, sensitivity, quantitation, depth and lateral resolution.

II. SENSITIVITY AND QUANTITATION

The popularity of SIMS stems from its ability to commonly detect elemental concentrations at levels $\leq$1 ppm. Therefore to be competitive with the high sensitivity of SIMS, post-ionization methods must provide efficient ionization as well as efficient detection of ions. The overall efficiency of any mass spectrometric surface analysis method is often referred to as the "useful yield" of the instrument. Useful yield is defined as the ratio of detected ions to sputtered atoms for a monoisotopic sample and is equal to the product of the transmission of the instrument (T) and the fraction ionized (α^+) of sputtered or desorbed neutral species. Obviously the useful yield should be as high as possible in order to produce the desired sensitivity (<1 ppm) in the quickest possible measurement and consuming the least amount of material. The technique called surface analysis by laser ionization (SALI-described in section IV) currently has the highest useful yield (10^{-3}) of any of the nonselective post-ionization techniques [19].

The quantitative ability of any of the postionization methods relies on the desorption (sputtering) and ionization processes. Under steady-state sputtering conditions, or to a large degree under quasi-steady-state sputtering for nonuniform profiles where the mixing depth is much larger than the depth needed to reach a steady-state limit, the relative flux of a particular element is proportional to the atomic concentration of that element in the sample due to conservation of mass. Therefore under these conditions the relative intensities of the observed ion signals can potentially be used for quantitative determination of relative concentrations. Of course no sputtering technique is free from the effects of preferential sputtering before steady-state or quasi-steady state sputtering is reached; this can cause depletion/enhancement of some species at the surface and in the sputtered flux of a multicomponent target. Note however that even after steady-state sputtering conditions are reached this surface depletion/enhancement will be reflected in the relative signals observed for each component by direct surface monitoring methods such as Auger electron spectroscopy and X-ray photoelectron spectroscopy, but not when observing the sputtered material [20].

When preferential sputtering is not an issue or when steady-state sputtering is reached, observed deviations from the true concentrations reflect the relative elemental ionization probabilities which may differ depending on the method of ionization used. With the exception of species with very high ionization potentials (i.e., N, O, F, H, and rare gases) it is generally true that the relative ionization probabilities for most elements are well within a factor of ten of each other when nonselective ionization methods are used [3,21]. Of course one can determine concentrations more accurately (within 10%) by using reference materials rather than the semiquantitative raw data alone. In general, the reference materials need not be very closely matched to the target, thus making quantitation of the analysis of surfaces, interfaces, and extremely thin films possible, which is typically much more difficult in SIMS analyses.

III. POST-IONIZATION BY ELECTRON IMPACT

A. Low Pressure Plasma Ionization

Early work of Honig involved sputtering by ions produced in a low voltage plasma. Neutral species emitted from the sample surface were subsequently ionized by impact from the plasma electrons and analyzed by a 180° magnetic sector mass spectrometer [1]. The method has been further developed by Woodyard and Cooper [10], and by Oechsner [3,4]. This now commercially available technique produces low energy ions in a low pressure high frequency discharge and is best known for its high depth resolution potential. The depth profile of a multilayer Ta-Si sample with approximately 5.0 nm single layerthickness is displayed in Figure 1. The depth resolution represented by the peak to valley intensity ratio degrades very little over the entire eroded depth of more than 100 nm. This high depth resolution is due to the low impact energies of the primary ions (as low as 20 eV) when the sample is placed in the plasma which reduces their penetration depth and can limit knock-on mixing to ~2 atomic layers. Under proper plasma conditions a flat bottomed crater is formed eliminating the need for a focused raster/gated primary ion source. However, for depth profiling in this direct bombardment mode (DBM), flat samples are required with relatively large sizes (typically at least 5 millimeters in diameter) making small spot analysis impossible. Of course, optimum sensitivity (≥10 ppm) is achieved with these larger size samples but sample areas as small as 10^{-2} cm^2 can be analyzed in cases where high sensitivity is not required. To address these problems the commercial instrument is equipped with an ion gun which produces a focused ion beam for sputtering in investigations of smaller samples (SBM-separate bombardment mode).

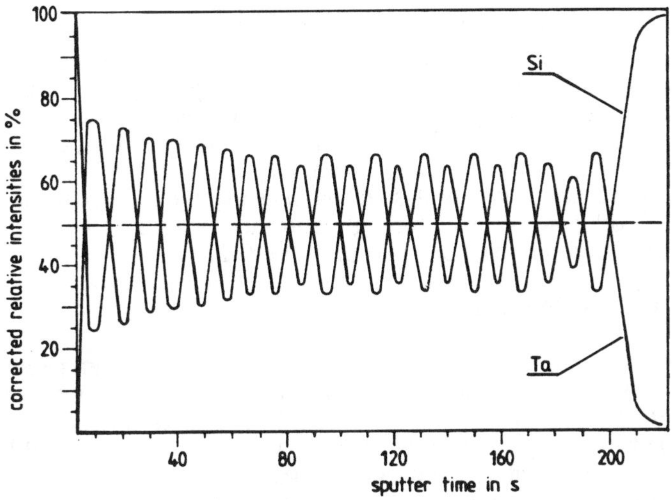

Figure 1. Depth profile of a Ta-Si multilayer system with ~5.0 nm single layer thickness; taken with low energy ion bombardment (150 V). Figure reprinted from Ref. 22.

Insulating materials can be studied using either direct bombardment mode (DBM) or separate bombardment mode. This is because in favorable cases using high sample bias the plasma electrons or secondary electrons emitted from the sample may be used to neutralize the charging surface of the sample [24]. Generally a conductive grid or mask is also needed although this can degrade depth resolution significantly.

The useful yield of the low pressure plasma instrument is ~10^{-9} [5] and in order to detect low level constituents a significant amount of material must be removed from the sample. Therefore optimum depth resolution cannot be achieved simultaneously with optimum sensitivity. It has also been shown that at low primary ion bombardment energies (< 1 keV) the relative sensitivity factors are a function of matrix dependent angular distributions especially in samples containing low mass elements (H, Be, C, etc.) [23]; this is an important issue because of the small solid angle of neutrals accepted for the apparatus' mass spectrometer.

B. Electron Beam Post-ionization

Another method first used by Honig also involves electron impact of neutrals subsequent to sputtering by an ion beam [2]. The electron source in this method is an electron beam emitted from a filament rather than plasma electrons as described in the previous section. Due to the short dwelling time of the neutrals in the relatively small volume of the electron beam, the post-ionized fraction (~10^{-4}) is ~10^2 smaller than possible for ionization in a plasma. However, the collection efficiency of the instrument is higher by ~10^2 because there is no plasma chamber separating the sample from the extraction optics [14,15]. Therefore this technique also has a useful yield of ~10^{-9}, is capable of providing detection limits in the ppm range, and can be used with a small spot primary ion beam. Both types of instruments discussed so far can be used in a SIMS mode fairly easily although the sensitivity will be higher in the electron beam instrument due to the more efficient ion optics. Another advantage of this technique is that the relative sensitivities are all within a factor of twelve even for high ionization potential elements (i.e., C and N) when using electron impact energies of ~40 eV [25].

IV. LASER POST-IONIZATION

Laser post-ionization simply involves passing a laser beam parallel to and close to but above a surface undergoing stimulated desorption, often by ion beam bombardment. The use of a pulsed desorption laser results in a narrow distribution of ion creation times and thus easily interfaces with time-of-flight (TOF) mass analysis. This is important in that the entire mass spectrum may be recorded for every laser pulse, in contrast to conventional quadrupole or magnetic sector mass spectrometers. Transmission is very high (often $\gtrsim$50%). Useful yields (which do not include the multiplex advantage in mass) in these high transmission instruments have been reported to be as high as 10^{-1} with achievable elemental detection limits of 1 ppb when resonantly-enhanced multi-photon ionization (MPI) is employed for postionization [18]. Furthermore, the observable mass range is essentially unlimited when using TOF instruments with a mass resolution >2000 at ~300 amu, a figure which may soon exceed 10,000.

A. REMPI

Resonantly-enhanced multi-photon ionization has the highest sensitivity of any of the post-ionization techniques currently in use [17,18]. This is largely due to the high efficiency of the resonance process which allows saturation of the ionization for most elements with laser beams not tightly focused (except for high I.P. elements such as C, N, O, As, P, F, Cl, and rare gases). The method is not generally applicable to survey analysis because the laser wavelength must be tuned to resonances. Quantitative information for relative amounts of different species is problematic, as is also true of matrix effects caused by excited state and molecular formation [17,26]. However, the technique has several distinct advantages in addition to its great sensitivity. By selective ionization, isobaric interferences generally can be eliminated to fairly significant dynamic ranges. It is also quite simple to vary the time delay between the ion beam and focused laser beam for studies of sputtered neutral velocity distributions [27]. With the laser beam focused into a ribbon shape a position sensitive detector can be used to study angular distributions of the sputtered neutrals [27]. This technique thus provides an excellent means of making detailed investigations of the sputtering process (though the other approaches have capabilities in this regard too) and has many other applications which have not yet been fully explored.

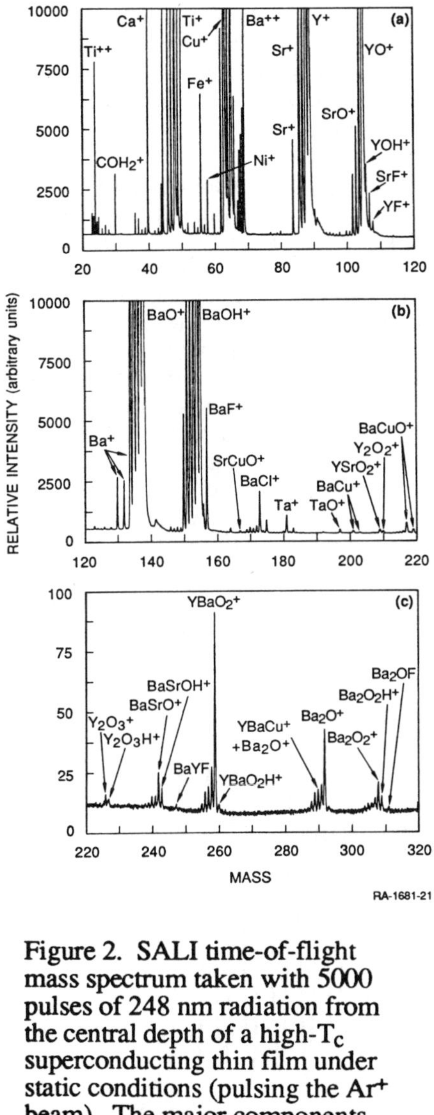

Figure 2. SALI time-of-flight mass spectrum taken with 5000 pulses of 248 nm radiation from the central depth of a high-T_c superconducting thin film under static conditions (pulsing the Ar^+ beam). The major components are off scale and from the signal to noise of the 0.1% Ba isotope the dynamic range is determined to be $>10^6$. Figure reprinted from Ref. 28.

B. SALI

Surface analysis by laser ionization (SALI) employs an untuned high power uv laser for nonresonant photoionization of sputtered neutrals [6,8,9,19]. Commonly the output of a standard excimer laser (KrF, 248 nm or ArF, 193 nm) is focused to produce $\sim 10^{10}$ W/cm^2 in an area of 10^{-3} cm^2. This is generally sufficient power to saturate the ionization for two-photon ionizable species within the laser focal volume. Higher I.P. species requiring three of four photons are not efficiently ionized with a standard excimer laser, i.e., O, N, F, He, Ne, Ar. Higher powered lasers, becoming more readily available, can overcome this limitation. The laser beam passes about 1 mm above the sample surface intersecting ~10% of the sputtered neutrals solid angle and ~5% of the velocity distribution for primary ion pulse widths of ~1 μs. Although the intersected fraction is rather low ~0.5%, the high transmission of the reflecting TOF mass spectrometer gives a useful yield of $\sim 10^{-3}$. The useful yield thus is still $\sim 10^{-6}$ higher than that of any other nonselective postionization surface analysis method. Therefore, 1 ppm detection limits are achieved with combined submonolayer and fairly small area sampling (as low as 10^{-4} cm^2), which is highly desirable for studying sharp interfaces in depth profiling. Furthermore, the mass multiplex advantage allows rapid data accumulation for multicomponent systems so that complete mass spectra can be obtained with a dynamic range $\sim 10^6$ about 10 sec.

As a typical example, the spectrum obtained at the surface of a bulk $YBa_2Cu_3O_7$ superconducting thin film is shown in Figure 2 [28]. This spectrum represents signal averaged over 5000 laser pulses and was obtained in less than one minute (100 Hz laser). It is estimated that ~4×10^{-2} monolayer was removed from the sample surface during the course of this measurement.

Even though the primary ion beam impact energy is typically of several keV, high depth resolution is still achievable. Typically an ~80 µm diameter ion beam is rastered over a 1 mm^2 area and the laser is triggered to fire only when the ion beam is sampling from the center of the flat bottomed crater. Figure 3 displays the depth profile of a sample consisitng of a 280 nm thick GaAs layer on top of 20 layers of InGaAs (5.8 nm)/GaAs (21.5 nm) where the In^+ signal intensity is plotted as a function of depth (sputter time). 5 nm depth resolution (16-84%) is measured at the onset of the layered structure at ~280 nm from the surface. After this point the depth resolution degrades slightly due to the natural sputter tail but each layer is still easily resolvable throughout the entire layered structure between 280 nm and 820 nm. Of course this depth resolution is also available with any of the techniques previously discussed which use a rastered/gated well collimated ion beam for sputtering as well as with bombardment by low pressure plasma ions where no raster is required. Note that when comparing the depth resolution of different instruments one must take into consideration the ion beam energy as well as the mass of the primary ions relative to the target atoms. The penetration depth of the primary ions (and therefore the depth resolution) will depend on all of these parameters.

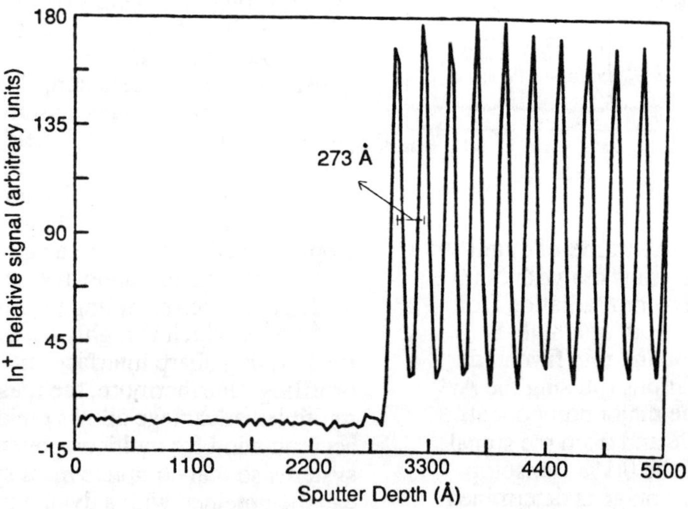

Figure 3. Depth profile monitoring In^+ signal in a layered structure containing 0.28 µm thick GaAs on top of 20 layers of InGaAs (5.8 nm)/GaAs (21.5 nm). Taken with SALI using a 7 keV Ar ion beam rastered over 1 mm^2 and 248 nm radiation.

V. CONCLUSION

Table I summarizes the optimal useful yields and typical detection limits of the various techniques presently used for studying secondary neutrals by mass spectrometry. All of the nonselective ionization techniques have the advantage of providing sensitive semiquantitative raw data which is not available with SIMS. Each method has its own advantages and difficulties and must be chosen for its simplicity, versatility and applicability. While these techniques can provide detection limits in the ppm-ppb range, only laser postionization can do so without consumption of a significant amount of material due to very high useful yields. Most of the methods are very versatile in that one can in principle use a variety of stimulated desorption processes other than ion beam bombardment (i.e., laser, electron beam or thermal desorption) in conjunction with the selected postionization method.

Table I

	Optimal Useful Yield	Typical Concentration Limit*
low pressure plasma		
plasma ionization	10^{-9} [5]	≥10 ppm
electron beam	10^{-9} [5]	≥10 ppm
REMPI	10^{-1} [18]	~1 ppb
SALI	10^{-3} [19]	~30 ppb
SIMS	10^{-2}-10^{-9} range	ppm-ppb

*Not considering amount of material consumed.

REFERENCES

1. R. E. Honig, "The Application of Mass Spectrometry to the Study of Surfaces by Sputtering," in Advances in Mass Spectrometry, J. D. Waldron, Ed. (Pergamon Press, London, 1959) p. 162.

2. R. E. Honig, "Sputtering of Surfaces by Positive Ion Beams of Low Energy," J. Appl. Phys. 29(3), 549 (1958).

3. H. Oechsner, "Recent Applications of Secondary Neutral Mass Spectrometry for Quantitative Analysis of Homogeneous and Structured Samples," Nuclear Instruments and Methods in Physics Research, B33, 918 (1988).

4. H. Oechsner and E. Stumpe, "Sputtered Neutral Mass Spectrometry (SNMS) as a Tool for Chemical Surface Analysis and Depth Profiling," Appl. Phys. 14, 43 (1977).

5. W. Reuter, "Post-ionization of Sputtered Particles: A Brief Review," in Secondary Ion Mass Spectrometry-SIMS V (Springer Verlag, 1986), p. 94.

6. C. H. Becker and K. T. Gillen, "Surface Analysis of Contaminated GaAs: Comparison of New Laser-Based Techniques with SIMS," J. Vac. Sci. Technol. A3, 1347 (1985).

7. B. Schueler and R. W. Odom, "Nonresonant Multiphoton Ionization of the Neutrals Ablated in Laser Microprobe Mass Spectrometry Analysis of GaAs and Hg 0.78 Cd 0.22 Te," J. Appl. Phys. 61(9), 4652 (1987).

8. J. B. Pallix, U. Schühle, C. H. Becker, and D. L. Huestis, "The Advantages of Single-Photon Ionization over Multiphoton Ionization for Mass Spectrometric Surface Analysis of Bulk Organic Polymers," Anal. Chem., submitted for publication.

9. J. B. Pallix, U. Schühle, and C. H. Becker, "Analysis of Organic Solids by Combined Stimulated Desorption and Single-Photon Ionization," Proceedings of the 36th ASMS Conference on Mass Spectrometry and Allied Topics, 1988, p. 1203.

10. J. R. Woodyard and C. B. Cooper, "Mass Spectrometric Study of Neutral Particles Sputtered from Cu by 0 to 100-eV Ar Ions," J. Appl. Phys. 35(4), 1107 (1964).

11. W. W. Harrison, K. R. Hess, R. K. Marcus, and F. L. King, "Glow Discharge Mass Spectrometry," in Secondary Ion Mass Spectrometry-SIMS V (Springer-Verlag, 1986) p. 75.

12. P. K. Chu, J. C. Huneke and R. J. Blattner, "Application of Glow Discharge Mass Spectrometry and Sputtered Neutral Mass Spectrometry to Materials Characterization," J. Vac. Sci. Technol. A5(4), 295 (1987).

13. C. H. Becker and K. T. Gillen, "Nonresonant Multiphoton Ionization as a Sensitive Detector of Surface Concentrations and Evaporation Rates," Appl. Phys. Lett. 45, 1063 (1984).

14. D. Lipinsky, R. Jede, O. Ganschow, and A. Benninghoven, "Performance of a New Ion Optics for Quasisimultaneous Secondary Ion, Secondary Neutral, and Residual Gas Mass Spectrometry," J. Vac. Sci. Technol. A5(5), 2007 (1985).

15. H. Gnaser, J. Fleischhauer, and W. O. Hofer, "Analysis of Solids by Secondary Ion and Sputtered Neutral Mass Spectrometry," Appl. Phys. A37, 211 (1985).

16. G. Blaise and R. Castaing, "Analyse des Surfaces par Ionisation Thermique des Produits de Pulverisation Cathodique," C. R. Acad. Sci., Paris Series B 284, 449 (1977).

17. F. M. Kimock, J. P. Baxter, and N. Winograd, "Detection of Sputtered Neutrals by Multi-Photon Resonance Ionization," Nuclear Instruments and Methods in Physics Research, 218, 287 (1983).

18. C. E. Young, M. J. Pellin, W. F. Calaway, B. Jorgensen, E. L. Schweitzer, and D. M. Gruen, "Laser Based Secondary Neutral Mass Spectroscopy: Useful Yield and Sensitivity," Nuclear Instruments and Methods in Physics Research, B27, 119 (1987).

19. J. B. Pallix, N. Newman, and C. H. Becker, "Surface Analysis by Laser Ionization," Materials Research Society Bulletin, Vol XII, 52 (1987).

20. C. M. Stahle, D. J. Thomson, C. R. Helms, C. H. Becker and A. Simmons, "Quantitative Measurements of the Stoichiometric of Ariodic Oxides Grown on $Hg_{0.78}Cd_{0.22}Te$," Appl. Phys. Lett., 47(5), 521 (1985).

21. C. H. Becker and K. T. Gillen, "Surface Analysis by Nonresonant Multiphoton Ionization of Desorbed or Sputtered Species," Anal. Chem. 56, 1671 (1984).

22. R. Jede, H. Peters, G. Dunnebier, O. Ganschow, U. Kaiser, and K. Siefert, "Quantitative Depth Profile and Bulk Analysis with High Dynamic Range by Electron Gas Sputtered Neutral Mass Spectrometry," J. Vac. Sci. Technol. A6(4), 2271 (1988).

23. J. F. Geiger, M. Kopnarski, H. Oechsner, and H. Paulus, "SNMS-Analysis of Insulators," Microchim. Acta I, 497 (1987).

24. A. Wucher, F. Novak, and W. Reuter, "Relative Elemental Sensitivity Factors in Secondary Neutral Mass Spectrometry," J. Vac. Sci. Technol. A6, 2265 (1988).

25. O. Ganschow, "Electron Beam Postionization of Sputtered Neutrals," Secondary Ion Mass Spectrometry-SIMS V (Springer-Verlag, 1986) p. 79.

26. W. Husinsky, P. Wurz, B. Strehl, and G. Betz, "Cr Atoms Sputtered from Different Matrices," Nucl. Instrum. Meth. Phys. Res. B18, 452 (1987).

27. G. A. Schick, J. P. Baxter, J. Subbiah-Singh, P. H. Kobrin, and N. Winograd, "Multi-Photon Resonance Ionization of Emitted Particles," Secondary Ion Mass Spectrometry-SIMS V (Springer-Verlag, 1986), p. 90.

28. J. B. Pallix, C. H. Becker, N. Missert, K. Char, and R. H. Hammond, "Depth Profiling of Superconducting Thin Films Using Rare Gas Ion Sputtering with Laser Postionization," AIP Proceedings on Thin Film Processing and Characterization of High-Temperature Superconductors, 413 (1988).

ANALYSIS OF CERAMICS BY LASER ABLATION FOURIER TRANSFORM MASS SPECTROMETRY

Sahba Ghaderi
IBM Corporation
Materials Analysis Laboratory
General Products Division
5600 Cottle Rd, San Jose, CA 95193

INTRODUCTION

Mass spectrometry is one of the most powerful analytical techniques. Beside providing detection sensitivities in the low pico gram range, the advantages of mass spectrometry include the capability to detect all elements with isotopic information, the ability to provide molecular ion as well as fragmentation pattern information for organic compounds, and the flexibility to select either positive or negative ions generated by a variety of methods.

Several means of ion generation are available to mass spectrometry, each suitable for different applications. Electron Impact (EI) and Chemical Ionization (CI) are, of course, the the traditional techniques. Other ionization methods include the use of energetic ions (or Secondary Ion Mass Spectrometry (SIMS), Fast Atom Bombardment (FAB), Field Ionization and Field Desorption (FIMS, FDMS), Inductively Coupled Plasma (ICPMS), and the more recent one, Laser Desorption/Ionization (LDMS/LIMS)[1-5].

In the past few years lasers have become an important mass spectrometry tool. Laser ablation, desorption and ionization of non-volatile compounds have greatly expanded mass spectrometry applications especially in the area of surface analysis.

INSTRUMENTATION

Bulk and solid surface analyses are of vital importance in material science and related areas. Most laser mass spectrometry experiments, thus far, have been performed with Time of Flight (TOF) and quadrupole type instruments[6,7]. A new type of mass instrument, called Fourier Transform Mass Spectrometer (also known as Fourier Transform Ion Cyclotron Resonance or FTICR), has proven itself to be most suitable with laser applications[8-14].

The pulsed nature of FTMS operation, along with its capability for accurate mass measurement, high mass and ultra-high resolution ion detection, allows the best use of a pulsed laser as an ion generating source. A single laser pulse can be used for ablation and ionization of non-volatile substances, eliminating complex arrangements necessary for a separate volatilization and ionization system.

Mass analysis in a FTMS instrument is performed by detection of the cyclotron frequencies of ions. A mass spectrum measured by this procedure is the result of resolving different frequencies of ions of different mass, rather than resolving different ion trajectories. This frequency measurement provides FTMS with advantages analogous to those obtained by FTNMR and FTIR over the older continuous-wave (CW) and dispersive spectrometers.

Microprobe is the desirable method in surface analysis and it requires fine focusing of the laser beam, as well as the capability for viewing and manipulation of the specimen. Because commercially available FTMS instruments lack laser microprobe capability, a Nicolet FTMS-2000 mass spectrometer was modified in our laboratory, for the first time, to perform such task[15-18].

In the modified instrument, the beam of a pulsed CO_2 laser is directed to the FTMS cell region through a specially designed interface system. The laser is focused on the tip of a motor driven sample introduction probe, providing the means for ablation and ionization of samples.

Viewing of the magnified image of the sample is possible by a solid state video camera through the same optical path as for the laser beam. A fiber optic bundle and an external light source are used to illuminate the sample. Magnification is about 60X. Figure 1 shows a simplified schematic diagram of the system.

A Zinc selenide window (Z) and lens (L1) are used in the laser path to permit transmission of both the 10.64μm wavelength CO_2 laser light and visible light. The germanium window (G) passes the laser beam while reflecting the visible light to the optics for the video camera. This allows uninterrupted viewing of the sample during the experiment. Sample positioning is accomplished by rotational and axial movement of the probe.

The mechanism of ionization with a laser beam in the far IR region is a thermal process. Surface and near surface ion-molecule reactions, such as cationization, producer ionic products. Ions formed by these processes are trapped within the source cell (S) boundaries due to the magnetic and electrostatic fields present. The tubular vacuum chamber, housing the dual cell and its accessories, is surrounded by a wide bore superconducting magnet operating at 3 tesla. This high field, homogeneous magnetic field causes ions in the cell to move in an orbital motion, restricting them in the x/y plane (plane perpendicular to the magnetic flux lines). A low voltage positive or negative potential (depending on whether positive or negative ions are to be analyzed) is applied to the trapping plates, restricting the ions in the Z direction (parallel to the magnetic flux lines). The combination of these fields create a very efficient ion trap which makes high mass resolution detection possible.(see Fig.1)

The frequencies of the orbital motion of trapped ions (known as ion cyclotron frequency) depend upon their mass to charge ratios and on the strength of the applied magnetic field. At a constant field of the superconducting magnet, detection of these ion frequencies can be easily translated to ion masses. Ions are excited using a rapid wide band frequency sweep to bring them into larger orbits and coherent motions. Image currents induced on the receiver plates by orbiting ions are amplified and digitized. A fast Fourier transformation routine is used to unravel the composite frequencies of all detected ions and convert them to masses. The result is simultaneous detection of all excited ions.

A key obstacle in obtaining high sensitivity and high resolution mass spectra with FTMS is the loss of ion coherencies due to collisional damping effects. This is because of the presence of neutral molecules, either from sample outgassing or from background pressure in the cell, which leads to ion molecule collisions. A differentially pumped, dual cell arrangement helps to resolve that problem by reducing the pressure in the ion detection environment. In high resolution experiments, ions are transferred from the source region (S) to the analyzer region (A) through the center aperture in the common trap plate, prior to excitation and detection. The analyzer cell operates in the low 10^{-9} torr range. An electron gun (F) is also available for analysis of gasses and also for certain experiments involving post electron ionization of the laser ablated material. Results reported in this paper are, however, obtained by laser ablation/ionization only.

A single laser shot is often sufficient to produce a complete mass spectrum. Multiple shots are also used when signal averaging or depth profiling is desired. The typical time to obtain a mass spectrum varies from a few millisecond for low resolution, to several seconds for high resolution.

The major problem in interfacing the laser microprobe to the FTMS instrument is the limited accessibility to the cell region. The distance between the cell and the port flanges, as well as a crowded internal geometry, are the consequences of the magnet surrounding the vacuum chamber. These are limiting factors in achieving minimum laser spot size and maximum image magnification. Using a 10.64μm wavelength laser, a minimum spot size of 30μm has been obtained. This does not mean, however, that smaller spots are excluded from analysis. In most cases very good results can be obtained from an area as small as 5-10μm. A background subtraction method can reveal minute mass spectral differences between spots on the same surface.

RESULTS AND DISCUSSION

The majority of the samples analyzed by this system are related to magnetic disk storage technology. Other applications include identifications of organic and inorganic contaminants on various electronic components, mass analysis of polymers and lubricants, and bulk and surface analysis of ceramics.

An example for bulk analysis of a alumina/titanium carbide ceramic is given in figure 2. One laser shot resulted in ablation and ionization of very small quantity of material (in the low picogram range). The spectrum shows the elemental composition of the ceramic. It is important to point out that the laser ionization efficiencies for different elements can vary drastically. This variability has to be taken in account and correction must be made if any quantitative measurement is to be considered.

Contaminations on surfaces can be easily differentiated from bulk material by repeated firing of the laser. Mass spectra obtained from the same spot by the subsequent laser shots should differ from the first as the laser removes the surface material with each pulse. An example is given in figures 3a and 3b. Niobium is present as a surface contaminant in 3a. In 3b, which is the spectrum from the second shot, the niobium signal is diminished substantially. A change in the potassium peak intensity is also due to surface contamination. Both spectra are normalized to the most intense peak.

Figure 4a shows positive ion mass spectrum of another type of ceramic used in heads for magnetic storage devices. In addition to the elemental peaks from the bulk material, there are peaks at higher masses which are due to contamination. Negative ion mode (4b) showed a series of peaks which extended beyond 2000 amu with 166 amu repeating units. From these two spectra, the contaminant was identified as a perfluorinated polyether with a Y type backbone structure, ie; $X\text{-}(CF_2CF(CF_3)O)_k\text{-}(CF_2O)_l\text{-}X$.

Laser/FTMS can be used to analyze the organic materials in "green body" ceramics. Rapid and direct analyses of these organic substances, used as binders, fillers, plasticizers, conditioners, etc., are valuable in optimizing properties of ceramics. Figures 5a and 5b show positive and negative ion mass spectra of a green body ceramic.

Analyses of superconducting ceramics are also of great interest. Characterization of the materials and of their stoichiometry are the keys to obtaining desired properties. Figure 6a shows a laser ablation mass spectrum of a 1-2-3 type superconductor. Metal oxides contribute to the most intense peaks. An expanded plot (6b) of the peaks between 500 to 800 amu shows that most predominant ions are from combinations of barium, copper and oxygen. In fact, this is true for the entire spectrum. Most intense peaks are separated by 154 amu,the mass of barium oxide. The high abundance of the peaks is due to the high ionization efficiency of barium relative to the other elements in the ceramic.

Another type of ceramic, the highest temperature superconductor known to date, is a thallium type with elemental composition $Tl_1Ca_3Ba_1Cu_3O_x$. In this case thallium peak dominated the spectrum, in figure 7, again because of a very high ionization efficiency (low ionization potential). Vertical expansion of the spectrum shows combinations of metal oxides up to 1000 amu.

Further improvements in this instrument are underway. They include the use of a Nd/YAG laser and modifications in the laser delivery and optical imaging system. These changes are designed to reduce the laser spot size to below 10μm and to improve signal reproducibility. Design and interfacing of an external ion source is also being considered in order to improve sample handling and analysis of larger specimens.

REFERENCES

1. D. F. Torgenson, R. P. Skowronski, and R. D. Macfarlane, Biochem. Biophys. Res. Commun., 60 (1974) 616.

2. A. Benninghoven and W. Sishtermann, Org. Mass Spectrom., 12 (1977) 595.

3. H. Grade, N. Winograd, and R. G. Cooks, J. Am. Chem. Soc., 99 (1977) 7725-77.

4. M. A.Posthumus, P. G. Kistemaker, H. L. C. Meuzelaar, and M. C. Ten Noever de Brauw, Anal. Chem., 50 (1978) 985-991.

5. C. R.Blakly, J. J. Carmody, and M. L. Vestal, J. Am. Chem. Soc., 102 (1980) 5931-5933.

6. R. J. Cotter, and J. C. Tabet, Anal. Chem., 56 (1984)1662-1667.

7. H. J. Neusser, U. Boesl, R. Weikauf, and E.W. Schlag, Int. J. Mass Spectrom. Ion Proc., 60 (1984) 147-156.

8. M. B. Comisarow and A. G. Marshall, "Fourier Transform Ion Cyclotron Resonance Spectroscopy," Chem. Phys. Lett. Vol. 25 pp. 28-283 (1974).

9. M. B. Comisarow and A. G. Marshall, "Frequency Sweep Fourier Transform Ion Cyclotron Resonance Spectroscopy," Chem. Phys. Lett. Vol. 26 pp. 489-490 (1974).

10. A. G. Marshall, "Fourier Transform Ion Cyclotron Resonance mass spectrometry," Accounts of Chemical Research Vol.18 pp. 316-322 (1985).

11. R. E. Hein, and R. B. Cody, Anal. Chem., 57 (1985).

12. S. Ghaderi, P. S. Kulkarni, E. B. Ledford, Jr., C. L. Wilkins, and M. L. Gross, Anal. Chem. 53, 428-437 (1981).

13. Fourier Transform Mass Spectrometry; Evolution, Innovation, Applications, ed. M. V. Buchanan, American Chemical Society, Washington, DC (1987).

14. D. A. Laude, Jr., C. L. Johlman, R. S. Brown, D. A. Weil and C. L. Wilkins, "FTMS; recent instrumentation developments,"Mass Spec. Reviews Vol. 5 pp. 107-166 (1986).

15. Sahba Ghaderi, "Applications of laser microprobe FTMS," Proceedings of the 36th ASMS Conference on Spectrometry and Allied Topics, June 1988, San Francisco, CA, pp. 1126.

16. J. T. Brenna, W. R. Creasy, W. McBain, and C. Soria, Rev. Sci. Instruments, Vol. 59(6), pp. 873-879 (1988).

17. M. G. Sherman, J. R. Kingsley, J. C. Hemminger, and R. T. McIver, Jr., Anal. Chem. Acta Vol.178 (1) pp. 79-89 (1985)

18. M. Pelletier, G. Krier, J. F. Muller, R. M. Johnston, and D. A. Weil, Rapid Comm. Mass Spec., Vol. 2 pp. 146-150 (1988).

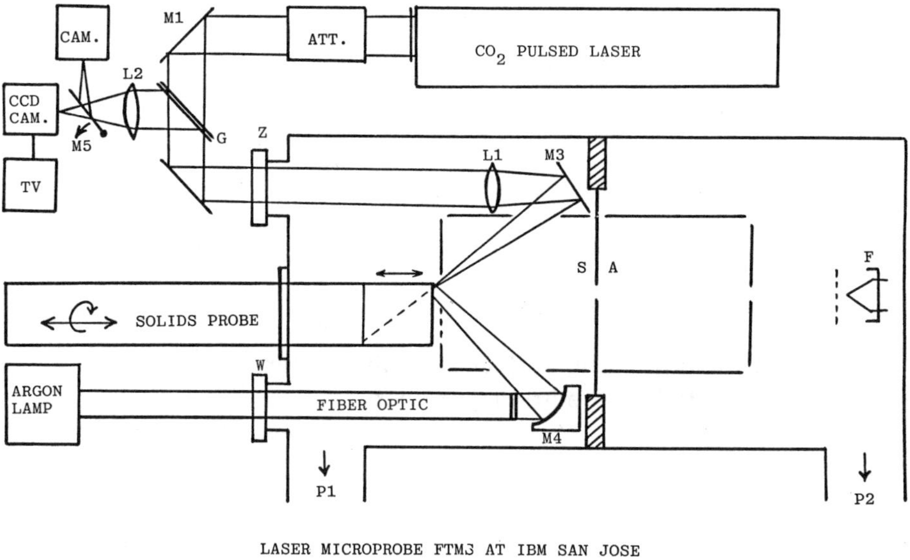

Figure 1) Schematic diagram of the laser microprobe FTMS system.

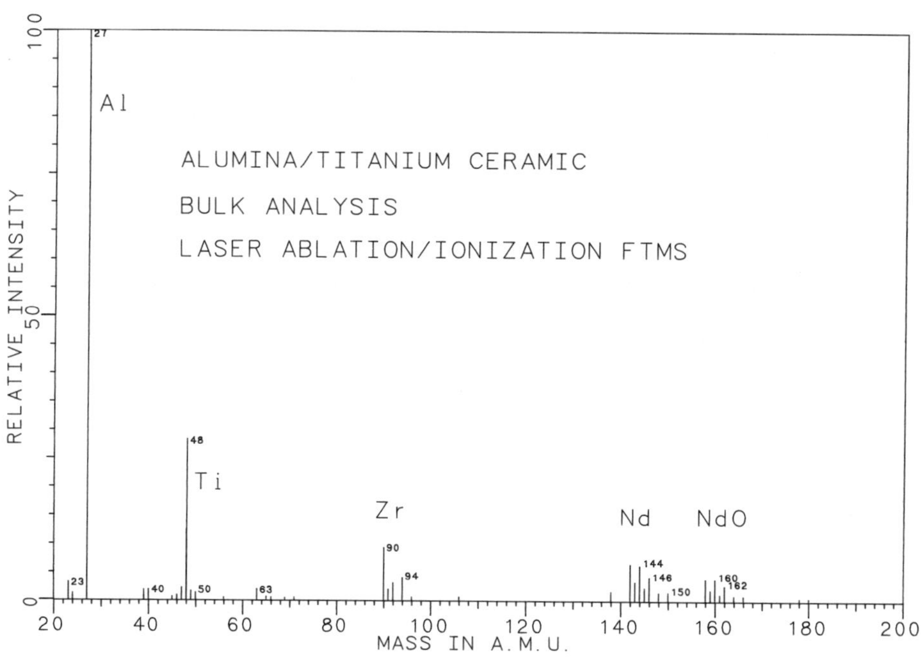

Figure 2) Bulk analysis of an alumina/titanium ceramic.

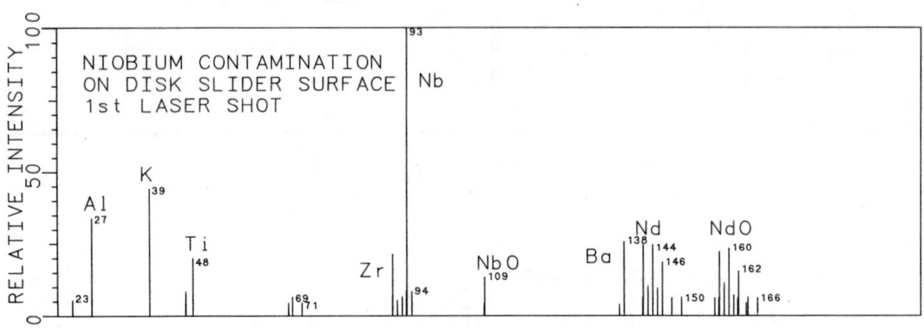

Figure 3a) Niobium on the surface of a disk slider, detected by the first laser shot.

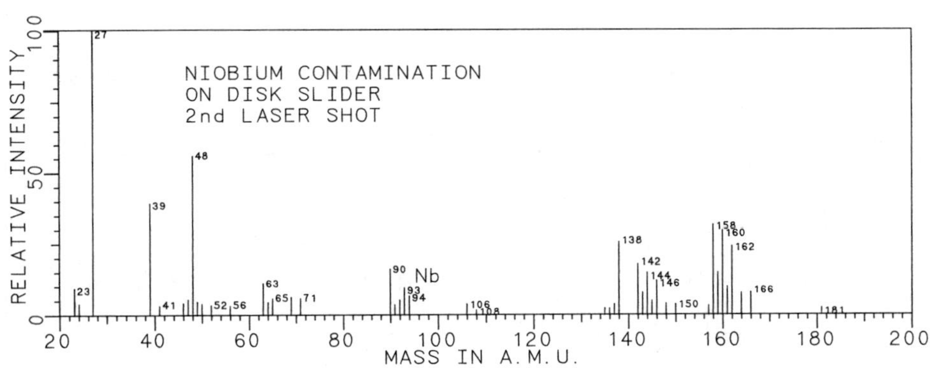

Figure 3b) Niobium signal is diminished in the second laser shot.

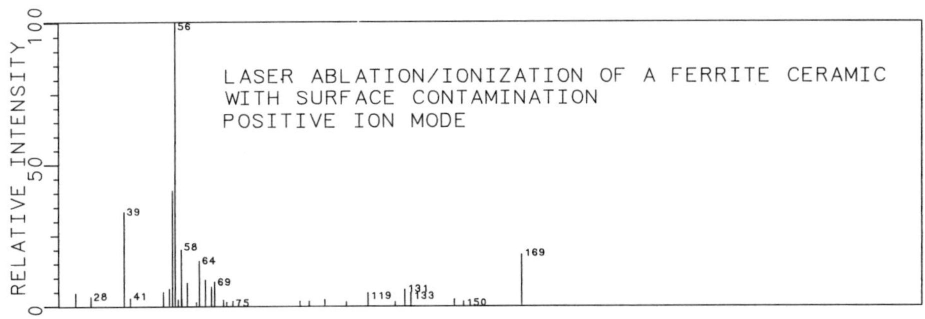

Figure 4a) Laser ablation/ionization of a ferrite ceramic with surface contamination.

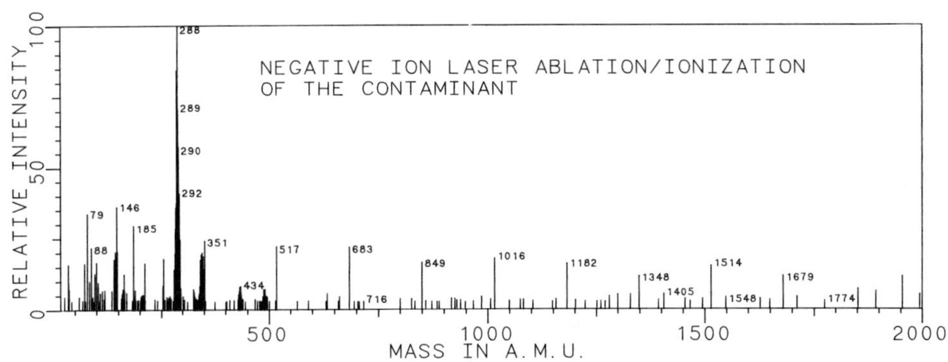

Figure 4b) Negative ion mode laser mass spectrum of the surface contaminant.

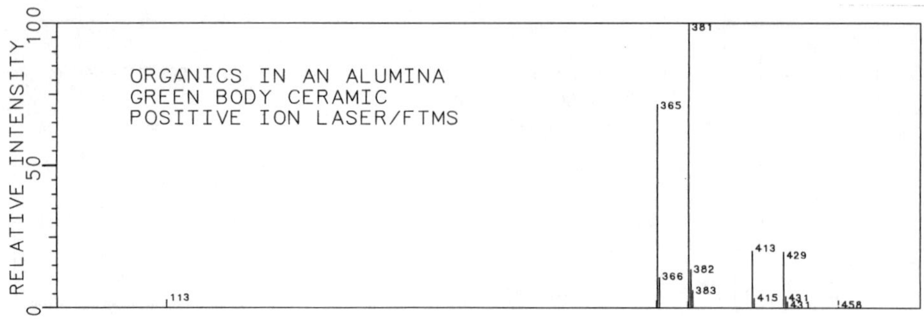

Figure 5a) Organics in a green body alumina ceramic - positive ions.

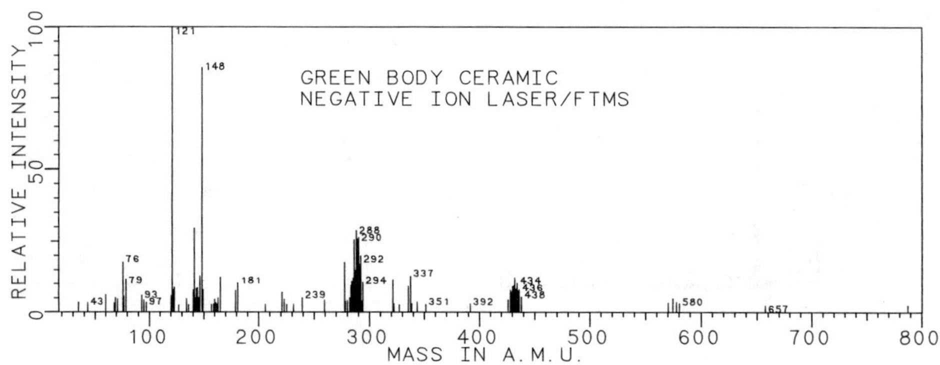

Figure 5b) Organics in a green body alumina ceramic - negative ions.

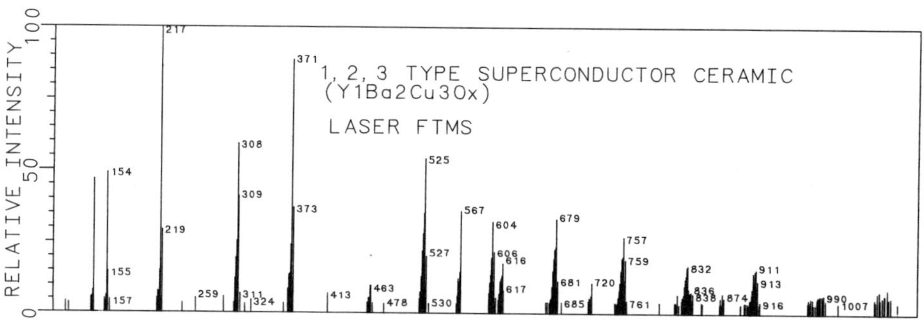

Figure 6a) Laser ablation/ionization of a 1-2-3 type superconductor.

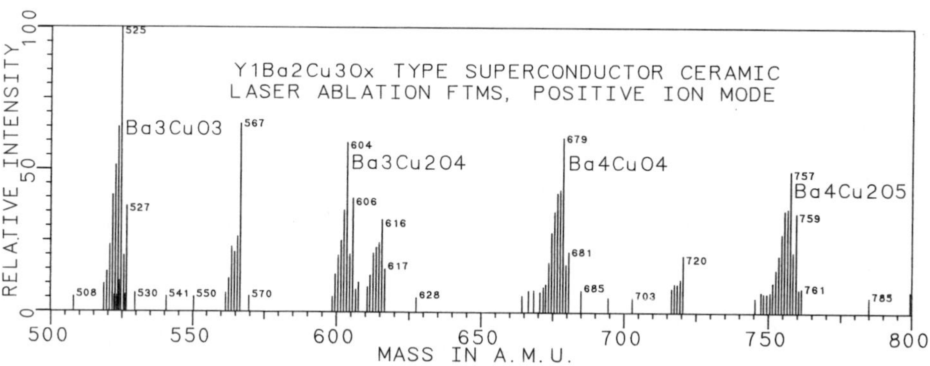

Figure 6b) Expanded region, 500 to 800 amu, showing metal oxide clusters.

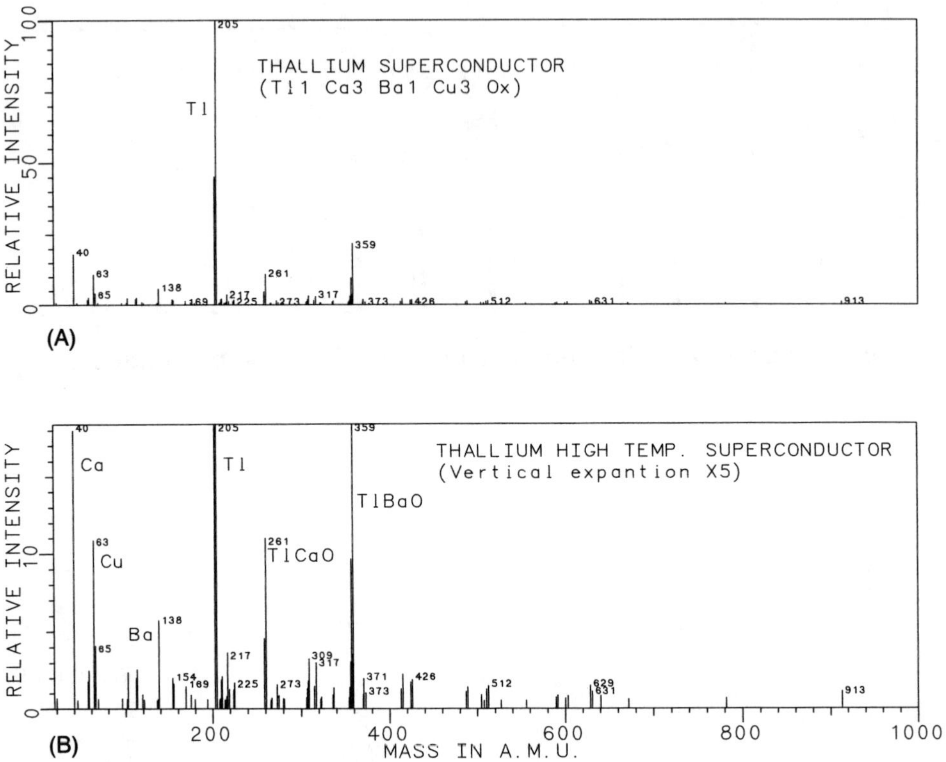

Figure 7) Vertically expanded mass spectrum of a thallium type superconductor.

ANALYSIS OF MATERIALS BY LASER ABLATION INDUCTIVELY COUPLED PLASMA MASS SPECTROMETRY (ICP-MS)

Peter Arrowsmith, E41/013, IBM General Products Division, 5600 Cottle Road, San Jose, CA 95193. Tel (408)256-4749.

ABSTRACT

A laser ablation sample introduction system has been developed for plasma source mass spectrometry (MS) for direct analysis of materials. Ablated particles, produced by focusing a pulsed laser onto the surface of the sample, are transported by gas flow to an inductively coupled plasma (ICP) for subsequent vaporization and ionization. The technique has several useful features for bulk elemental analysis of solids. Sample preparation is usually not required, a wide range of materials including metals, ceramics and glasses can be analyzed and rapid semi-quantitative determinations over the concentration range 0.1 ppm to 1% can be performed for many elements. Although the amount of material ablated depends on the sample matrix, the ion yield in the ICP does not appear to be matrix dependent. The separation of laser ablation and plasma ionization allows independent optimization and simplifies the approach to quantitative analysis. An internal standardization technique is used to compensate for any variation in ablation yield and analytical results are reported for a set of ceramics.

INTRODUCTION

Since commercial introduction five years ago, ICP-MS has become an accepted technique for trace elemental analysis of solutions. Compared to traditional methods of solution analysis, ICP-MS has similar sensitivity to atomic absorption spectroscopy (AAS) combined with the multielement analysis capability of sequential atomic emission spectroscopy (AES). However, in contrast to optical spectra, mass spectra are much simpler since all the stable isotopes of all the elements give only ~200 peaks at unit mass resolution. The isotopic selectivity of ICP-MS also allows isotope ratios to be determined for application to tracer and isotope dilution analysis. A review of ICP-MS is given in Ref. 1.

Fig. 1 shows a schematic of a typical ICP mass spectrometer. A solution of the material to be analyzed (0.1-1% maximum dissolved solids) is flowed through a nebulizer and a small fraction (~1%) of the resulting aerosol droplets enter the

ICP. The ICP is driven by an rf generator (1-2 kW at 27 MHz) and typically runs at a gas kinetic temperature of 7000 K in Ar gas at atmospheric pressure. The aerosol solvent (generally water) is quickly vaporized and any solid material is atomized and ionized within the ICP. Many elements have first ionization potentials <9 eV, substantially below that of Ar (15.8 eV), and these are >90% ionized by the ICP, resulting in a uniform response to different elements. The central zone of the plasma containing the analyte ions is sampled by means of a differentially pumped sampler and skimmer mounted on a water cooled flange. The nozzle diameters are sufficiently large (~1 mm) to produce continuum flow and the resulting jet of gas rapidly cools without significant ion-neutral recombination as it expands into the low pressure (10^{-5} Torr) region containing the mass spectrometer. Hence, the ions sampled by the mass spectrometer are representative of the plasma. Both the currently available commercial ICP-MS systems (developed by Sciex Perkin-Elmer and VG Elemental) use quadrupole mass spectrometers. The quadrupole acts as a mass filter, transmitting ions of a single mass/charge ratio (m/z) to the detector, usually a continuous dynode electron multiplier operating in pulse counting mode. Although these are single channel instruments, the mass range of 0-250 dalton can be scanned in a short time (typically 100 ms) or alternatively, the signal at selected m/z values can be rapidly sampled by peak-hopping the quadrupole under computer control. The measured count rate for a given isotope is proportional to the concentration of that element in the solution and detection limits are in the range 0.01-1 ng/ml (ppb) for most elements.

Sample dissolution is often time consuming (particularly for ceramics), prone to contamination and cannot give information on the spatial variation in composition over the sample. Hence, methods for direct introduction of solids to the ICP are currently of great interest. Laser ablation sample introduction has been previously used with AES and AAS (for a review see Ref. 2) but because of the sensitivity and flexibility of ICP-MS the new combination of laser ablation ICP-MS appears to be particularly useful.[3,4]

In the laser ablation technique, sample preparation, i.e. conversion of the solid material to suitable form for transport to the ICP, is performed by the laser. Ablated particulate material (a so-called solids aerosol) is transported by gas flow to the ICP for secondary vaporization, atomization and ionization. The method has several advantages. Little, or in most cases, no sample preparation is required and the focused beam of a medium energy (100-500 mJ) pulsed laser is capable of ablating most materials. Short duration (~10 ns) laser pulses, such as those from a Q-switched Nd:YAG laser, appear to give minimal fractional vaporization. The laser beam can be readily focused down to spot diameters of 10-100 μm, allowing moderate lateral resolution. In the absence of solvent, loading of the ICP is relatively light for laser ablation sample introduction (particularly for short duration laser pulses) and any variation in ion yield with sample matrix is expected to be less than that for solution nebulization (see below). In contrast, single-step sputter/ablation-excitation techniques such as spark source AES and MS, secondary ion MS (SIMS) and laser microprobe (LAMMA and LIMA) tend to give excited atom or ion yields that depend strongly on the ablation conditions and the nature of the sample matrix. Finally, since ablation, transport and secondary ionization occur sequentially and are separated in space, each is relatively independent, allowing separate optimization and simplification of the method for quantitative analysis.

This paper describes the development of a practical laser ablation ICP-MS system and presents some of the results obtained in the ~1 year the system has been working.

EXPERIMENTAL DEVELOPMENT

The work described in this paper was performed with a Sciex Perkin-Elmer ICP-MS (model ELAN 250) and a pulsed Nd:YAG laser (Lumonics model HY 200) operating at the fundamental wavelength of 1.06 μm. The 180 mJ laser beam was focused to a diameter of ~50 μm. The laser can be operated either Q-switched or free-running giving 10 ns and 180 μs duration pulses respectively, at repetition rates from single shot to 20 Hz.

The sample housing, shown in Fig. 2, comprises a gas tight outer box (side 12 in.) and an inner small ablation cell, shown separately in Fig. 3. Ar gas flows into the outer box at 4-5 Torr above atmospheric pressure and the resulting pressure gradient along the 80 cm-long transfer tube connecting the cell to the sample inlet of the ICP torch (see Fig. 1), serves to transport ablated material to the ICP. The inlet flow is usually fixed and higher than the transport gas flow which can be optimized for maximum signal by means of two outlet flow valves that control the pressure within the outer box. The motorized XYZ stage is under computer control, allowing the sample to be translated up to 10 cm in any direction with respect to the fixed laser beam and optical viewing axis at a resolution of 3 μm per step.

The gas inlet for the first two cell designs (Fig. 3a and 3b) did not lead directly to the cell and these cells were found to exhibit a long washout time, as demonstrated by the slow decay (half-life ~2 min) of the signal on switching off the laser. The signal decay arises from the reentry of material that is ablated sideways and escapes from the cell to the large volume outer box. Any memory effect is undesirable since it prevents rapid sequential sampling of different areas of a sample or depth profiling of a single area. The problem was overcome by flowing inlet gas directly into the outer part of an annular cell (Fig. 3c). This design cannot prevent high velocity ablated particles from escaping but it does stop material from reentering the cell and the signal decays to the background level within 1-2 s for this design. Cell 3c was used for all subsequent work. A single laser pulse transient signal obtained with cell 3c is shown in Fig. 4, the half-life of the decay is 180 ms. The transient signal is sufficiently long to give a continuous signal at laser repetition rates >10 Hz and the signal can be maintained constant for extended periods by simultaneously translating the sample to avoid defocusing the laser beam by hole-drilling. This sampling mode is used to acquire complete mass scans and for quantitative analysis (see below). The rate of sample translation is usually chosen to place 1-20 overlapping laser shots within the beam diameter at the sample surface. The time dependent response of the continuous signal to switching the 20 Hz laser on and off is shown in Fig. 5. This data was obtained by peak-hopping the mass spectrometer between several m/z values (isotopes) of interest, the rapid rise and decay in signal over 5 orders of magnitude is clearly seen.

By comparing the signal from transported material and that obtained by sampling the outer box immediately after ablation, the efficiency of entrainment of ablated material by the cell was estimated to be 90%. A mass balance measurement was performed by weighing a small piece of Mo metal before and after ablation and collecting transported particles on a filter placed at the end of the transfer tube for subsequent solution analysis. For this metal the transport efficiency was found to be ~50%. The major loss mechanism is probably gravitational deposition within the transfer tube and an efficiency of 50% appears reasonable for a dense material.

The final cell design (Fig. 3c) has a number of useful features. Since the cell does not form a seal to the surface of the sample and the width of the gap at the base of the cell is not critical over 1-3 mm, there is no need for the sample to be flat. The cell has been found to work well with both nonuniform samples and those with a cylindrical cross-section. Also, the excess flow of gas around the base of the cell is sufficient to prevent air from being entrained in the transport gas flow and thereby reaching the air sensitive ICP. Hence there is no need to purge the outer box when a new sample is introduced.

It should be pointed out that the overall sensitivity of the system is determined by the convolution of three particle size distributions; the size distribution of the ablated particles, the transfer function for entrainment within the cell and transport through the transfer tube, and the response function of the ICP. The first two functions together determine the transport efficiency. The response function falls off for particles larger than 10-20 μm, depending on the physical properties of the ablated material and the temperature distribution and residence time in the ICP, because of the inability of the plasma to completely vaporize large particles.[5] Further details of cell design, entrainment and transport efficiency and measurement of an ablated particle size distribution can be found in Ref. 6.

RESULTS AND DISCUSSION

Typical background spectra for laser ablation ICP-MS, obtained by running an Ar-only plasma with the laser off, are shown in Figs. 6 and 7. Major background species include the isotopes of Ar at m/z 36, 38 and 40 and the molecular ion Ar_2. Other background species such as C, N, O and their molecular ions probably arise from impurities in the Ar gas, entrainment of air by the ICP and outgassing from surfaces. All these ions can potentially interfere with analyte ions, for example, the isotopes of Ar_2 mask the major isotopes of Se at m/z 76, 78 and 80. However, the intensities of molecular ions, especially those containing H and O, are typically two orders of magnitude lower with the dry Ar ICP, compared to aqueous solution sample introduction. This is particularly evident for oxides of analyte ions and the mixed species ArC, ArN and ArO, that give rise to interferences with the isotopes of Cr and Fe.

The mass ablated with the short duration (10 ns) Q-switched laser is typically 1-10 ng per pulse. At these high irradiances ($\sim 10^{10}$ W/cm^2) heating is extremely rapid, plasma breakdown occurs and there is intense emission at the focus of the laser. A significant fraction of the incident laser energy is absorbed by the plasma, resulting in inefficient coupling of the laser to the sample material, although the plasma itself may cause ablation. This mechanism is supported by several observations. Craters formed by the Q-switched laser are relatively shallow but significantly wider than the laser beam spot, and there is little increase in the amount ablated when the pulse energy is increased beyond ~200 mJ. Also, ablation yields appear to be similar, within an order of magnitude, for many materials, such as metals and ceramics. Electron microprobe X-ray analysis performed on the rim and sides of craters formed by Q-switched ablation of Al_2O_3-TiC-ZrO_2 ceramics showed similar composition to the bulk. This indicates that (at least for these elements) fractional vaporization during ablation is not a serious problem, presumably because the depth of the heated zone around these craters is small. In terms of solids content, ICP loading with Q-switched laser ablation at 10 Hz is equivalent to running a solution of ~100 ppm through a nebulizer. This is a factor of 3-10 times lower concentration than that found to give significant matrix effects for solution sample introduction. Matrix effects in the ICP may arise from space-charge interaction as plasma gas expands through the sampler and skimmer nozzles.[7] It is possible that for laser ablation, matrix effects become significant only at relatively high levels of ICP loading because of the absence of solvent.

In contrast, ablation with the long-pulse duration free-running laser produces narrower, deeper craters and 10-100 times more material is ablated. This is described by a different ablation mechanism. In this case there is sufficient time for equilibrium to be established, thermal conduction is significant and material is removed by steady-state boiling and evaporation since the ablation zone is relatively cool. Free-running ablation of metals is observed to sputter large molten droplets. For many materials, significant amounts of solidified melt material are deposited around the crater rim. Since there is increased probability of fractional vaporization from molten material, Q-switched laser ablation is preferred for quantitative analysis. In addition, the large amount of material ablated by the free-running laser can cause loading of the ICP and blockage of the sampler and skimmer, particularly for refractory samples. The various ablation mechanisms have been discussed by Ready.[8]

Semi-quantitative information is conveniently obtained for an unknown sample by scanning the complete mass range with continuous ablation, i.e. 10-20 Hz laser pulses and simultaneous sample translation. Examples of this type of mass spectrum are shown in Figs. 8 and 9 for a ceramic material. The major elements Al and Ti are both saturated and a signal of 10^6 counts/s is observed for Zr, present at ~1% in the ceramic. This demonstrates the excellent dynamic range of 5 orders of magnitude and sensitivity that can be obtained with ICP-MS. Since the background is at ~10 counts/s, the detection limit is ~0.1 ppm, in terms of a background equivalent concentration. Assuming 1-10 ng of material is ablated per pulse, then 10^{12}-10^{13} atoms/s of Zr are ablated and the efficiency of ICP-MS is 10^{-6}-10^{-7}.

For comparison to Fig. 9, the 100-250 dalton region was rescanned with free-running laser ablation. The spectrum, shown in Fig. 10, shows improved signal because a larger amount of material is ablated. The higher levels of oxide ions, particularly NdO, indicates the ICP is running cooler compared to Q-switched ablation, possibly because of sample loading.

The ease with which comparative analyses can be performed is demonstrated by Fig. 11 which shows the mass spectrum a Zr-free Al_2O_3-TiC ceramic from a different vendor. Nd and other rare earth elements are absent, indicating that the rare earth elements observed in the previous spectra (Fig. 9) may be impurities in the Zr, or more likely the Nd, found in these materials.

The usual mode for performing quantitative analysis is continuous mode Q-switched ablation in conjunction with a peak-hopping mass spectrometer, to give data similar to that of Fig. 5. The advantages of this method are the improved duty cycle per isotope compared to mass scanning (where considerable time may be spent at m/z values of no interest) and ease with which the data can be extracted by computer since no peak recognition is required. A disadvantage is that the elements of interest must be preselected, requiring a preliminary mass scan for analysis of an unknown sample.

The approach we have developed for quantitative analysis involves use of an internal standard element and response factors. Since the ablation yield is unknown and varies with ablation conditions and the physical properties of the sample, the measured intensities of the unknown elements E are corrected by dividing them by the intensity of an element of known concentration in the same sample (the internal standard IS). Use of internal standardization corrects for any systematic variation in ablation yield, such as change in laser focus, that affects the signals for the unknown elements and the internal standard in a similar manner. The internal standard method has a further advantage that it also corrects, to some extent, for short term fluctuations in signal when intensities are correlated. The internal standard may be a major element, even if the signal for that element would normally be saturated, since software in the ICP-MS enables the signal at selected m/z values to be attenuated by say 1000 fold. Ideally, a matrix-matched standard containing the same elements (and internal standard) as the unknown samples is ablated to determine response factor ratios R(E)/R(IS) that can be used to convert the intensity ratios I(E)/I(IS) obtained for the unknowns to concentrations. The validity of internal standardization for laser ablation ICP-MS has been previously tested by analysis of a set of NBS microprobe steel standards.[4] Plots of measured versus certified concentration for several elements were found to be linear over 4-5 orders of magnitude, precision and accuracy were estimated to be $\pm 5\%$.

Recently a similar method was used to quantify several Al_2O_3-TiC-ZrO_2 ceramics. In this case X-ray fluorescence (XRF) was used to create a reference standard (Sample A) which was subsequently analyzed by laser ablation ICP-MS to obtain response factors that could be applied to the other samples. The results are shown in Table 1 (see below), together with the XRF data for comparison. Some of the obvious differences between the ICP-MS and XRF results may arise from sample heterogeneity.

Table 1. Analysis of Al_2O_3-TiC-ZrO_2 ceramics by laser ablation ICP-MS. Mean of two determinations. Element concentrations are in ppm or percent by weight (*).

	Sample A		Sample B		Sample C	
	Ablation	XRF	Ablation	XRF	Ablation	XRF
Fe	2230	2000	4070	660	460	400
Ni	230	210	260	360	5	
Zr	1.68*	1.67*	1.88*	1.69*	2.14*	1.61*
Ba	210	190	70		50	
Nd	2860	2700	3370	2500	3640	2600
Hf	460	440	500	670	560	450
W	440	420	1.08*	6500	260	
Th	21	20	15		25	
U	44	40	50		50	

We are currently exploring a method that does not require matrix-matched standards. Unfortunately, response factor ratios cannot be obtained from solution analysis because there is evidence that the ion energy, and therefore the instrument response for a given element, differs for ablation, presumably because of modified ICP conditions in the absence of solvent. However, it appears that response factors obtained by ablating a standard having a different matrix can be used. Since the elements of interest are unlikely to be in the standard it is necessary to perform a fit to the response factors over the mass range so that a factor may be calculated for any element. Therefore the reference standard should contain elements covering the mass range and the same internal standard as the unknown samples. In practice it is found that R(E)/R(IS) ratios can be fitted as a product of two element (or mass) dependent terms; a relative ionization efficiency, given by the Saha equation for ionization in a plasma (with the ICP temperature as a single adjustable parameter) and a smoothly varying (polynomial) instrument response function. These response ratios appear to be applicable to different sample matrixes, suggesting that ion intensities are matrix independent under the conditions of plasma loading encountered with Q-switched laser ablation. Hence, atomization and ionization in the ICP appears to be insensitive to the chemical and physical nature of the ablated material.

CONCLUSIONS

Laser ablation sample introduction extends the technique of ICP-MS, originally developed for solution analysis, to direct analysis of solids. The typical detection limits of 0.1-1 μg/g (ppm) compare favorably to established techniques for analysis of solids. Laser ablation ICP-MS separates sample preparation/ablation from plasma atomization/ionization and it appears to have better reproducibility than some single-step methods such as spark source AES and MS. Although it is unlikely to displace other techniques such as XRF, laser ablation ICP-MS is complimentary in nature since it offers rapid analysis with moderate lateral resolution.

1. R. S. Houk, *Mass Spectrometry of Inductively Coupled Plasmas*, Anal. Chem. **58**, 97A-104A (1986).
2. K. Dittrich and R. Wennrich, *Laser Vaporization in Atomic Spectroscopy*, Prog. Anal. Atom. Spectrosc. 7, 139-198 (1984).
3. A. L. Gray, *Solid Sample Introduction by Laser Ablation for Inductively Coupled Plasma Source Mass Spectrometry*, Analyst **110**, 551-556 (1985).
4. P. Arrowsmith, *Laser Ablation of Solids for Elemental Analysis by Inductively Coupled Plasma Mass Spectrometry*, Anal. Chem. **59**, 1437-1444 (1987).
5. B. Raeymaekers, T. Graule, J.A.C. Broekaert, F. Adams and P. Tschopel, *Characteristics of nebulized suspensions of refractory oxide powders used for the production of ceramics and their evaporation behaviour in an inductively coupled plasma*, Spectrochim. Acta **43B**, 923-940 (1988).
6. P. Arrowsmith and S. K. Hughes, *Entrainment and Transport of Laser Ablated Plumes for Subsequent Elemental Analysis*, Appl. Spectrosc., **42**, 1231-1239 (1988).
7. G. R. Gillson, D. J. Douglas, J. E. Fulford, K. W. Halligan and S. D. Tanner, *Nonspectroscopic Interelement Interferences in Inductively Coupled Mass Spectrometry*, Anal. Chem. **60**, 1472-1474 (1988).
8. J. F. Ready, *Effects of High-Power Laser Radiation* (Academic Press, New York, 1971).

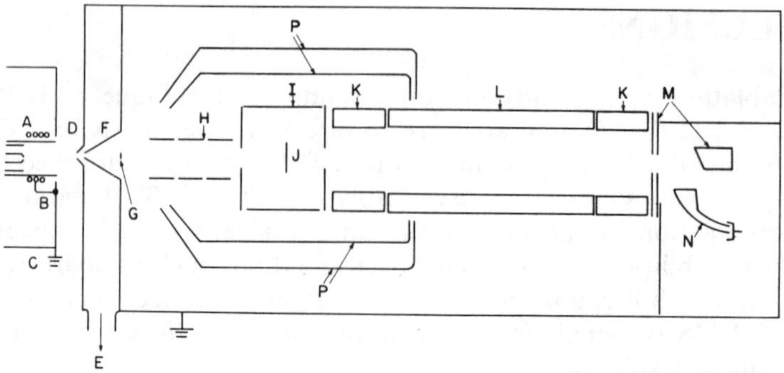

Fig. 1. Schematic of the ICP mass spectrometer. The sample flow is introduced through the innermost tube of the ICP torch A, surrounded by the center-tapped load coil B supplying the rf power via a matching circuit C. D and F are the sampler and skimmer nozzles respectively, with the intermediate region E evacuated by mechanical pump. The ion optical elements that focus and transmit the ion beam consist of light baffles (photon stops) G and J, a three element Einzel lens H, bezel box I and ion deflectors M at the exit of the quadrupole mass filter L. The short entrance and exit rods K serve as additional focusing elements. Ions that reach the detector N are converted to voltage pulses and counted electronically.

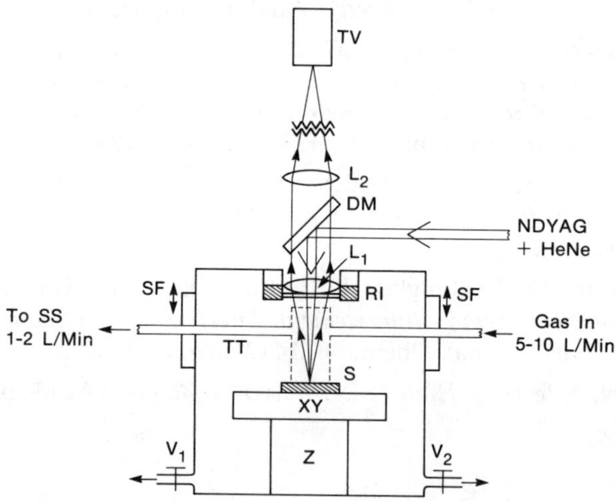

Fig. 2. Schematic of the laser ablation system. TT is the transfer tube connecting the ablation cell (located within the dashed area) to the secondary source SS (the ICP). SF are sliding flanges, DM is a dichroic mirror, RI is a ring illuminator, V_1 and V_2 are flow control valves, L_1 (behind a window) and L_2 are lenses. The sample S sits on the XYZ translation stage.

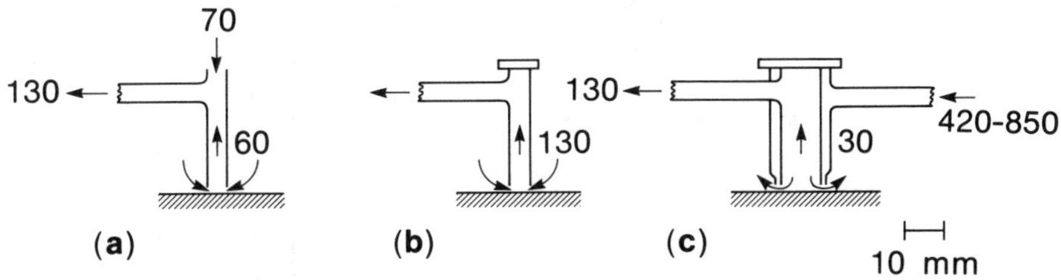

Fig. 3. The three cell designs evaluated for entrainment of ablated material, cells are to scale and approximate average flow velocities (in cm/s) are indicated.

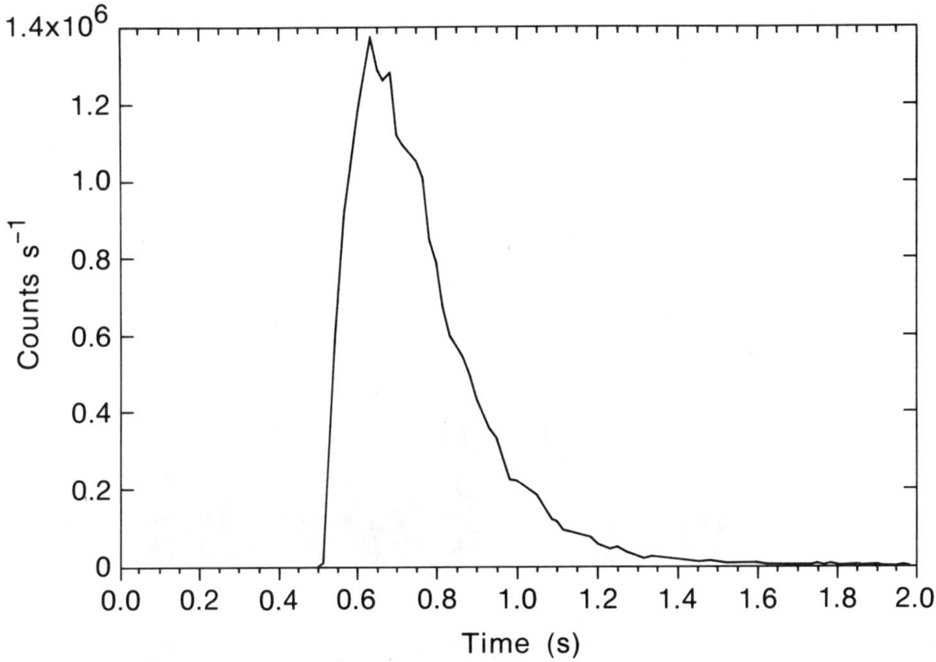

Fig. 4. The single laser pulse transient response of cell 3c, showing the Mg signal from NBS aluminum alloy (4.54% Mg). The measurement time was 5 ms.

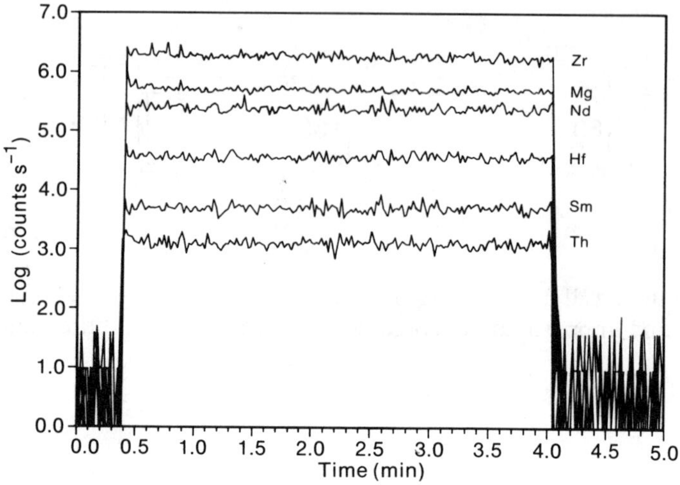

Fig. 5. Temporal response of cell 3c obtained by ablating an Al_2O_3-TiC-ZrO_2 ceramic (~1% Zr) with the 20 Hz laser. The laser was switched on and off at approximately 0.4 and 4.0 min respectively. The mass spectrometer was peak-hopped between the isotopes with a measurement time of 40 ms.

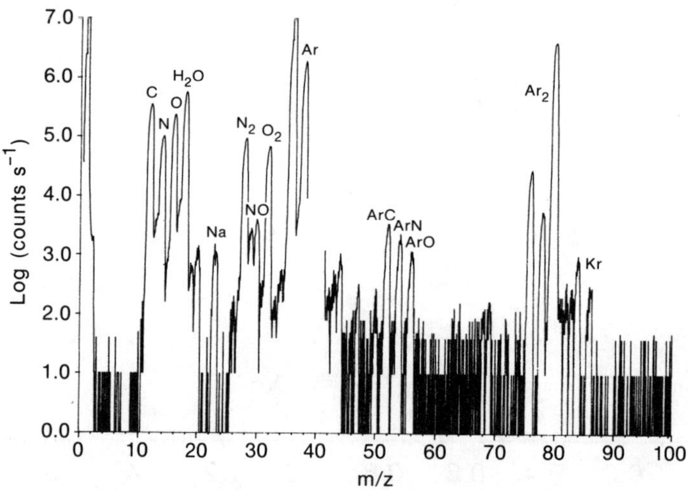

Fig. 6. Background mass spectrum over 0-100 dalton of the dry (laser-off) Ar ICP. Taken at 100 ms measurement time with 20 points per dalton and resolution of unit mass at 10% of the peak height. The intense Ar peak at m/z 40 was skipped.

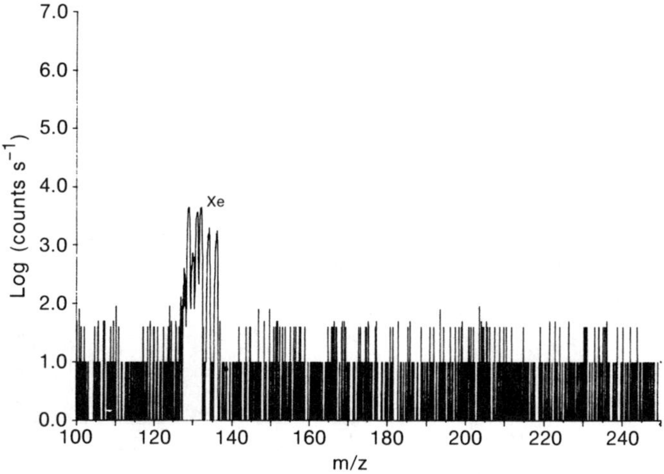

Fig. 7. Background mass spectrum over 100-250 dalton. Obtained under the same conditions as Fig. 6.

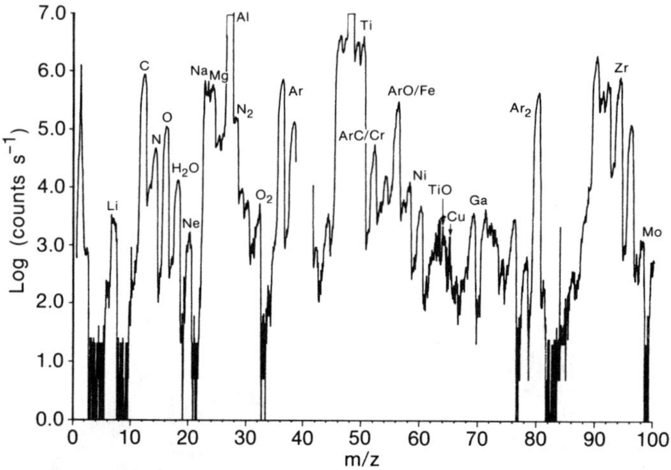

Fig. 8. Mass spectrum over 0-100 dalton of Al_2O_3-TiC-ZrO_2 ceramic obtained by 10 Hz Q-switched laser ablation with simultaneous sample translation. Approximately 15 laser pulses were overlapped within the diameter of the focused beam. The data acquisition time was ~3 min. Other conditions as for Fig. 6.

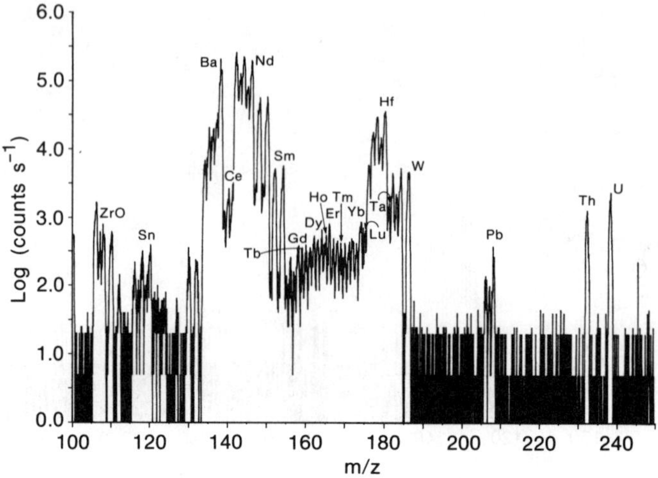

Fig. 9. Mass spectrum over 100-250 dalton of Al_2O_3-TiC-ZrO_2 ceramic obtained under the same conditions as Fig. 8.

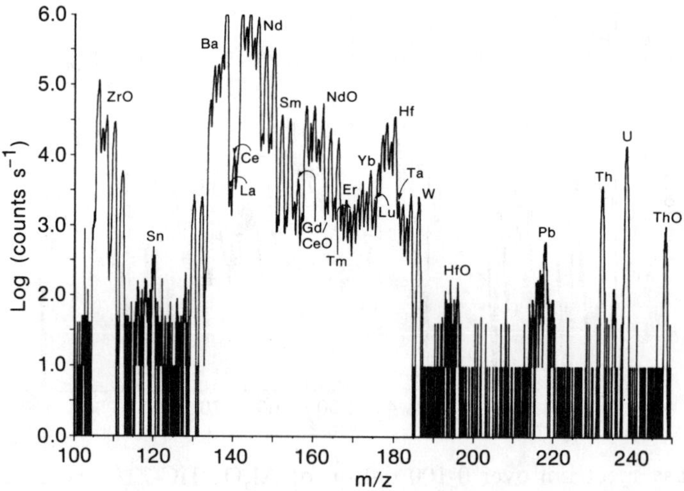

Fig. 10. Mass spectrum of Al_2O_3-TiC-ZrO_2 ceramic obtained by free-running laser ablation at 10 Hz.

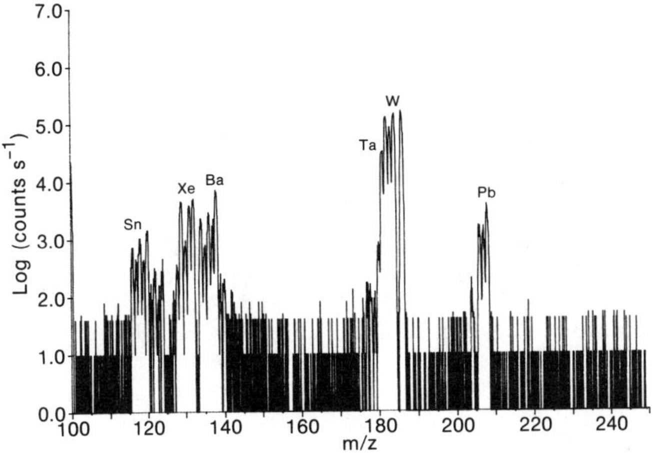

Fig. 11. Mass spectrum of a Zr-free Al_2O_3-TiC ceramic. Taken under the same conditions as Fig. 8.

CHARACTERIZATION OF MATERIALS USING SIMS IMAGE DEPTH PROFILING

Steven W. Novak, Charles Evans & Associates, 301 Chesapeake Dr., Redwood City, CA 94063 USA; Hilda Kanber, Hughes Aircraft Company, 3100 W. Lomita Blvd., Torrance, CA 90509 USA; and Richard Lowden, Oak Ridge National Laboratory, P.O. Box X, Oak Ridge, TN 38831 USA.

ABSTRACT

The SIMS image depth profiling technique consists of sequentially recording ion images as a sample is sputtered. Once recorded, the imaged volume can be examined sequentially, as cross sectional ion images, or as a selected area depth profile. The use of a conductive metal coating and electron bombardment during sputtering compensate for sample charging, allowing relatively insulating ceramic materials to be examined by this technique. Lateral resolutions of ion images is typically 1 μm whereas the depth resolution can be as low as 5-10 nm. We demonstrate the use of this SIMS technique to characterize disordering of GaAs/AlAs heterostructures, Ti diffusion in $LiNbO_3$, and fiber-matrix interfaces in SiC-fiber reinforced SiC composites.

INTRODUCTION

The secondary ion mass spectrometry (SIMS) technique is now widely used to characterize electronic materials because of its great elemental sensitivity and excellent depth resolution but has been less commonly applied to ceramic materials (1, 2). This is because many bulk ceramics are highly insulating. Therefore, during SIMS ion bombardment analysis, a charge develops on the sample surface which severely alters emitted secondary ion energies, or in the most severe cases, prevents the primary ion beam from sputtering the sample. Several techniques have been developed to counteract sample charging, however, allowing SIMS to be used to characterize virtually any material. Because the sample is sputtered, i.e. eroded, during a SIMS analysis, depth profiling is the natural mode of analysis. When the SIMS instrument is capable of recording ion images sequentially as a sample is sputtered, a volume within the sample can be examined. This combination of high sensitivity and high spatial resolution gives us a powerful method to examine elemental distributions in the extremely fine scale structures now being produced in many manufactured materials. The purpose of this paper is to discuss this image depth profiling technique and to show several applications to modern analytical problems.

SIMS TECHNIQUE

The dynamic SIMS technique utilizes a well focussed (<100 μm) moderate energy (1-20 keV) ion beam of 10's of nA to 10μA to sputter a sample (3). Secondary ions ejected during sputtering are extracted into a mass spectrometer for mass filtering and detection. An O_2^+ or O^- primary ion beam is generally used to enhance positive secondary ion yields, whereas Cs^+ is commonly utilized to enhance negative ion yields. The beam is rastered during analysis so that the secondary ions are extracted from a flat bottomed crater. Although ions are emitted only from the surface atomic layer, the primary beam impact produces a collisionally mixed zone which typically limits the depth resolution of this technique to 5-10 nm. Any pre-existing surface topography or preferential sputtering effects that cause sputter roughing will degrade these depth resolution limits significantly.

Bombarding an insulating sample ($\rho > 10^7$ Ωcm) with a positively charged ion beam will cause a net positive charge to buildup on the sample surface, because the sample resistance prevents electron flow within the sample (4, 5). This problem is especially acute if the sample is negatively biased to aid extraction of negative secondary ions since secondary electrons are repelled as well. To alleviate sample charging, a conductive metal coating (typically Au) can be used to conduct electrons to the vicinity of the primary beam. An electron gun can also be used to flood the sample surface with electrons (6, 7). These charge compensation techniques allow insulating films up to 1 μm thick to be routinely profiled (8) and have also been used to profile ion implanted single-crystal sapphire (9), as well as the single crystal $LiNbO_3$ discussed later (10). For extremely insulating samples, a negatively charged primary ion beam (generally O^-) may be required (4), although this technique was not needed for the materials examined in this work.

The mass spectrometer design determines the mode of ion imaging for a particular instrument. Most imaging SIMS instruments utilize the scanning imaging method in which the pulse-counting detector output is used to modulate the X-Y raster signal (11). This technique is directly analogous to that used for scanning Auger microscopy (12) and X-ray mapping (13). An alternative imaging method involves utilizing a stigmatic focussing mass spectrometer which directly forms an ion image (1, 3). These latter instruments operate in a direct imaging mode. For the SIMS scanning imaging method, as in Auger mapping, the beam size determines the resolution of the resulting image. In order to obtain lateral resolutions <1 μm, primary beam currents must be kept quite low, in the picoampere range. Such low beam currents significantly reduce the sputtering rate. It has been calculated that the scanning imaging technique requires 450 times longer than direct imaging to collect images with similar resolution on the same mass spectrometer (14). Thus, although the scanning imaging technique has attained the smallest lateral resolutions available by SIMS, it is limited to imaging surfaces because of the low sputtering rates involved. To examine significant volumes by imaging, it is necessary to employ a direct imaging

SIMS instrument. It may be possible to utilize a second high current ion beam to alternate with the imaging beam on a scanning instrument and therefore profile the sample in depth but such an application has yet to be demonstrated.

The CAMECA IMS-4f microanalyzers used in this study have a stigmatic double focussing mass spectrometer capable of forming ion images with lateral resolutions of <1 μm (15). Because the mass spectrometer determines the image resolution, high primary beam currents, and therefore high sputtering rates (>10 nm/sec), can be used to profile samples while acquiring ion images. We have utilized a position-sensitive pulse counting detector (16), which, in conjunction with a PC-based instrument control system, allows us to sequentially record ion images in digital form as a sample is sputtered. Thus, a three dimensional ion image record of a sample can be rapidly recorded and is available for subsequent analysis using standard image processing techniques.

Once recorded, the imaged volume can be examined in several ways. The images can be sequentially displayed to show a "movie" of the sample as it is sputtered. A thin slice through the imaged volume can be displayed as a cross sectional ion image. This representation is useful for examining small scale structures in depth, as the vertical resolution of the cross section is defined by the depth resolution of the profiling technique. Finally, a box can be defined within the imaged area and the integrated counts for each image can be plotted as a function of sputter time to yield a selected area depth profile.

APPLICATIONS

One example of the use of image depth profiling is for the examination of optical wave guides formed by selective area disordering of III-V compound laser heterostructures. Waveguides are formed by introducing a dopant into a heterostructure sample through surface diffusion or ion implantation in selected areas. During subsequent annealing, diffusion of the dopant atoms causes disordering of the active regions of the heterostructure by inducing layer interdiffusion (17). The result is optically confined channels of undisturbed material surrounded by disordered material. Figure 1 shows an Al cross-sectional ion image of such a sample. The structure consists of alternating layers of 6.5 nm of GaAs and 6.5 nm of AlAs with a 20 nm GaAs cap. A pattern of stripes was defined on the sample surface using photoresist and the sample was then implanted with ^{28}Si to a dose of 1×10^{15} atoms/cm^2 at an energy of 150 keV. The implant disorders the layers in the implanted regions by recoil mixing as shown by the reduced contrast between layers in the ion cross-sectional image. To quantify the amount of layer mixing caused by implantation, a depth profile was reconstructed from an implanted stripe and compared with one from an unimplanted region (Fig. 2a and 2b).

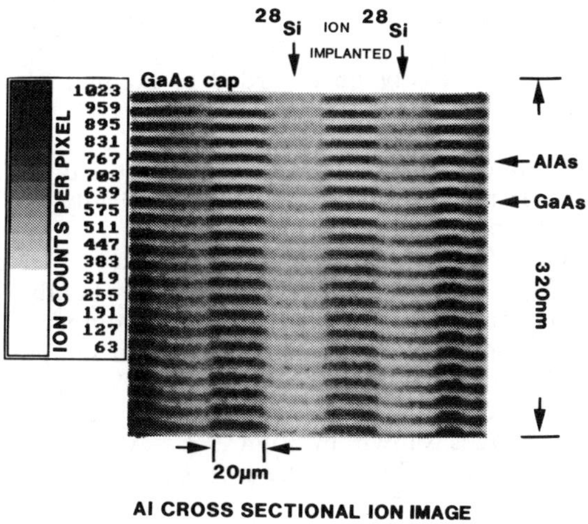

Figure 1. Cross-sectional Al^+ ion image of a ^{28}Si ion implanted AlAs-GaAs heterostructure.

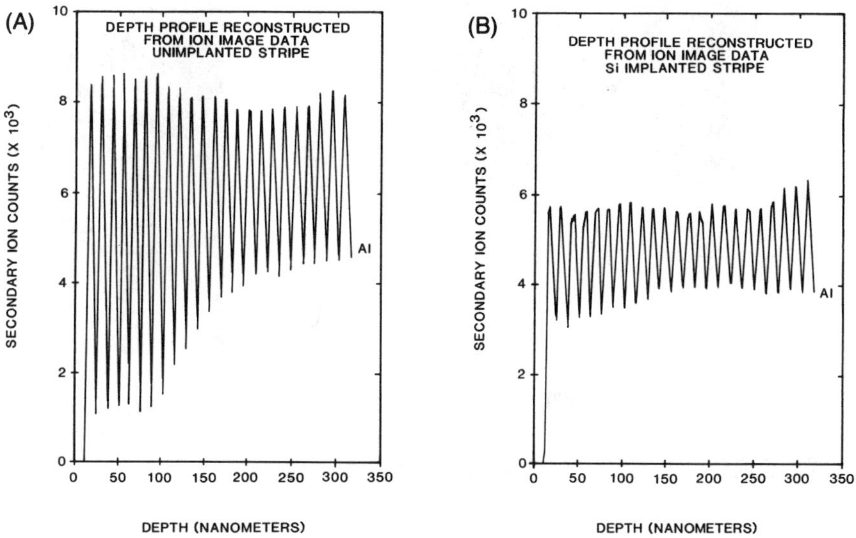

Figure 2a. Selected area Al^+ depth profile from an unimplanted 20 μm wide stripe reconstructed from the image data of Figure 1.

Figure 2b. Selected area Al^+ depth profile from an implanted 20 μm wide stripe on Figure 1.

The peak to valley ratio is reduced in the implanted region and can be used to infer the amount of layer interdiffusion (18). Although not shown, the dopant atom profile can easily be recorded to compare its behavior before and after annealing.

Image depth profiling can also be used to examine the diffusion of Ti or H in $LiNbO_3$ to form optical wave guides (19, 20). Typically, Ti metal stripes are deposited on single crystal $LiNbO_3$ and the Ti is driven in by a high temperature anneal. In contrast to semiconductors, $LiNbO_3$ is relatively insulating and a Au coating together with electron bombardment were used during SIMS depth profiling. Uniform charge compensation is especially critical during image depth profiling as variations in sample voltage as low as 10-20V can severely distort the ion images. A 9kV 30μA normal incidence electron beam focussed to 150μm provided suitable charge compensation to obtain the Ti ion images shown in Figure 3. A slight charging transient distorted the images within 0.5 μm of the surface.

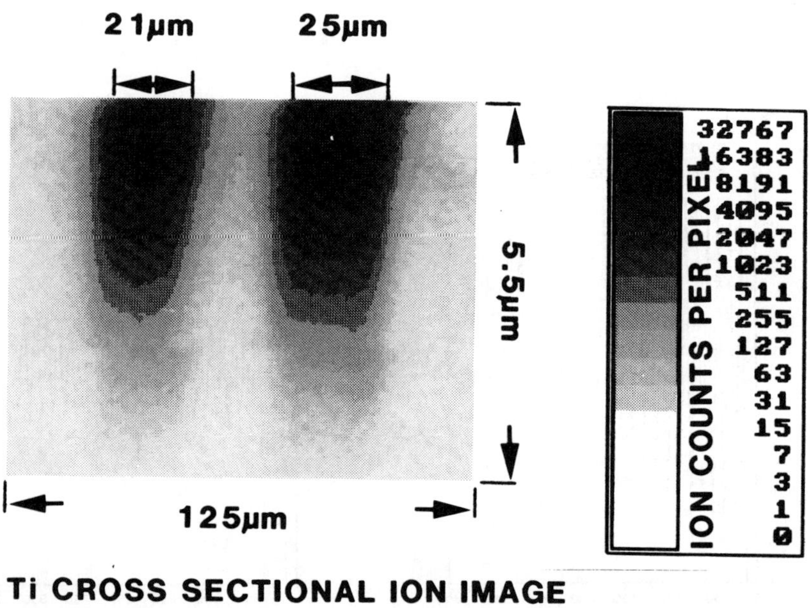

Figure 3. Ti^+ cross-sectional ion image of diffused stripes in $LiNbO_3$.

This cross sectional image is useful for measuring the lateral diffusion of Ti. A quantitative selected area depth profile showing Ti concentrations versus depth in the 21 μm wide stripe is given in Figure 4.

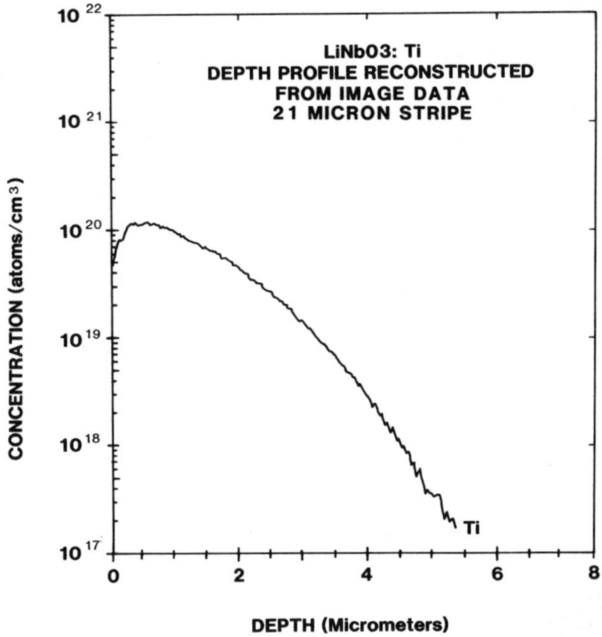

Figure 4. Selected area depth profile of Ti^+ reconstructed from the image data of Figure 3.

Quantitation of the Ti signal was accomplished by the use of an ion implanted standard. The depth scale was established by measuring the analytical crater depth with a stylus profilometer.

Characterization of the fiber-matrix interface in SiC-fiber reinforced ceramics is well suited to image depth profiling. We have examined a SiC composite that consists of 10-20 μm diameter SiC fibers woven into cloth that has been chemically embedded in a SiC matrix (21). In order to tailor the strength of the fiber interface, different precoat materials were deposited on the SiC fibers before the SiC matrix was grown. SIMS image depth profiling can determine whether B from the precoat material (BN) has diffused into the fiber or matrix, the uniformity of the fiber coating, and the compositional differences between fiber and matrix. The image depth profile analysis was arranged so that the fibers dipped gently below the surface at the analysis point (Fig. 5a and 5b).

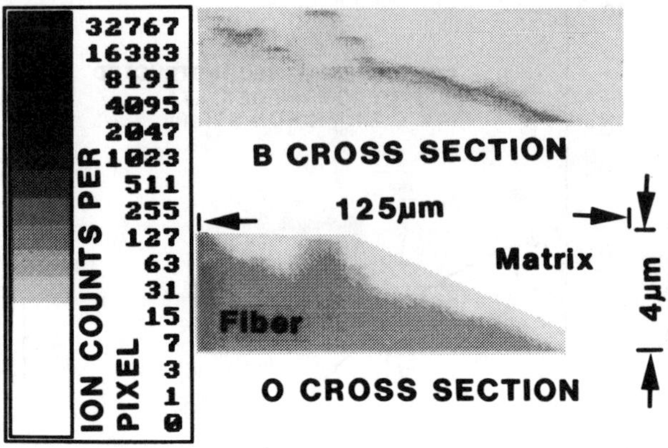

Figure 5a. Cross-sectional ion images of B and O in a longitudinal fiber section from a SiC composite. The fiber slopes from left to right and has a 1 μm thick BN coating.

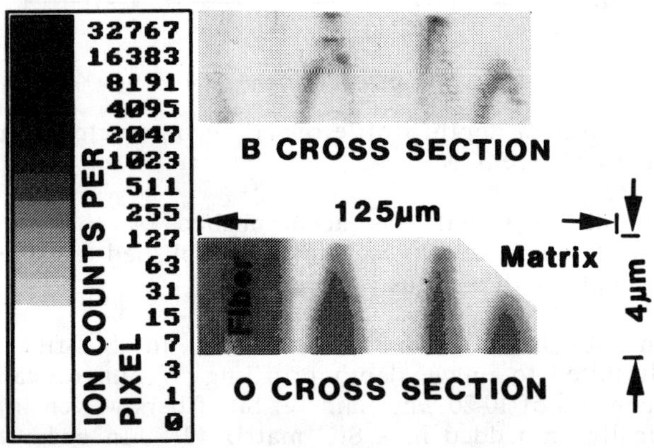

Figure 5b. Transverse fiber sections from the same composite at 90° to Figure 5a.

By recording ion images from such an area as the sample is sputtered, we can see successive section from the matrix down into the fibers passing through the fiber-matrix interface at a shallow angle. Using a 125 μm image field, a 45 minute profile examines an imaged volume 4 μm thick, or ~50000 μm^3. Figure 6 shows a selected area depth profile into one fiber in the SiC composite.

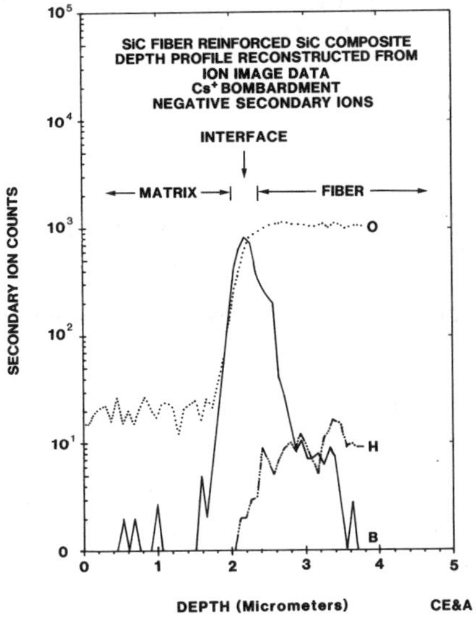

Figure 6. Selected area depth profile of fiber-matrix interface in a SiC composite. The fiber was coated with BN before the matrix was grown.

From this profile, we can determine the O and H contents of the fiber and matrix, and that B present at the fiber interface has diffused approximately 2 μm into the fiber. No charge compensating methods were required for analysis of this material. Bccause of the large O content of the fiber, measured ion count rates may not be a simple function of concentration. However, ion implantation provides a convenient method to "manufacture" in situ standards for virtually any materials system. By profiling ion implanted composites, we can determine the local variation in ion yield due to compositional differences, and therefore calculate accurate elemental contents in localized areas.

CONCLUSIONS

The SIMS image depth profiling technique provides a means to rapidly examine the three dimensional major and trace elemental distributions within the top tens of micrometers of a sample. Structures as small as 1μm laterally and 10nm in depth can be examined. Conductive samples are the most straightforward to analyze, although charge compensation techniques exist that allow profiling of even the most insulating materials. Many multicomponent materials traditionally thought of as insulators, such as the SiC composite examined in this study, may in fact contain enough impurities and internally conductive paths that the bulk sample is not an insulator. Thus, the SIMS technique may be useful for examining many ceramic materials.

REFERENCES

[1] E. Zinner, Depth Profiling by Secondary Ion Mass Spectrometry, Scanning, 3(1980)57-78.

[2] P. Williams, Secondary Ion Mass Spectrometry, Ann. Rev. Mat. Sci. 15(1985)517-548.

[3] A. Benninghoven, F. G. Rudenauer, and H. W. Werner, *Secondary Ion Mass Spectrometry,* John Wiley and Sons (1987).

[4] H. W. Werner and A. E. Morgan, Charging of Insulators by Ion Bombardment and Its Minimization for Secondary Ion Mass Spectrometry (SIMS) Measurements, J. Appl. Phys. 47(4)(1976)1232-1242.

[5] H. W. Werner and N. Warmoltz, Beam Techniques for the Analysis of Poorly Conducting Materials, J. Vac. Sci. Tech. A2(2)(1984)726-273.

[6] K. Wittmack, Primary-Ion Charge Compensation in SIMS Analysis of Insulators, J. Appl. Phys. 50(1)(1979)493-497.

[7] G. Slodzian, M. Chaintreau, and R. Dennebouy, The Emission Objective Lense Working as An Electron Mirror: Self-Regulated Potential at the Surface of an Insulating Sample, *Secondary Ion Mass Spectrometry, SIMS V,* A. Benninghoven, J. Colton, D. S. Simons, eds. (Springer Verlag), (1986)158-160.

[8] P. K. Chu and S. L. Grube, Quantitative Determination of Boron and Phosphorous in Borophosphosilicate Glass by Secondary Ion Mass Spectrometry, Anal. Chem. 57(1985)1071.

[9] R. G. Wilson and S. P. Smith, Depth Profiling and Implantation Profiles in Crystalline Al_2O_3, *Secondary Ion Mass Spectrometry (SIMS VI),* A. Benninghoven, A. M. Huber, H. W. Werner, eds. (John Wiley 1988)525-528.

[10] F. A. Stevie, V. V. S. Rana, A. S. Harris, T. H. Briggs, and P. Skeath, Summary Abstract: High Sputter Rate Secondary Ion Mass Spectrometry Analysis of Insulators Used in Microelectronics and Lightwave Applications, J. Vac. Sci. Tech. A., 6(3)(1988)2082-2084.

[11] H. Liebl, SIMS Instrumentation and Imaging Techniques, Scanning, 3(1980)79-89.

[12] A. Joshi, L. E. Davis, and P. W. Palmberg, Auger Electron Spectroscopy, *Methods of Surface Analysis*, A. W. Czanderna, ed. Elsevier (1975)159-222.

[13] D. E. Newbury, D. C. Joy, P. Echlin, C. E. Fiori, and J. I. Goldstein, *Advanced Scanning Electron Microscopy and X-Ray Microanalysis*, Plenum (1986).

[14] M. Schuhmacker, B. Rosser, H.-N. Migeon, and G. Slodzian, Ion Microscope and Ion Microprobe Images Under Oxygen and Cesium Bombardment, *Secondary Ion Mass Spectrometry (SIMS VI)*, A. Benninghoven, A. M. Huber, and H. W. Werner, eds., (John Wiley)(1988).

[15] M. Lepareur, Le Micro-Analyseur Ionique de Second Generation CAMECA, Modele 3f, Revue Technique, THOMSON-CSF 12(1)(1980)225-268.

[16] R. W. Odom, B. K. Furman, C. A. Evans Jr., C. E. Bryson, W. A. Peterson, M. A. Kelly, and D. H. Wayne, Quantitative image acquisition system for ion microscopy based on the resistive anode encoder, Anal. Chem. 55(1983)574-578.

[17] T. Venkatesan, S. A. Schwartz, D. M. Hwang, R. Bhat, M. Koza, H. W. Yoon, P. Mei, Y. Arakawa, and A. Yariv, Dose-Dependant Mixing of AlAs-GaAs Superlattices by Si Ion Implantation, Appl. Phys. Lett. 49(1986)701.

[18] S. A. Schwartz, T. Venkatesan, R. Bhat, M. Koza, H. W. Yoon, Y. Arakowa, and P. Mei, Depth Dependant Mixing of an AlAs-GaAs Superlattice by Ion Implantation, Proceedings of the Materials Research Society 56(1986)321.

[19] M.N. Armenise, C. Canali, M. DeSario, A. Carnera, P. Mazzoldi, and G. Celloti, Characterization of TiO_2, $LiNb_3O_8$, and $(Ti_{0.65}Nb_{0.35})O_2$ compound growth observed during Ti:$LiNbO_3$ optical waveguide fabrication, J. Appl. Phys. 54(1983)6223-6231.

[20] J. L. Jackel, C. E. Rice, and J. J. Veselka, Proton Exchange for High-Index Waveguides in $LiNbO_3$, Appl. Phys. Lett. 41(1982)607.

[21] D. P. Stinton, A. J. Caputo, and R. A. Lowden, Synthesis of Fiber-Reinforced SiC Composites by Chemical Vapor Infiltration, Ceram. Bull. 65(1986)347.

THE APPLICATION OF RUTHERFORD BACKSCATTERING AND CHANNELING TECHNIQUES TO THE CHARACTERIZATION OF CERAMICS

L. A. Boatner and C. W. White
Oak Ridge National Laboratory
Solid State Division
Oak Ridge, TN 37831-6056

ABSTRACT

Rutherford backscattering spectroscopy (RBS) is a versatile and powerful technique for investigating the near-surface properties of polycrystalline ceramics, glasses, and single crystals. In the case of macroscopic single-crystal samples, the variant of RBS known as "channeling spectroscopy" can be used to obtain detailed information regarding the near-surface structural, impurity, and defect characteristics of ceramic materials.

INTRODUCTION

The techniques of Rutherford backscattering and channeling spectroscopy have previously been applied to numerous detailed investigations of the near-surface region of important semiconducting materials such as silicon and germanium, and these methods have also been extensively used in studies of metal and alloy systems. In particular, RBS and channeling have been used to investigate the effects of ion implantation on the properties of semiconductors and metals. In the case of ceramic materials, however, the application of these methods to characterizations of the near-surface properties (i.e., to a depth of a few microns) has only occurred relatively recently. Nevertheless, these more recent investigations have firmly established the utility of RBS and channeling in determining the structural, chemical, and physical properties of ceramic materials. Examples of the versatility and power of these techniques are provided by previous RBS investigations of the corrosion of glass, channeling studies of the amorphization and epitaxial regrowth of ion-beam-damaged surface layers on crystalline oxides, and RBS/channeling determinations of ion-implantation-induced chemical composition alteration and recovery effects in the $YBa_2Cu_3O_x$ family of superconducting ceramics. These three somewhat diverse examples will be used to illustrate the general nature of the applicability of RBS/channeling methods to different types of ceramic materials in the following sections. It should be emphasized, however, that the utility of these methods is not bounded by the present examples and that RBS-based analytical techniques can be applied to essentially any type of ceramic (i.e., insulating, conducting, polycrystalline, or amorphous) and to ceramics in any form (i.e., as bulk materials, thin films, or buried subsurface layers).

RUTHERFORD BACKSCATTERING AND CHANNELING SPECTROSCOPY

The physical arrangement typically employed in carrying out RBS analysis is illustrated in Fig. 1.[1] The sample to be analyzed (in the present case a ceramic material) is mounted in a vacuum chamber at nominally 10^{-6} to 10^{-8} Torr. A well-collimated beam of energetic particles from a Van de Graaff or other accelerator impinges on the sample surface. The example in Fig. 1 shows 2.0 MeV ^{4}He particles striking the sample, and while these conditions have become somewhat of a standard in RBS analysis, other scattering particles or other energies may be employed.

While most of the incident He particles simply penetrate the sample and are effectively "implanted," a small fraction interacts with the nuclei of the atoms comprising the sample via nuclear Coulomb (or Rutherford) scattering. This scattering process is essentially elastic so that both energy and momentum are conserved between the He projectile and the target (sample) atom. Since the probability for scattering of the projectile nuclei is small, it is reasonable to assume that the detected scattered alpha particles are not associated with multiple scattering events.

The probability with which a single collision takes place between the incident He particle and a constituent atom in the sample to yield a scattering event that will be detected at the scattering angle θ is related to a quantity known as the differential scattering cross section $d\sigma/d\Omega$,

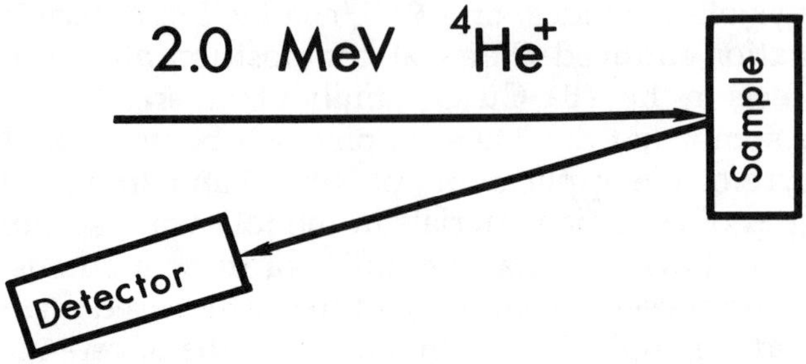

Fig. 1. The configuration typically used in RBS analysis is shown. A beam of 2.0 Mev ${}^4He^+$ ions is scattered from the component nuclei of the ceramic sample (typical beam diameters range from 0.5 to 2.0 mm). The ${}^4He$ particles lose energy as they penetrate the sample, and hence, the scattered energy depends on the depth at which the scattering event takes place.

where $d\sigma$ represents the differential solid angle subtended by the detector. With respect to the center-of-mass system of the incident and target particles, the formula derived by Rutherford for the differential scattering cross section is given by:[1]

$$(d\sigma/\,d\Omega) = \left[\frac{Z_1 Z_2 e^2}{4E\sin^2(\theta/2)}\right]^2 \tag{1}$$

where Z_1 and Z_2 are the atomic numbers of the projectile (e.g., He) atom and the target atom in the sample, respectively; e is the electronic charge; E is the energy of the projectile when the scattering event occurs; and θ is the scattering angle in the center of mass system (i.e., the angle between the forward direction of the projectile and its scattered direction). In the laboratory frame of reference, the scattering angle is generally large as the term "backscattering" implies, and a scattering angle of 160° or more is frequently used. As indicated by Eq. 1, the scattering cross section depends on the square of the atomic number of the target atom so that relatively high scattering yields will be associated with high Z elements.

When an incident particle with mass m and energy E_{in} is scattered near $\theta = 180°$ from an atom with mass M on the surface of the sample, the energy of the backscattered particle E_s can be approximated by the expression:

$$E_s \approx \left[\frac{M - m}{M + m}\right]^2 \cdot E_{in} \tag{2}$$

The quantity E_s, therefore defines the "leading edge" of an RBS spectrum for the target element with mass M, since it represents the highest energy scattered particle that will be detected for a given sample element. For the frequently employed case of 2.0-MeV ^{4}He backscattered from surface atoms, examples of leading edge energies are ~1.87, 1.43, and 0.72 MeV for U, Ti, and O atoms, respectively.

As the incident projectile penetrates the sample, it progressively loses energy either through electronic interactions with the atoms traversed in the target or through small-angle collisions with the target nuclei.[1] For the case of light projectiles such as He or H, the dominant contribution to the energy loss of the projectile arises from electronic interactions with the target atoms, and the contribution from nuclear interactions is only significant near the end of the projectiles' path when its energy is very low. Taking the case of 2.0 MeV ^{4}He ions as an example, those particles that penetrate the sample a distance x before being scattered will lose energy prior to (and after) the scattering event. These particles will, therefore, be scattered with an energy whose value is less than E_s. The energy loss is essentially proportional to x so that the energy of the scattered particle is approximated by $E \approx E_s - S \cdot x$ where S is the proportionality factor. Accordingly, the RBS spectrum from, for example, a single element such as uranium will consist of a number of counts representing the detection of particles scattered from U atoms on the sample surface (i.e., the detection of scattered He particles with an energy $E_s \approx 1.87$ MeV) plus counts at successively lower energies representing He particles scattered from different depths within the uranium sample. The number of scattered particles detected (or counts) at a given energy is proportional to the concentration of the target nuclei either at the surface or at some corresponding depth within the sample. The RBS spectrum for the present example of a polycrystalline sample of uranium is represented schematically in Fig. 2. From Eq. 1 it is known that the scattering cross section varies as $1/E^2$ so that the number of counts or yield will not be perfectly flat with respect to the backscattered energy since the energy of the scattered particles decreases with increasing depth in the sample. Accordingly, a characteristic $1/E^2$ "tail" will occur at the lowest detected energies as illustrated in Fig. 2. The idealized uranium spectrum shown in Fig. 2 essentially represents a "map" of uranium concentration (as represented by the number of backscattered particles or counts) as a function of depth of uranium below the surface (as represented by the energy of the backscattered particles with the most energetic particles representing scattering from uranium atoms on the surface of the sample). For the case of 2-MeV ^4He particles, the concentration can generally be profiled to a depth of ~2–3 μm with an optimum depth resolution of ~50 Å.

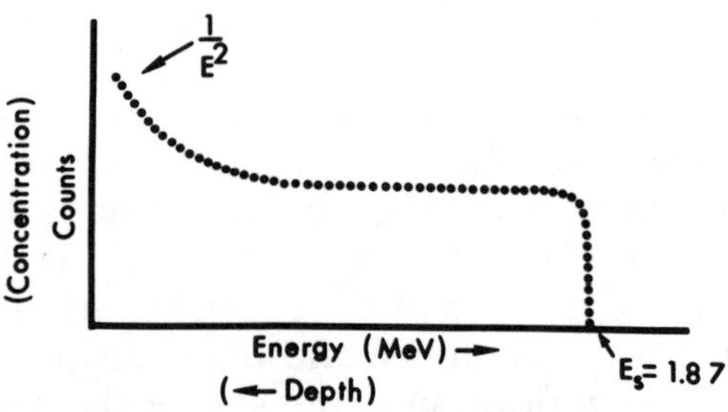

Fig. 2. Representative RBS spectrum for 2.0-MeV $^4He^+$ particles incident on a polycrystalline sample of uranium. The spectrum is effectively a map of uranium concentration vs depth.

The example shown in Fig. 2 is, of course, only representative of the RBS spectrum from a simple, single-component specimen. In such a case there are no concentration variations, and the backscattered yield simply increases monotonically and slowly until the $1/E^2$ tail becomes very apparent at low-scattering energies. A schematic spectrum that illustrates the situation corresponding to a multicomponent sample with a concentration of the constituents that is unform in depth is shown in Fig. 3. This spectrum is effectively a cumulative superimposition of the RBS spectra corresponding to the four different components in this case. The differences in the "step" heights occurring at the varying E_S values of the components would correspond to differences in the relative concentration of the components. It is important to note that in a multicomponent system, variations in the concentration of any given component in the near-surface region of the sample will be reflected by either increases or decreases in the backscattered yield associated with that component. It may be difficult to interpret such variations, however, if the leading edges of different components are very close in energy.

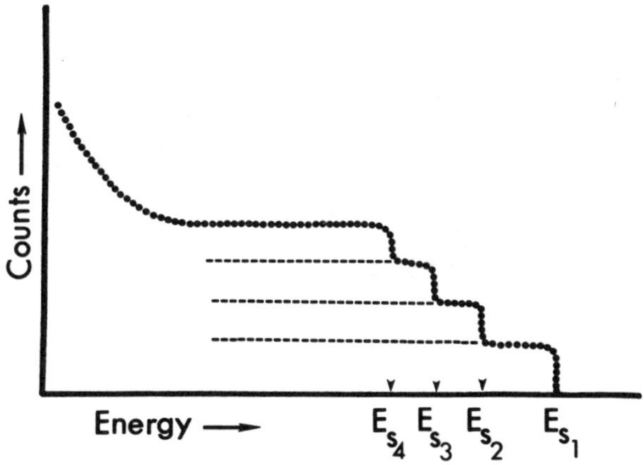

Fig. 3. Idealized RBS spectrum from a random polycrystalline-multicomponent specimen. The energies, E_S, correspond to scattering from atoms on the surface of the sample for the components 1 to 4. As shown in the figure, the counts from the various components are progressively cumulative for energies below the surface energy of each component.

The above brief discussion has introduced the basic concepts of RBS as applied to random polycrystalline or amorphous materials. In the case of macroscopic single-crystal samples, an additional effect referred to as "channeling" can be observed under certain conditions. These effects arise as a result of correlated interactions with the ordered planes or rows of atoms in the structure of a single crystal. In this case the incident projectile particles are effectively guided down "open" directions or "channels" in the crystal structure with the net effect of significantly reducing the number of sample or target atoms available for a scattering event. In this scenario, the incident particles travel a relatively large distance in the crystal before they are scattered so that the backscattered yield is greatly reduced relative to that corresponding to a random polycrystalline material with the same composition. Since channeling effects occur when the single-crystal specimen is oriented relative to the incident beam such that the atoms in the sample are effectively lined up in rows along the beam direction, the atoms on the surface present a relatively large cross section and their associated yield produces a so-called "surface peak" in the channeling spectrum. The nature of the effects of channeling on RBS spectra is shown in the example of Fig. 4.

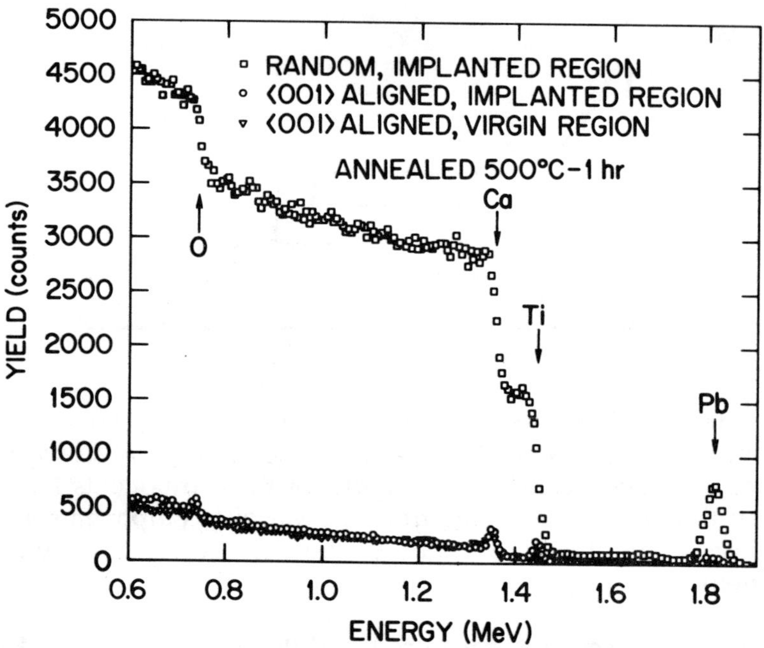

Fig. 4. A comparison of the "random" spectrum of a $CaTiO_3$ single crystal implanted with lead and two "channeling" spectra representing the aligned yield from an unimplated region of the crystal and the aligned yield from the implanted region after the implantation damage was removed by a thermal anneal. The small Ti, Ca, and O "surface peaks" can be seen in the channeling spectra.

The "random" spectrum shown in Fig. 4 for the $CaTiO_3$ single crystal was obtained by rotating the crystal in the He beam while the spectrum was accumulated. This rotation effectively randomizes the incident beam with respect to various directions in the single crystal and is performed in a manner that avoids channeling directions or planes. The peak in the random yield at an energy of about 1.8 MeV in Fig. 4 is due to the presence of the implanted lead impurity (^{208}Pb, 250 keV, $4 \times 10^{15}/cm^2$). This random yield can be compared with that at the same backscattered energy in the channeling or aligned spectrum for the implanted, thermally annealed crystal. The low yield from the lead implant in the channeling spectrum indicates that most of the lead occupies substitutional sites in the $CaTiO_3$ lattice following the thermal anneal. The use of channeling techniques to investigate implantation effects in single crystals will be discussed in more detail in a subsequent section. As discussed in detail in Ref. 1, channeling effects are very important in backscattering analysis since they can provide information related to the degree of distribution in depth of disorder in the crystal lattice, they can be used to determine the thickness and composition of disordered surface layers on crystals, and they can be used to determine where impurity atoms are located in the host crystal lattice.

RBS INVESTIGATIONS OF GLASS CORROSION

As discussed above, RBS can be used to determine the compositional variations that occur in the near-surface regions of multicomponent materials—that is, depth profiles of the elemental composition of materials including glasses can be obtained. This ability has been utilized in a number of studies of corrosion processes in nuclear waste glasses.[2,3] Nuclear waste glass corrosion represents a topic of considerable technological importance since the corrosion process can result in the release of radioactive waste elements into ground waters. The RBS spectra obtained from a uranium-doped borosilicate glass are illustrated in Fig. 5.[3] The spectrum shown at the bottom of the figure was obtained for the as-prepared borosilicate glass. The surface yield steps for the various glass components are indicated at the top of the figure, and the spectrum qualitatively resembles the idealized example shown in Fig. 3. The sharp peak that occurs near the silicon leading edge is due to a thin aluminum film that was deposited on the sample to prevent any effects due to charging of the glass sample in the He ion beam used in the RBS analysis. This spectrum (marked 0.0 h) is indicative of uniform concentrations of the glass components as a function of depth below the glass surface. When the glass is exposed to water at 90°C for only 30 min, the significant compositional alterations shown in the spectrum marked 0.5 h occur. The spectrum shows that the U, Ti, and Ca concentrations are significantly enhanced at the surface of the glass after the short "leaching" period. When a U-doped glass specimen is exposed to 90°C distilled water for 3 h, the associated RBS spectrum denoted as 3.0 h in Fig. 5 is obtained. In this case, the RBS results show a large enhancement in the near-surface concentration of uranium that now extends deeper into the leached glass surface. Significant enhancements in the Ti and Ca concentrations are also evident. The yield corresponding to the silicon surface edge is seen to be significantly lower in this case indicating a loss of Si from the leached surface. A conversion of the backscattering energy into a depth scale using an established computational algorithm

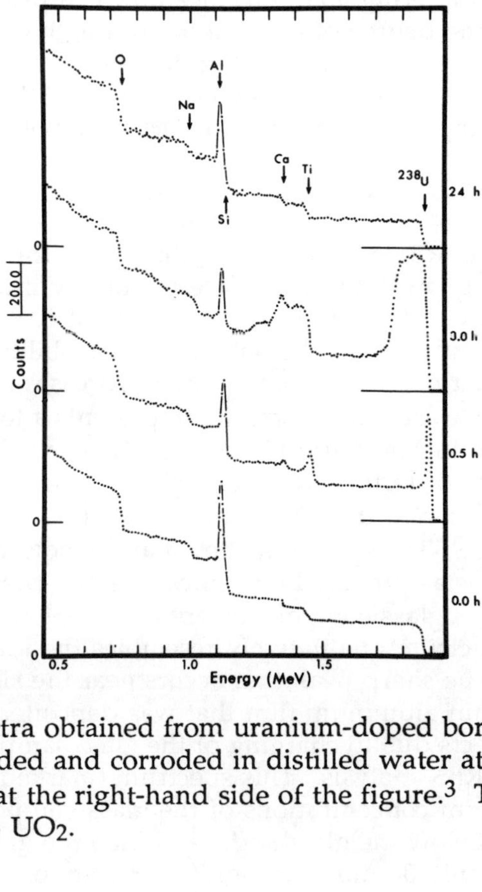

Fig. 5. RBS spectra obtained from uranium-doped borosilicate glasses that are uncorroded and corroded in distilled water at 90°C for the times indicated at the right-hand side of the figure.[3] The glass contained 5 wt% UO_2.

shows that the leached glass surface layer extends to a depth of over 2000 Å. The top RBS spectrum shown in Fig. 5 corresponds to the surface of a glass specimen that was leached in 90°C H_2O for a period of 24 h. A comparison of this spectrum with that obtained for the unleached glass shows that the near-surface properties are essentially identical for the two cases. This result was obtained because the pH of the fixed volume of leaching solution increased during the 24-h time period to a pH value of about 10. At this elevated pH value, the glass surface is simply dissolved congruently, and the altered surface layers associated with the RBS results shown for 0.5 and 3.0 h are not stable. RBS results of this type proved to be instrumental in clarifying the processes and mechanisms associated with the corrosion of glasses in aqueous media.

RBS techniques combined with ion implantation have been applied to direct determinations of the kinetics of glass corrosion.[4] This method is illustrated by the results shown in Fig. 6. In this case, a "marker" represented by the implant profile of a heavy element (i.e., bismuth) is

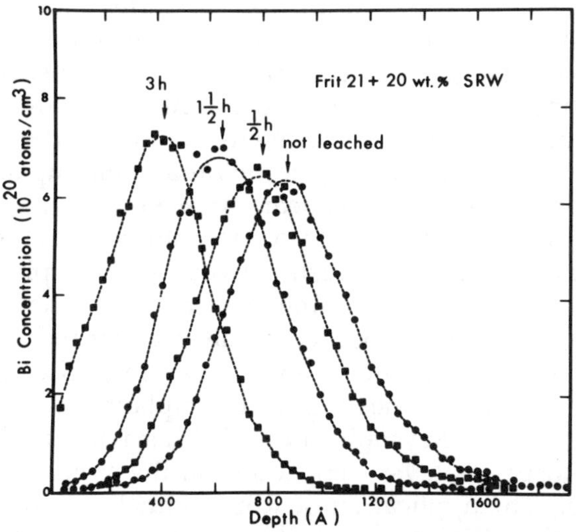

Fig. 6. RBS-determined depth profiles for bismuth implanted at 250 keV into borosilicate glass samples. The Bi-profile peaks at a depth of 850 Å in the as-implanted glass. When the glasses are exposed to H_2O for 0.5, 1.5, and 3.0 h, the surface dissolves away and the peak in the profile moves toward the glass surface.

placed below the surface of the glass sample by ion implantation. RBS is then used to depth profile the implanted Bi concentration and determine the depth of the peak of the profile below the glass surface. This profile is labeled "not leached" in Fig. 6. The implanted glass specimens are then exposed to the H_2O leachate for 0.5, 1.5, and 3.0 h at 90°C. As the glass surface is removed by dissolution, RBS can be used to track the movement of the peak in the implant profile toward the surface as a function of leaching time. These measurements provide a direct determination of the glass dissolution kinetics. For the example shown in Fig. 6 (i.e., a borosilicate frit 21 glass containing 20 wt% of a simulated defense nuclear waste),[4] the matrix dissolution rate was found to be 1.1 g/m^2d by using this combined implantation and RBS analysis method.

CHANNELING STUDIES OF ION BEAM DAMAGE AND REGROWTH

RBS channeling techniques have been used to study the effects of heavy ion damage resulting from the implantation of various elements into single crystals of insulating oxides such as $CaTiO_3$, $SrTiO_3$, and Al_2O_3.[5] These investigations have shown that it is possible to recover the crystallinity of the ion-damaged near-surface region of these crystalline oxides by a suitable thermal annealing procedure. The conceptual outline of these experiments is illustrated in Fig. 7. As shown at the top of Fig. 7, a low-to-moderate ion implantation dose is used to produce an amorphous near-surface layer on a principal cyrstallographic face of the single-crystal sample. (It should be noted that RBS alone cannot distinguish between an amorphous layer and a completely random polycrystalline layer and the amorphous nature of the ion-damaged surface is best established by means of the examination of a cross-sectional sample by transmission electron microscopy.) Thermal annealing of the implanted sample can be used to produce epitaxial recrystallization of either part or all of the originally amorphous layer depending on the annealing times and temperatures. The application of RBS channeling, rather than simply obtaining random RBS spectra in this situation, is important since the "channeling" yield will only decrease in the thermally regrown portion of the surface layer if the regrowth takes place epitaxially—that is, if the atoms in the regrown region are in exact crystallographic register with those of the underlying single-crystal "substrate" as represented by the undamaged portion of the single crystal.

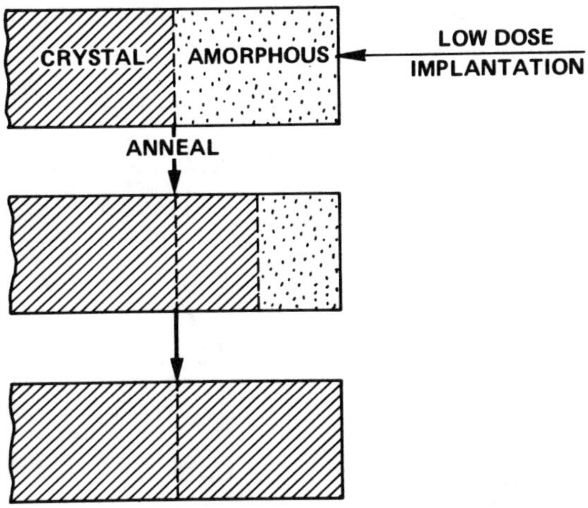

Fig. 7. Ion implantation of some single-crystal complex oxides can be used to form an amorphous near-surface layer (top). A suitable choice of thermal annealing time and temperature can produce epitaxial regrowth of part of this layer (middle) or eventually complete epitaxial recrystallization of the layer (bottom).

Figure 8 shows a comparison of the channeling spectrum obtained from a single crystal of Pb-implanted $CaTiO_3$ (250 keV, 1×10^{15} ions/cm^2) with a rotating random spectrum of the same region. The channeling spectrum clearly shows the effects of the disorder produced by the implantation, and the channeling yield in the titanium sublattice reaches the random value—indicating complete disorder in this sublattice. Similarly, the yield from the Ca and O sublattices shows complete disorder. The channeling spectrum obtained following a thermal anneal for 30 min at 350°C is given by the open triangles in Fig. 8. The reduction in backscattered yield in this channeling spectrum, together with the movement of the low-energy edge of the damage peaks to higher energy, shows that slightly over half of the originally amorphous ion-damaged region has regrown by a process of epitaxial crystallization. The spectrum corresponding to the inverted open triangles was obtained by channeling into an unimplanted region of the sample. Complete epitaxial regrowth can be achieved—as shown previously in the example provided by Fig. 4.

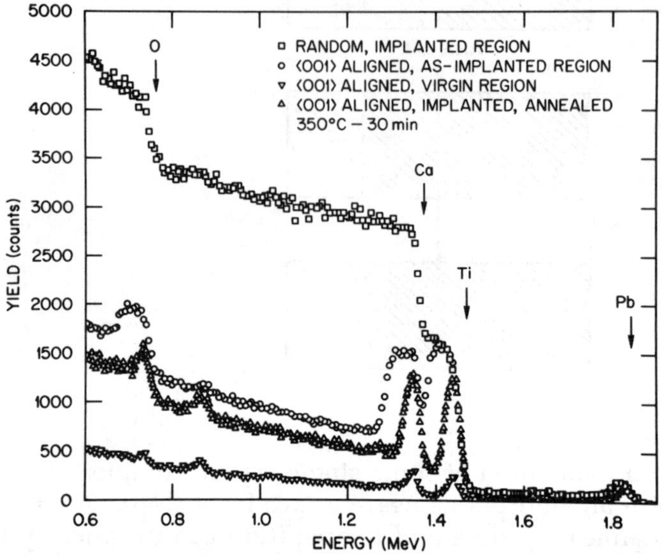

Fig. 8. The channeling spectra obtained for an as-implanted $CaTiO_3$ crystal (1×10^{15} Pb ions/cm^2 at 250 keV) (open circles) can be compared with that following an anneal for 30 min at 350°C. Random (implanted) and aligned (unimplanted or virgin) results are shown for comparison.

By carrying out carefully controlled thermal anneals of implanted single crystals, it is possible to determine the epitaxial regrowth velocity as a function of annealing temperature and thereby determine the activation energy for the epitaxial regrowth process.[5] An example of a determination of this type for regrowth in ion-damaged single crystals of $CaTiO_3$ is shown in Fig. 9. $CaTiO_3$ is an orthorhombic crystal with a so-called pseudo-cubic (or high-symmetry) direction and two orthogonal low-symmetry directions. As shown in Fig. 9, the data provided by channeling analysis have established that the activation energy for regrowth is different for the two types of symmetry directions.

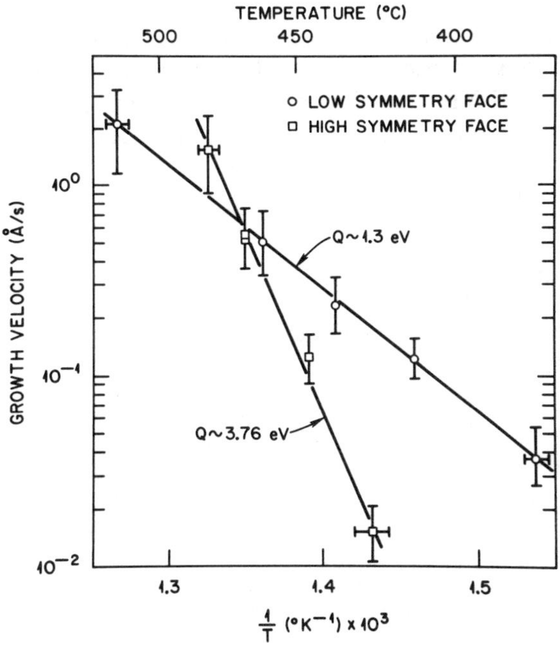

Fig. 9. Channeling techniques have been used to determine the crystallization kinetics of ion-beam-damaged surface layers on the high- and low-symmetry directions of $CaTiO_3$.

RBS ANALYSIS OF IMPLANTATION AND ANNEALING OF $YBa_2Cu_3O_x$

Both RBS and channeling techniques have recently been applied to investigations of the effects of ion implantation and thermal annealing on the properties of high-temperature ceramic oxide superconductors.[6] These efforts have followed a course that is somewhat similar to that described above for the case of complex insulating ceramic crystals, and attempts have been made to remove ion-implantation-induced damage in these superconducting materials by carrying out thermal anneals. An example of some of the results of these studies is provided by Fig. 10. This figure illustrates the combined effects of ion implantation of $YBa_2Cu_3O_x$ and subsequent thermal annealing at 500°C for 4 h in an atmosphere of pure oxygen. The figure compares the random spectra of an as-grown crystal (open circles) with that of an implanted-annealed sample (closed circles). The combined effects of the implantation and annealing have obviously resulted in large alterations in the chemical composition of the near surface of the $YBa_2Cu_3O_x$ sample. Subsequent studies have shown that it is necessary to anneal implanted $YBa_2Cu_3O_x$ ceramics at temperatures in the range of 850°C in O_2 in order to remove the implantation-induced damage.

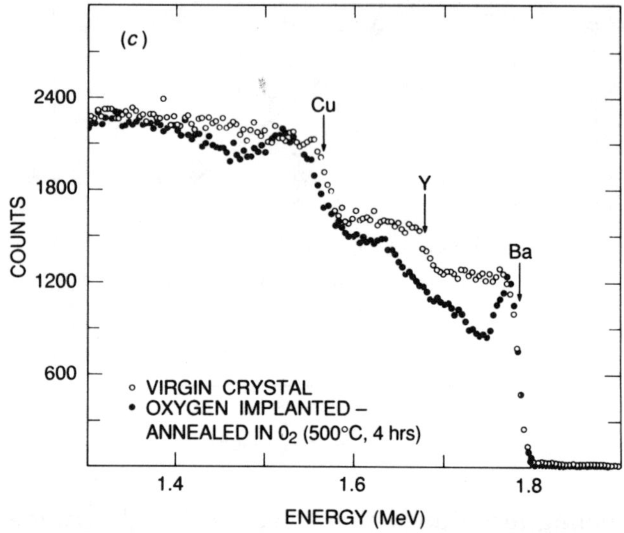

Fig. 10. The random spectrum for an as-grown single crystal of $YBa_2Cu_3O_x$ is shown (open circles) along with the random spectrum for a crystal that has been implanted with 3×10^{16} O ions/cm^2 at 80 keV. Compositional alterations are clearly evident in the implanted and annealed specimen.

SUMMARY

The techniques of Rutherford backscattering and channeling spectroscopy have been introduced and shown to be versatile and powerful methods for the study of a variety of ceramic materials. Examples of the applications of these methods to the study of the corrosion of nuclear waste glasses and to the investigation of ion-beam-induced damage and thermal regrowth of complex oxides have been presented along with an additional example of the use of RBS techniques to investigate implantation and annealing effects in the ceramic oxide superconductors of the type $YBa_2Cu_3O_x$.

Research sponsored by the Division of Materials Sciences, U.S. Department of Energy, under contract DE-AC05-84OR21400 with Martin Marietta Energy Systems, Inc.

REFERENCES

1. W. -K. Chu, J. W. Mayer, and M. -A. Nicolet, *Backscattering Spectrometry*, 1st Ed., Academic Press, Inc., New York, 1978.
2. B. C. Sales, C. W. White, G. M. Begun, and L. A. Boatner, "Surface Layer Formation on Corroded Nuclear Waste Glasses," *J. Non-Cryst. Solids* **67**, 245–64 (1984).
3. B. C. Sales, L. A. Boatner, H. Naramoto, and C. W. White, "Rutherford Backscattering Investigations of the Corrosion of Borosilicate Nuclear Waste Glass," *J. Non-Cryst. Solids* **53**, 201–26 (1982).
4. B. C. Sales, C. W. White, and L. A. Boatner, "A Comparison of the Corrosion Characteristics of Synthetic Monazite and Borosilicate Glass Containing Simulated Nuclear Defense Waste," *Nucl. and Chem. Waste Manage.* **4**, 281–89 (1983).
5. C. W. White, L. A. Boatner, P. S. Sklad, C. J. McHargue, J. Rankin, G. C. Farlow, and M. J. Aziz, "Ion Implantation and Annealing of Crystalline Oxides and Ceramic Materials," *Nucl. Instrum. and Methods Phys. Res. Sect. B* **32**, 11–22 (1988).
6. J. C. McCallum, C. W. White, and L. A. Boatner, "Ion Implantation and Thermal Annealing of Single Crystals of the Type $YBa_2Cu_3O_x$," *Mater. Lett.* **6**, 375–78 (1988).

Section II. Bulk Techniques

DYNAMIC X-RAY DIFFRACTION:A TECHNIQUE FOR FOLLOWING SOLID STATE REACTIONS

W. J. Thomson, Department of Chemical Engineering, Washington State University, Pullman, Washington

ABSTRACT

The hardware and software associated with the use of "Dynamic X-Ray Diffraction" (DXRD) to quantitatively monitor the progress of solid state transformations and reactions is described. The technique, which is compatible with temperature programs as high as 1700 K and gas pressures up to 125 kPa, has been applied to a number of solid-solid and gas-solid systems. In particular, DXRD results are presented for phase transitions in zirconia as well as the "123" superconductor, the formation of mullite from sol gel precursors and the firing of thick film copper conductors.

INTRODUCTION

High temperature powder x-ray diffraction is a technique which has been in use for decades[1]. Its primary use has been in the determination of high temperature solid state phase diagrams; in other words, for equilibrium measurements. What is described here is a powder x-ray diffraction technique which is useful for kinetic studies of solid state reactions and transformations. Of course, in order to follow kinetic phenomena, the measurements themselves must have a dynamic capability and thus the technique has been termed - "Dynamic X-Ray Diffraction," or DXRD. Although the technique is capable of following solid state reactions in general, because of its unique ability to monitor solid-solid reactions, there are many applications for DXRD in ceramics research.

First of all, DXRD is not a new instrument; it merely combines the high resolution of modern-day diffractometers with the revolution which has occurred in rapid digital data acquisition. In fact most of our effort over the past three years has been devoted to the development of the necessary software and to overcoming the obstacles which stood in the way of obtaining reasonably accurate and repeatable data. Another important point is that DXRD is not a "stand-alone" measurement or characterization device. For example, amorphous events during the course of solid state transformations will go undetected by DXRD. Thus it is always best to run parallel experiments using DTA or DSC (TGA if appropriate) in order to corroborate the DXRD measurements. However, for crystalline events, the distinct advantage of DXRD over DSC or DTA is its ability to obtain phase specific information and thus discriminate between simultaneous solid state reactions and/or transformations.

THE DXRD TECHNIQUE

Basic Strategy

The best way to describe the basic strategy of obtaining kinetic data from DXRD measurements is to consider the following solid-solid reaction scheme

$$A_{(s)} + B_{(s)} \longrightarrow C_{(s)} + D_{(g)} \qquad (1)$$

$$A_{(s)} + C_{(s)} \longrightarrow E_{(s)} \qquad (2)$$

What we have here is a series/parallel reaction network involving two original solid phase reactants (A & B), a solid phase intermediate (C), a solid phase product (E) and one gas phase product (D). IN DXRD we would have a prior knowledge of which species to expect (by pre-reaction if necessary) and would have already chosen the diffraction peaks we wish to monitor. Usually these are the major reflections but it is sometimes necessary to choose minor reflections in the event of peak overlapping. Upon initiation of the reaction, the computer is programmed to sequentially move to the peaks of interest (A, B, C, and E) and scan at a given resolution and at a predetermined rate. The latter is dictated by a desirable signal-to-noise ratio and the rapidity of the solid state reaction. The diffraction data along with time and temperature, are dumped to hard disks and then retrieved at the end of the experiment for data reduction and analysis. In our early work[2] it was necessary to spend some time deciding on the minimum resolution required for reliable peak integration in order to be able to follow fast reactions. However, with the addition of a Position Sensitive Detector (PSD), we can easily obtain accurate integrated data with high resolution (.01 deg. 2θ) every 15-20 seconds. In fact, for very fast reactions such as the tetragonal-orthorhombic phase transition in the "123" superconductor (which will be discussed below), we can use DXRD in a "fast scan" mode. That is, the PSD "window" can be opened to 6 degrees, and diffraction data collected without moving the goniometer. If there are multiple peaks within that window, three for example, then we can collect data at a maximum rate of three integrated peaks every 5 seconds. Thus we are able to discriminate between the three reactions listed above. Note that TGA would only record the extent of reaction (1) and, if one or more of the reactions occurred at the same time, neither DSC or DTA would be able to discriminate between them.

Description of the Hardware

The salient features of the DXRD system are shown in Figures 1 and 2. The heart of the system is a Siemens D500 diffractometer which is equipped with a numerically controlled goniometer under instructions from a DACO microprocessor. Diffracted x-rays are counted by a Position Sensitive Detector (PSD) which is capable of scanning at rates as high as 100 deg/min. Diffraction data are passed through the DACO unit to a PDP 11-23 (plus) computer which is interfaced with the campus main frame. When operated in the "fast scan" mode, the output of the multi-channel analyzer is sent directly to a PC/XT in order to eliminate unnecessary data handling associated with the software which interfaces the DACO to the PDP.

Figure 1 shows a sketch of a "flow through" hot stage, which is an HTK-10 unit manufactured by Anton-Parr, but modified to accomodate an optical pyrometer. The gas atmosphere can be controlled, but maximum pressures are limited to about 125 kPa. A metallic strip serves as both a sample holder and a heater and we typically use Inconel which is good to about 1200 K in air. Platinum is used when it is necessary to run at higher temperatures in an oxygen environment (< 1400 K) and tantalum is used for high temperature work (up to 1750 K in an anoxic environment). However, it is here where the most development was necessary. The problem is two-fold: the prevention of solid movement during goniometer movement and the measurement of accurate solid temperatures. For most systems the former can be solved by applying the solid sample as an acetone slurry. The difficulty in obtaining accurate sample temperatures arises from the large temperature gradient across the sample bed (as much as 150 K at high temperature operation). Since the x-rays usually penetrate to about 100 μm, and samples this thin do not generate a sufficiently high count rate for fast reactions, we have opted for a long wave length pyrometer (12 μm) in order to measure the exposed sample temperature. While this works very well, it is of course necessary to have accurate emissivity data for the solid sample. This is no easy task since temperatures in a given experiment can range over hundreds of degrees and reaction products may not have the same emissivity as the starting materials. We have found it best to obtain these measurements *in-situ*, and the details are discussed by Anderson[3].

Description of the Software

The details of the three years of software development which has been put into the DXRD technique cannot be presented here but it is instructive to mention its capabilities and to summarize the approach which has been taken. The basic software is capable of doing the usual things to diffraction data: baseline subtraction, smoothing (by fast Fourier transforms), peak integration and three-dimensional graphical output of the DXRD data. Examples of the output are shown below. However, because our prime interest is kinetics, we have spent a considerable effort at developing a methodology to do quantitative DXRD analysis; that is, to be able to calculate phase specific solid concentrations as a function of time.

The approach which is used depends on whether we are dealing with a homogeneous reaction or with a reaction that is "layered;" i.e., when the reaction product forms a distinct outer layer as in the case of many oxidation reactions. There are two basic methods for dealing with homogeneous systems: the *Internal Standard Method* and the *External Standard Method*[4]. When using the internal standard method, a known concentration of an inert material is added to the sample and concentrations of the species of interest can be obtained by comparing their diffraction intensities to those of the inert. Of course it is necessary to have prior calibrations of the relative strengths of the diffraction intensities (Reference Intensity Ratios, RIR) and to be sure that there is no overlap of the peaks of interest with those of the inert. As long as the concentration of the inert remains constant (it may not in gas-solid reactions), this method has the distinct advantage of being independent of the quantity of amorphous material in the sample. However, when the internal standard method cannot be used, then it is necessary to employ the calculational methods corresponding to the external standard method. In this case a set of equations described by equation (3)

$$\sum_{i=1, i \neq j}^{n} v_i \mu_i + v_j \mu_j \left(1 - \frac{I_{jp}}{I_j}\right) = 0 \qquad (3)$$

must be solved for the concentrations of each species and these require a knowledge of the absorption coefficients (μi) of all the species as well as their RIR values. Anderson[3] describes how these equations can also be solved to account for the presence of amorphous materials. When dealing with a layered system it is more appropriate to measure the dynamic nature of the thickness of the product layer. Anderson and Thomson[5] describe an improved technique for making these calculations from DXRD data and show experimental results for the high temperature oxidation of zirconium.

Limitations

Naturally there are limitations with DXRD, just as with any experimental technique. Some of these have already been mentioned but it is useful to summarize them here. The primary limitation of DXRD is its inability to detect and follow amorphous events. While modern materials synthesis technology is heading in the direction of nanometer and molecular sized precursors[6]; processing temperatures are still high enough to result in crystallization. Indeed, DXRD is very useful for quantifying crystallization rates. Another limitation is inherent in powder diffractometry; viz., its inability to detect low concentrations (5 % is a "rule-of-thumb"). Finally, accurate temperature measurements are still the single most difficult measurements in a DXRD experiment. Although we have managed to overcome this problem by using optical pyrometry, the lack of accurate data on the emissivities of materials as a function of both temperature and wave length necessitate tedious, in-situ calibrations. Fortunately most ceramic materials have similar emissivities at high temperatures; otherwise one would have to also account for the time varying behavior of the sample emissivity as reactions occur.

APPLICATIONS

It should now be apparent that DXRD is useful in a wide variety of applications. Although we have only used it as a research tool, it is also a valuable aid in materials processing; particularly for process optimization. Many of the new ceramic materials are formed from two or more precursors which must react and/or transform in multistep sequences (the processing of the 123 superconductor is a good example). Failure to achieve complete reaction at any one of these steps can have serious consequences in terms of the properties of the final product. Since DXRD can mimic any set of temperature-time processing conditions , it can be used to determine whether or not the process conditions are optimal for a given objective. Nevertheless, the remainder of this manuscript describes a few of the research areas where we have successfully applied DXRD and serves to illustrate its utility in a research environment.

Phase Transitions

The Tetragonal-Monoclinic Zirconia Phase Transition

The importance of the tetragonal-monoclinic phase transition in partially stabilized zirconia is well known[7,8]. Since transformation toughening is really a matter of the *rate* of this transition, DXRD represents an ideal method by which to study the attendant phenomena. Figures 2 and 3 show the DXRD results for heating and cooling, respectively. This points out another advantage of DXRD; namely, that it can follow reversible transitions. This can be extremely important in those cases where the act of "quenching" a sample can result in unobserved transitions prior to subsequent analysis. Notice that there is a rather large hysteresis in this transition (approximately 175 K), characteristic of its martensitic nature[9]. This points to still another possible application of DXRD; i.e., the measurement of more accurate phase diagrams.

The Tetragonal-Orthorhombic Transition in The 123 Superconductor

The importance of the tetragonal-orthorhombic phase transition and its dependence on temperature and oxygen partial pressure, has been well documented[10,11]. In the early studies of this transition, indirect methods of analysis (such as TGA[12]) were employed. Later on, neutron diffraction studies was able to establish the position of oxygen in the lattice structure as a function of temperature[13]. Recently, however, we have applied DXRD to the problem and obtained unambiguous, in-situ measurements of the lattice constants as a function of both temperature and oxygen pressure[14], and an example of these results is shown in Figure 4.

Solid-Solid Reaction Kinetics

Mullite Synthesis

Mullite and its formation from silica and alumina has been the subject of study for a good number of years[15,16]. Traditional methods of synthesis start with finely ground powders and, after a series of phase transitions, the final product is formed by high temperature (1800 K or higher) solid state diffusional processes[17]. However, when the precursors are derived from sol-gels or inorganic polymers, the formation temperatures are considerably lower and the reaction rates may or may not be diffusion limited. We are currently using DXRD in an attempt to quantify the rates of all the steps leading to the ultimate formation of mullite from sol-gel derived powders. Some of these results are shown in Figure 5 which shows the alumina phase transitions as well as the formation of mullite in a non-isothermal experiment between 1275 and 1475 C. One interesting result here, is that separate isothermal experiments indicate that it is possible to have high mullite conversions without forming α-alumina. This is an example of how DXRD can aid in process optimization.

Gas-Solid Processing

The Firing of Thick-film Conductors

The final example to be presented here is an application in electronic materials processing; viz., the firing of thick-film conductor pastes. Thick-films are generally used to build up hybrid circuits in a multilayer configuration and the precursor "pastes" are deposited on a suitable substrate by screen printing. The pastes are carefully tailored with organic carriers in order that they have the necessary rheological properties[18]. Organic burnout, substrate binding, and sintering is then achieved by firing at temperatures in the neighborhood of 1100 K. In the case of copper conductor pastes, firing in air (to effect organic burnout) can seriously affect the final conductivity since the copper can be oxidized in the process. Firing in nitrogen, on the other hand, can produce a reducing environment which may lead to reduction of the binding agents (bismuth oxide, for example). The goal is then to control the atmosphere as well as the temperature-time history of the firing process in order to maximize organic burnout while minimizing copper oxidation. In other words, it is a kinetics problem. Figure 6 shows the DXRD results for an isothermal experiment conducted in air at 200 C. The oxidation of copper is clearly seen in the reduction of the (111) copper peak and the appearance of the (111) Cu_2O peak. It is interesting to note that thermodynamic considerations would dictate the formation of CuO, *not* Cu_2O. In fact, at higher temperatures we observe the subsequent oxidation of Cu_2O to CuO.

CONCLUSIONS

During our development of DXRD we have discovered many interesting materials processing problems to which we could apply the technique with considerable success. There is still some room for improvement in the software, particularly in the quantitative calculation of phase specific concentrations. The development of the "fast scan" mode of operation now allows for very rapid data acquisition and is probably adequate for most conceivable applications in materials processing. However, additional development of the hardware is definitely needed. In particular, better methods of temperature measurement and/or heater design would be a considerable improvement. The utilization of a theta-theta goniometer would not only solve some of the sample movement problems, but it could also lead to the extension of the technique to more fluid-like systems. The technique does suffer, of course, from the disadvantages inherent to powder diffractometry; the inability to detect amorphous events or low solid state concentrations. Nevertheless, DXRD opens the way for materials science to begin to concentrate on the *kinetics* of materials transformations in addition to traditional thermodynamic approaches.

REFERENCES

1. S.S. Iyengar, P.Engler, M.W. Santana, and E.R. Wong, "Advances in X-ray Analysis", 28, 331-338, C.S. Barret, P.K. Predecki and D.E. Leyden, Eds., Plenum: New York (1985)

2. D.E. Anderson and W.J. Thomson, "Solid State Kinetic Measurements Using Dynamic X-ray Diffraction", Ind. Eng. Chem. Res., 26, 1628-1632, (1987)

3. D.E. Anderson, "The Application of Dynamic X-ray Diffraction to Limestone Sulfation", PhD Dissertation, Washington State University, (1988)

4. B.D. Cullity, "Elements of X-ray Diffraction", 2nd ed., Addisson-Wesley: Reading, Ma. (1978)

5. D.E. Anderson and W.J.Thomson, "Layer Thickness Determination with X-ray Diffraction", in press, J. Appl. Crystal. (1988)

6. I.A. Aksay, in "Advances in Ceramics", 9, J.A.Mangels and G.L.Messing, eds., Am. Ceram. Soc., Columbus, Ohio, (1984)

7. G.M. Wolten, "Diffusionless Phase Transformations in Zirconia and Hafnia", J. Am. Ceram. Soc., 46, 418-422, (1963)

8. C.Grain and R.Garvie, "Mechanism of the Monoclinic to Tetragonal Transformation in Zirconium Dioxide", U.S. Bur. Mines Rept. Inv., No. 6619, (1965)

9. I.Chen and Y.Chiao, "Martensitic Transformations in ZrO_2 and HfO_2 - An Assessment of Small Particle Experiments with Metal and Ceramic Matrices", Adv. Ceram, 12, 33-45, (1984)

10. K.Kishio, J.Shimoyama, T.Hasegawa, K.Kitazawa and K.Fueki, "Detrmination of Oxygen Nonstoichiometry in a High T_c Superconductor $Ba_2YCu_3O_{7-}$ Jpn.J.Appl. Phys.,26,l1228-1230 (1987)

11. R.J.Cava, B.Batlogg, R.B. van Dover, D.W.Murphy, S.Sunshine, T.Siecrist, J.P.Remeika, E.A.Rietman, S.Zahurak and G.P.Espinosa, "Bulk Superconductivity at 91K in Single Phase Oxygen-Deficient Perovskite $Ba_2YCu_3O_{9-}$,Phys. Rev. Lttrs., 58, 1676-1679 (1987)

12. Y.Kubo, Y.Nakabayashi, J.Tabuchi, T.Yoshitake, A.Ochi, K.Utsumi, H.Igarashi and M.Yonezawa, "Determination of the Tetragonal-Orthorhombic $YBa_2Cu_3O_{7-}$ Phase Boundary in the δ - T Diagram", Jpn. J. Appl. Phys.,26,l1888-1891 (1987)

13. J.D.Jorgensen, M.A.Beno, D.G.Hinks, L.Sodeholm, K.J.Volin, R.L.Hitterman, J.D.Grace and I.K.Schuller, "Oxygen Ordering and the Orthorhombic-Tetragonal Phase Transition in $YBa_2Cu_3O_{7-x}$", Phys. Rev.B,36(7), 3608-3616 (1987)

14. H.Wang, D.X.Li and W.J.Thomson, "Transition Temperatures For the Tetragonal-Orthorhombic Phase Transition in the 123 Superconductor", in press, J.Am.Ceram.Soc. (1988)

15. W.L.de Keyser, "Reactions at the Point of Contact Between SiO_2 and Al_2O_3", Science of Ceramics,2, 243 (1965)

16. W.Staley and G.Brindley, "Development of Noncrystalline Material in Sibsolidus Reactions between SiO_2 and Al_2O_3", J.Am.Ceram.Soc.,52, 616-619 (1969)

17. I.Aksay and J.Pask, "Stable and Metastable Equilibria in the System Al_2O_3-SiO_2", Am.Ceram.Soc.Bull., 9, 710 (1973)

18. R.Vest, "Materials Science of Thick Film Technology", Ceram. Bull., 65, 631 (1986)

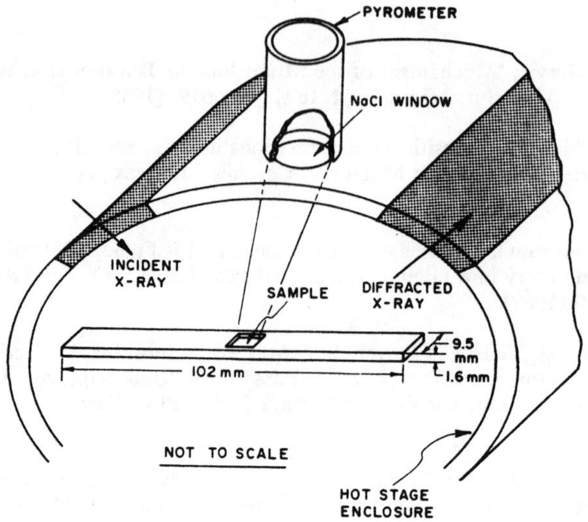

Fig. 1: Modified Hot Stage

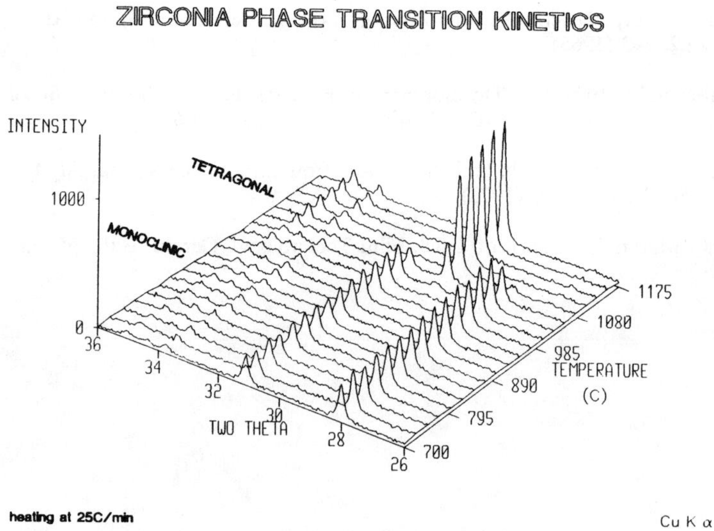

Fig. 2: Monoclinic-Tetragonal Phase Transition in Zirconia (Heating at 25 K/min)

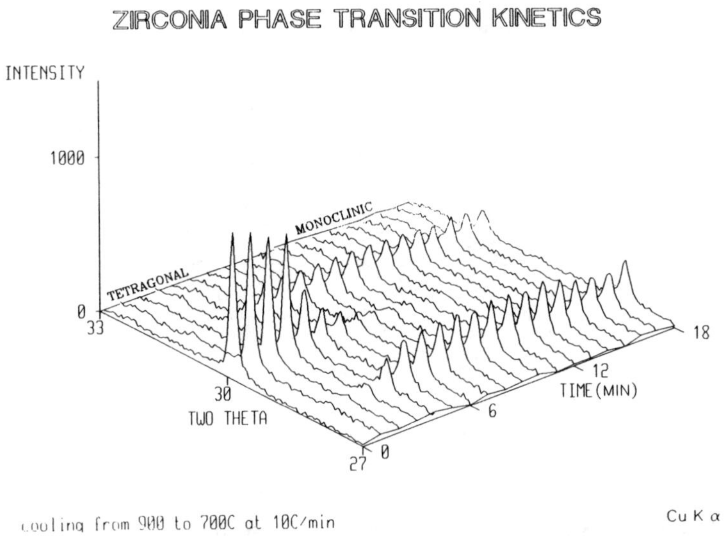

Fig. 3: Tetragonal-Monoclinic Phase Transition in Zirconia (Cooling at 20 K/min)

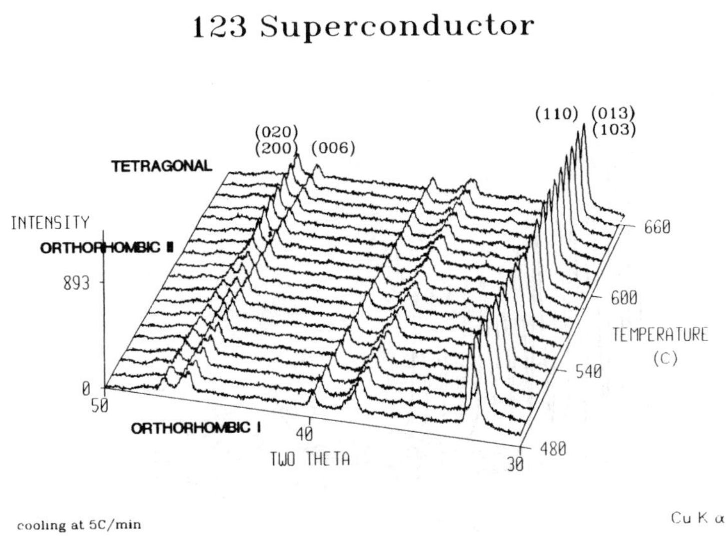

Fig. 4: Reversible Tetragonal-Orthorhombic Phase Transition in the 123 Superconductor (Cooling at 5 K/min)

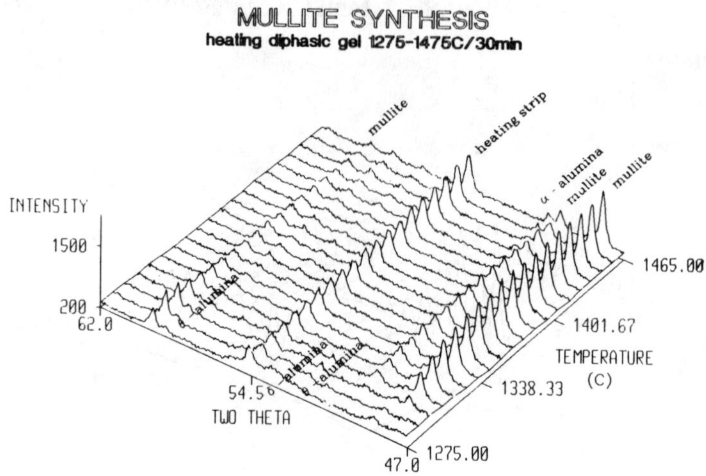

Fig. 5: Mullite Synthesis from a Diphasic Gel (Heating at 10 K/min)

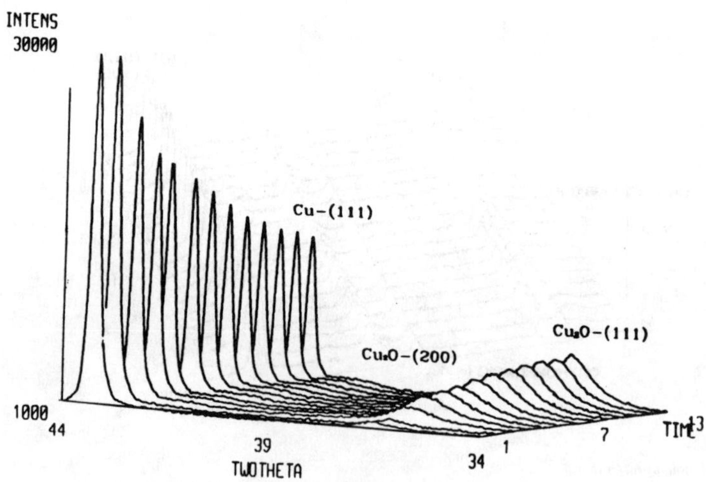

Fig. 6: Copper Oxidation in a Thick Film Conductor Paste (in air at 475 K)

MICROSTRUCTURAL CHARACTERIZATION OF $CuGaSe_2$ THIN FILMS BY X-RAY DIFFRACTION

D.Albin, R.Noufi, and J.Tuttle
Solar Energy Research Institute
1617 Cole Blvd., Bldg. 16/2
Golden, CO 80401

S.H. Risbud
Dept. of Materials Science and Engineering
University of Arizona
Tucson, AZ 85721

ABSTRACT

Photovoltaic devices utilizing Cu based chalcopyrite compounds fabricated by multi-source evaporation require a Cu poor/near stoichiometric bilayer structure for high efficiency.[1] XRD analysis of peak shifts for films deposited on Corning 7059 glass and polycrystalline alumina substrates provide conclusive evidence of a chalcopyrite to sphalerite disordering in films with Cu-poor compositions. Films deposited on molybdenum coated alumina substrates were chalcopyrite regardless of composition. Comparison of x-ray diffraction (XRD) data with a computer simulation of suspected defect chemistries suggests intra-granular phase separation of the cubic and tetragonal phases, with the latter retaining a near stoichiometric composition.

INTRODUCTION

Early photovoltaic devices were essentially single crystal or large grained due to the detrimental effects of grain boundaries. Today that trend continues in a variety of elemental (silicon) and compound (III-V) materials systems. Recently, in order to reduce costs associated with processing materials of such high structural perfection and purity, good efficiency devices have been fabricated in fine grained polycrystalline form. These devices are typically configured as heterojunctions, utilizing a wide bandgap window, such as CdS (Eg=2.4 eV) or ZnO (Eg=3.2 eV), coupled with a narrower bandgap absorbing material such as $CuInSe_2$ (Eg=1.0 eV) or $CuGaSe_2$ (1.7 eV). Thin film devices based on these latter materials have very recently exhibited efficiencies in excess of 11% (938 cm^2) and 14% (3.7 cm^2).[2]

In this paper we present a microstructural study of $CuGaSe_2$ thin films deposited by multi-source evaporation from the constituent elements. In doing so we shall concentrate on results obtained by XRD with emphasis on crystallographic structure, defect chemistry, and phase separation.

BACKGROUND

The tetragonal chalcopyrite structure of $CuGaSe_2$ is a derivative of the II-VI zincblende structure and consists of two interpenetrating face centered cubic lattices; one lattice consisting of Se and another comprised of an ordered array of Cu and Ga. The resulting superlattice typically has two structural effects for chalcopyrites in general: an internal displacement of the anion and a distortion of the c-axis from an ideal cubic value of c=2a. Lattice parameter measurements show a strong correlation between this latter distortion and the relative cation sizes.[3] An increase in atomic radii of the periodic group I atom and/or decrease in radii of the group III atom will generally result in larger distortion.

Phase studies of $CuGaSe_2$ [4] and the related material $CuInSe_2$ [5,6] have traditionally been restricted to studies involving binary systems with most emphasis placed on the $(I_2VI)_x(III_2VI_3)_{1-x}$ pseudobinary where x is defined as the molecularity. A characteristic trait of these materials is the room temperature existence of a one phase tetragonal region for values in x slightly less than stoichiometry (x=0.5). In the opposite sense, an increase in x should result in secondary phase formation, in this example, $Cu_{2-x}Se$. At higher temperatures, the cation lattice will disorder thus forming the sphalerite phase. This transformation temperature appears to be related to the internal anion position [7], which in turn reflects the relative strengths of the two cation-anion bonds.

X-ray diffraction studies for room temperature single crystal [8] and polycrystalline [9,10] $CuGaSe_2$ can be found in the literature with no studies of the high temperature disordered phase being reported. Theoretical intensity calculations incorporating the effect of suspected defect chemistries within the one phase chalcopyrite region have also been reported.[11] Allowed reflections for the chalcopyrite structure can be placed into any of three major classifications.[12] Group (i) reflections are sphalerite-like and have Miller indices given by (h,k,l/2) unmixed. Group (ii) reflections are dependent only upon the anion scattering factor and have indices given by (h,k) even (l/2) odd or vice versa. Group (iii) reflections involve both an anion term as well as a cation ionicity term related to the difference in cation scattering factors. Reflections of the latter variety are often referred to as superlattice peaks. In the case of ideal anion placement, group (ii) peaks become extinct as does the anion contribution to group (iii) reflections.

A typical powder pattern for chalcopyrite $CuGaSe_2$ is shown in Figure 1. The most obvious characteristic is the peak splitting of various group (i) peaks due to the c-axis distortion. The magnitude of this splitting increases with increasing tetragonal distortion and is the most readily visible observation of ordering in this material. Also present (although not visible in the scale used) is the (101) superlattice peak which is of of relatively low intensity in $CuGaSe_2$ due to the similar scattering power of Cu and Ga. In $CuInSe_2$ these trends are reversed; the dissimilar nature of Cu and In results in higher intensity group (iii) reflections while the larger group III atom, and consequently negligible tetragonal distortion results in barely resolvable peak splitting. Peak splitting and superlattice peaks have therfore formed a basis by which ordering in these materials have been studied in the past.

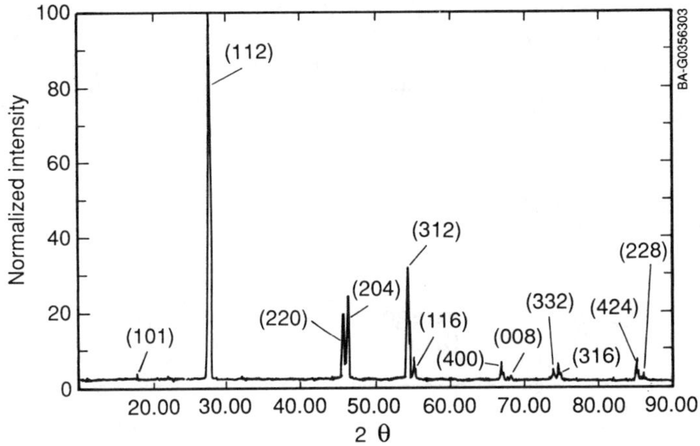

Fig. 1: Typical Chalcopyrite $CuGaSe_2$ X-ray Diffraction Pattern

EXPERIMENTAL PROCEDURES

Polycrystalline thin films of $CuGaSe_2$ were deposited by multi-source evaporation of the elements in vacuum using a previously described method.[11] Typical deposition fluxes for the Cu, Ga, and Se sources as referenced to cold substrate calibration were: 0.5-1.5, 2.8-6.0, and 12-14 Å/sec, respectively. Resulting film growth rates varied from 4.5 to 11.0 Å/sec. Rate control was obtained by an EIES (electron impact emmision spectroscopy) monitor for the Cu-Ga fluxes and a quartz crystal oscillator for the Se source. Substrates included Corning 7059 glass, polycrystalline alumina, and molybdenum coated (3 um) alumina, with substrate temperatures in the range 350 to 500°C. Film thickness was measured using an alpha-step surface profilimeter and varied from 200-300 nm for optical studies up to 1-2 um for general XRD measurements. XRD measurements were made on a Rigaku Dmax vertical goniometer and Cu rotating anode with computer control and data aquisition. Operating conditions of 40 kV and 60 mA in conjunction with a step width of .02 degrees and 1 sec/step counting were typical. Films were mounted on a special holder to eliminate substrate thickness effects.

Compositional analysis of films was obtained by electron probe microanalysis at 15-20 keV with a 20 um spot size and typically 50 nA beam current. Both polycrystalline and single crystal standards were used. Compositional measurements were accurate to ± 0.2 at.%.

RESULTS AND DISCUSSION

X-ray data for the high temperature, cubic sphalerite phase of $CuGaSe_2$ was not available. Consequently, it was necessary to calculate the cubic powder pattern. This was accomplished by replacing Cu and Ga with an atom of averaged atomic scattering and assuming that tetragonal distortion was limited to the <001> directions. It was also necessary to use the isotropic temperature factors of the tetragonal phase.[8] Anomalous dispersion data were taken from Cromer's compilation.[13] Table I compares the normalized powder intensities for representative group (i) and group (iii) reflections. Group (ii) reflections are absent due to an undistorted Se lattice.

Table I. Calculated XRD Normalized Intensities for the Cubic and Tetragonal Phases of $CuGaSe_2$

Tetragonal (hkl)	Cubic (hkl)	Tetragonal (I/Io)	Cubic (I/Io)	Tetragonal 2θ	Cubic 2θ
(101)	-	.147	0.00	17.73	-
(112)	(111)	100.0	100.0	27.69	27.52
(220)	(220)	21.8	64.3	45.71	45.71
(204)	(202)	42.2	64.3	46.14	45.71

Differences in XRD data, useful in phase studies, can be discussed in terms of whether the effect is dependent or independent of intensity. For instance, intensity differences clearly exist between all common (hkl) values and are especially important for group (iii) reflections where total extinction occurs in the cubic phase. An intensity independent difference is observed when comparing the values of 2θ for a given (hkl) pair. This shift in 2θ increases with increasing tetragonal distortion and can be determined for all reflections using the relationship given in reference [11]. Intensity independent comparisons are useful when dealing with thin films due to the likelihood of preferred orientation during film growth. Using the technique presented by Lotgering [14] for studying orientation effects, an orientation figure was constructed for $CuGaSe_2$ films deposited on glass substrates as a function of composition and substrate temperature and is shown in Figure 2. It is clear that a truly random powder orientation is never achieved and consequently phase analysis by intensity measurements may be impractical. The preferred orientation shown in Figure 2 can be expected to be even more severe when using polycrystalline substrates.

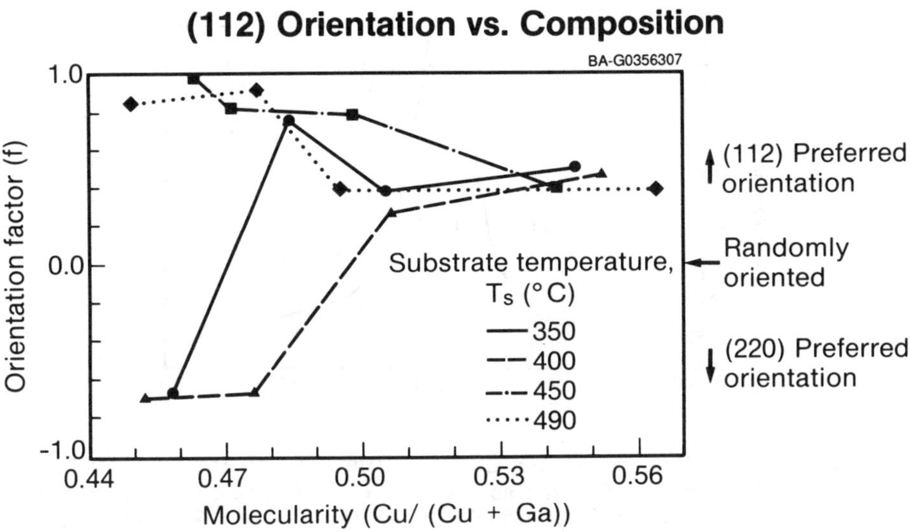

Fig. 2: Orientation Plot of $CuGaSe_2$ vs. Composition and Substrate Temperature Deposited on 7059 Glass.

Fortunately, intensity independent peak shifts could be used for studying the phase formation behavior of $CuGaSe_2$ as a function of composition, substrate type, and temperature. The magnitude of this shift is shown in Figure 3 for Cu-poor compositions (20-22 at.% Cu) deposited on various substrates. It was generally observed that Cu-poor films would tend to be disordered when deposited on either glass or alumina while films deposited on molybdenum coated alumina were always highly chalcopyrite. The effect of substrate temperature on this order-disorder existence is shown in Figure 4 for both thin and thick films deposited on glass. Higher substrate temperatures appear to sharpen the transition between the phases for thinner films on glass. Also shown in this figure, as well as the alumina sample of Figure 3, is the occurrence of 2θ values between the extremum points representing the pure phases. This effect is undoubtedly due to cubic-tetragonal phase mixing and is pronounced when using alumina substrates.

The delineation between Cu poor/cubic and stoichiometric-Cu rich/tetragonal material was also substantiated by optical measurements made on thin films deposited on glass.[15] Band splitting by crystal field and spin-orbit effects were observed at 1.741 and 1.932 eV in stoichiometric and Cu rich films with a fudamental optical direct gap of 1.662 eV. Cu poor material exhibited a single valence subband at 1.992 eV with a fundamental gap of 1.718 eV.

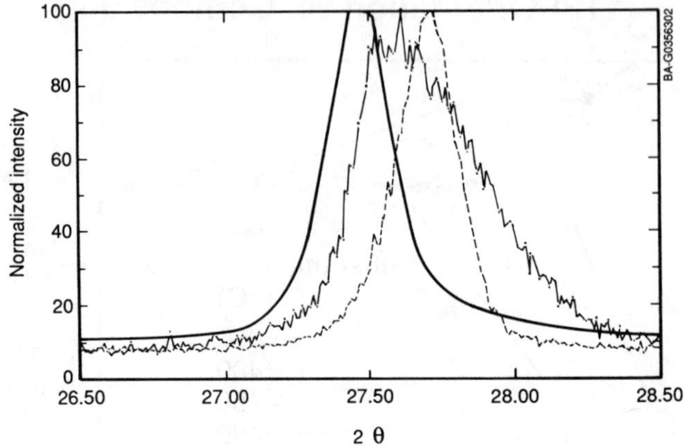

Fig. 3: (112) Peak Scans for Cu Poor Films (Substrate T ≃ 450°C) on Different Substrates (——— :21.96/25.43/52.61 glass; —·— :20.59/27.40/52.01 alumina; — — — :22.79/25.79/51.42 moly/alumina).

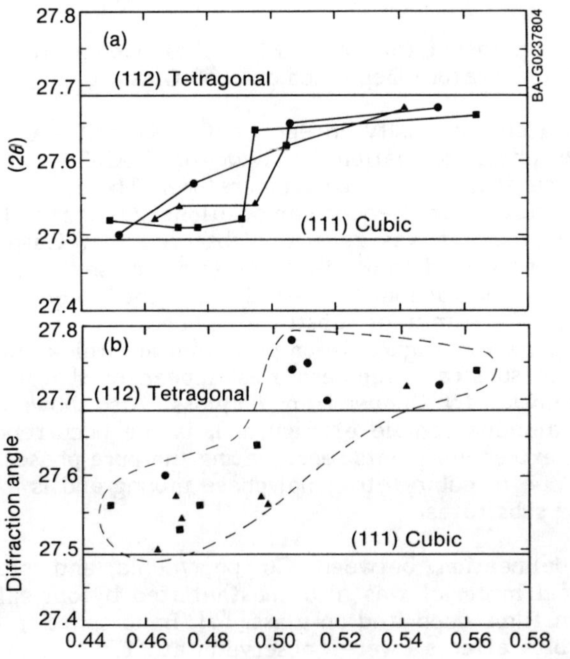

Fig. 4: (112) Peak Shift for $CuGaSe_2$ Deposited on 7059 Glass (a - 2000-3500 nm thick; b - 1-2 um thick; ● - 400°C, ▲ - 450°C, ■ - 490°C).

Useful information with regards to microstructure can further be obtained by comparison of single crystal and polycrystalline data. As mentioned previously, chalcopyrite $CuGaSe_2$ exhibits a wide existence range in the Cu poor region. By establishing a relationship between composition and structure within this one phase region, it was shown that the intensities of all group (iii) reflections increases as we make the material more Cu poor.[11] This result was supported by analysis of Cu deficient single crystals. However, in the case of our polycrystalline thin films, no group (iii) reflections were observed even though the films contained both the cubic and tetragonal phases as shown previously. If we assume that the absence of these peaks is not due to orientation effects, we must arrive at an interesting microstructure, i.e., that of compositional segregation between the phases: a tetragonal phase of near stoichiometric composition (therefore the absence of group (iii) peaks), and a cubic phase capable of compositionally adjusting to arrive at the final film composition. Possible variations of this are shown in Figure 5. It is believed that the uniform fluxes present during film growth will favor intragranular separation in this case.

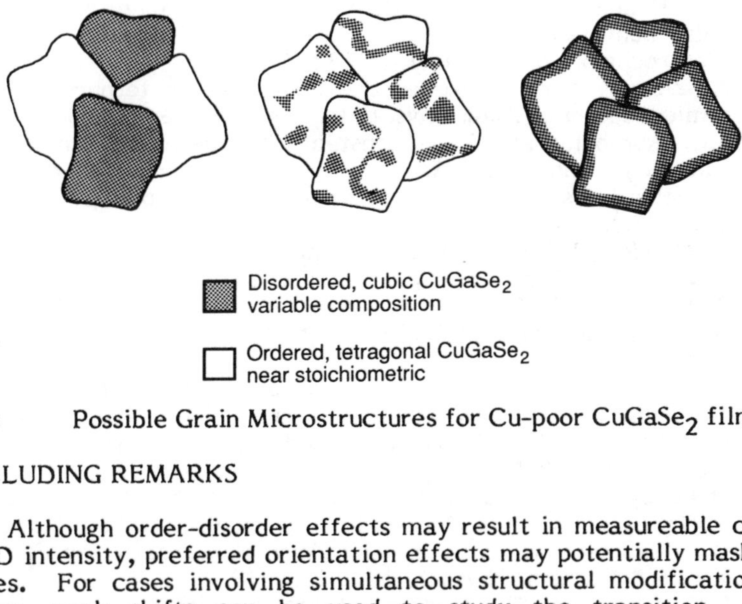

Fig. 5: Possible Grain Microstructures for Cu-poor $CuGaSe_2$ films.

CONCLUDING REMARKS

Although order-disorder effects may result in measureable changes in XRD intensity, preferred orientation effects may potentially mask these changes. For cases involving simultaneous structural modification with ordering, peak shifts can be used to study the transition. These observations are independent of intensity and consequently orientation.

ACKNOWLEDGEMENTS

This work was performed at SERI under contract DE-AC02-83CH10093 with the U.S. Department of Energy. Additional support was provided by Associated Western Universities, Inc. We also acknowledge the assistance of J. Dolan and R. Schwerdtfeger for both technical and experimental help.

REFERENCES

1. R. Noufi and J. Dick, "Compositional and electrical analysis of the multilayers of a CdS/$CuInSe_2$ solar cell," J. Appl. Phys., 58(10) 3884-3887 (1985).
2. ARCO Solar Company cell measured at SERI by K. Emery and D. Deblasio (1988).
3. J.L. Shay and J.H. Wernick, Ternary Chalcopyrite Semiconductors: Growth, Electronic Properties, and Applications, Pergammon Press, 1975.
4. J.C. Mikkelsen, Jr.,"Ternary Phase Relations of the Chalcopyrite Compound $CuGaSe_2$", J. Electron. Mater., 10(3) 541-558 (1981).
5. L.S. Palatnik and E.I. Rogacheva, "Phase Diagrams and Structure of Some Semiconductor $A_2^IC^{VI}$-B_2^{III}-C_3^{VI} Alloys," Dokl., 12(5) 503-506 (1967).
6. J.C. Folmer, J.A. Turner, R. Noufi and D. Cahen, "Structural and Solar Conversion Characteristics of the (Cu_2Se) (In_2Se_3) System, "J. Electrochem. Soc., 132(6) 1319-1327 (1985).
7. A. Zunger, "Order-disorder transformation in ternary tetrahedral semiconductors," Appl. Phys. Lett., 50(3) 164-166 (1987).
8. S.C. Abrahams and J.L. Bernstein, "Piezoelectric nonlinear optic $CuGaSe_2$ and $CdGeAs_2$: Crystal structure, chalcopyrite microhardness, and sublattice distortion," J. Appl. Cryst., 10 130-131 (1977).
9. L. Mandel, R.D. Tomlinson, and M.J. Hampshire, "Crystal data for $CuGaSe_2$," J. Appl. Cryst., 10 130-131 (1977).
10. B. Plenkovic, S. Popovic, B. Celustka, and B. Santic," Crystal data for $AgGa_xIn_{1-x}Se$ and $CuGa_xIn_{1-x}Se$," J. Appl. Cryst., 13 311-315 (1980).
11. D. Albin, R. Noufi, J. Tuttle, J. Goral, and S.H. Risbud, "Composition-Structure Relationships for Multi-Source Evaporated $CuGaSe_2$ Thin Films," to be published in J. Appl. Phys., Nov. 1988.
12. J.E. Jaffe and A. Zunger, "Electronic structure of the ternary chalcopyrite semiconductors $CuAlS_2$, $CuGaS_2$, $CuInS_2$, $CuAlSe_2$, $CuGaSe_2$, and $CuInSe_2$," Phys. Rev. B, 28(10) 5822-5847 (1983).
13. D.T. Cromer and D. Liberman, "Anomalous Scattering Factors for X-rays," J. Chem. Phys., 53 1891-1896 (1970).
14. F.K. Lotgering, "Topotactical Reactions with Ferimagnetic Oxides Having Hexagonal Crystal Structure - I," J. Inorg. Nucl. Chem., 9 113-123 (1959).
15. D. Albin, J. Tuttle, J.Goral, A. Mason, R.Noufi, and S.H. Risbud, "The Optical and Structural Properties of $CuGaSe_2$ Polycrystalline Thin Films," to be published in 20th IEEE Photovoltaics Specialists Conference (1988).

AN IN-SITU, HIGH-TEMPERATURE STUDY OF THE SUPERCONDUCTOR $YBA_{1.67}SR_{0.33}CU_3O_{6+X}$

John Faber, Jr., H. Shaked, B. W. Veal, R. L. Hitterman, A. P. Paulikas,
L. J. Nowicki and J. W. Downey
Argonne National Laboratory
Materials Science Division
Argonne, Il 60439

ABSTRACT

Neutron diffraction is a powerful scientific probe for the study of ceramic materials. Perhaps the most interesting structural studies of ceramics involve analyzing property behavior under in-situ conditions. This technique has been used to study the high-T_c, Sr-doped 123 material, $YBa_{1.67}Sr_{0.33}Cu_3O_{6+x}$ at high temperature. An order-disorder transition is observed at T=650C. Upon heating, the crystal symmetry changes from orthorhombic (Pmmm) below the transition to tetragonal (P4/mmm) above. The mechanism of the phase transition involves the redistribution of anions on the O(1) and O(5) sites. These sites become equivalent in the high temperature phase. In addition, the anions on these lattice sites do not exhibit temperature-dependent thermal vibration behavior, indicating that this behavior is dominated by static displacements due to nonstoichiometric disorder.

INTRODUCTION

Mechanisms responsible for superconductivity in the high T_c oxides are not well-understood. However, it is well-known that the oxygen stoichiometry and ordering of the associated anion vacancies strongly affect the transition temperature to superconductivity.[1] The so-called

123 compound, $YBa_2Cu_3O_{6+x}$, has been the subject of numerous experimental[2-7] and theoretical[4,7-11] investigations. The results of in-situ, neutron diffraction studies[4,8,9,11-12] have provided particularly lucid insight into this compound. The structure of $YBa_2Cu_3O_{6+x}$ can be represented[13] at x=1 by stacking layers of CuO, BaO, CuO_2,Y,CuO_2, BaO, CuO, along the c-axis of the orthorhombic unit cell, as illustrated in Fig.1. The model used to represent this structure assumes that Sr dopant ions substitute randomly on the Ba sites, that the cations vibrate isotropically and that the anion vacancies are localized, but distributed among the O(1) and O(5) sites.

An important approach for characterizing oxygen stoichiometry and associated ordering of anion vacancies is to study the effects caused by controlled substitution of dopant ions on different cation sites. A number of studies have been carried out to test the response of the structure to dopant additions. Surprisingly, a depression in T_c is rarely seen for rare-earth ion substitution on the Y sites in the lattice, even if the substitution is complete and the substituted ion is a rare earth with a large magnetic moment. In contrast to Y ion site behavior, substitutions on the Ba sites (although not as extensively studied) have a much more pronounced effect on T_c. In particular, for Sr-substituted material, there is a depression of T_c of approximately 0.45K/%Sr content.[14]

An orthorhombic to tetragonal phase transition is observed in the pure 123 material, $YBa_2Cu_3O_{6+x}$, with P_{O2}=1atm at T_{ot}=700C; however, the effects of Sr-doping on this transition are not known. In this paper, we will examine the behavior of $YBa_{1.67}Sr_{0.33}O_{6+x}$ in the temperature range 20<T<725C, using in-situ neutron diffraction techniques.

EXPERIMENTAL

The neutron scattering experiments were carried out with sintered, polycrystalline samples of $YBa_{1.67}$,$Sr_{0.33}Cu_3O_{6+x}$, using the high resolution, General Purpose Powder Diffractometer at Argonne's Intense Pulsed Neutron Source.[15-17] Details of the sample preparation and characterization are available elsewhere.[14] Constant scattering angle, time-of-flight (tof) methods are used to obtain the Bragg scattering information with a pulsed, polychromatic, white-beam source. In this case, the relationship between tof and interplanar d-spacings is given by

$$\lambda = hmt/L = 2d\sin\Theta. \quad (1)$$

In this expression, h is Planck's constant, m is the neutron mass, t is the time-of-flight, L is the total path length of the neutron from the source to the detector, d is the d-spacing of interest and Θ is the Bragg angle. For 2Θ=90 degrees, with λ in the range $0.5<\lambda<5$Å, the accessible d-space range is $0.5<d<3.95$Å. The resolution, $\delta d/d$, is 0.004 at this scattering angle. The data are analyzed using well-known Rietveld profile refinement techniques.

A unique, restricted-angle, high-temperature furnace provided a completely isolated sample environment. The restricted angle geometry eliminates interference effects due to scattering from the furnace components. This furnace is ideally suited for scattering experiments on polycrystalline materials. The high-temperature studies on $YBa_{1.67}Sr_{0.33}Cu_3O_{6+x}$ were carried out in the temperature range 20<T<725C, with an oxygen partial pressure, P_{O2}=1atm using a flowing gas over the sample. Temperature gradients at the sample position were <±2.5C.

RESULTS AND DISCUSSION

To compare the room temperature and high temperature structures, Rietveld profile refinement techniques were used. The results of a representative fit are illustrated in Fig. 2. There are approximately 440 Bragg reflections in the d-space range shown in the figure. There were 37 adjustable parameters in the scattering model used to calculate the intensity at 2625 data points, yielding a weighted residual, R_{wp}=0.06, with χ^2=1.9. The agreement between theory and experiment is clearly very good. The thermal expansion behavior of $YBa_{1.67}Sr_{0.33}Cu_3O_{6+x}$ at constant P_{O2}=1atm in the temperature range 550<T<725C is illustrated in Fig. 3. A phase transition is observed at T_{ot}=650C. The unit cell volume (not shown) is observed to be continuous through the transition; the transition is second-order. Although the atmosphere over the sample was maintained at P_{O2}=1atm, the stoichiometry of the sample shifted with increasing temperature. The occupation numbers for the anions on various lattice sites were varied during the structural refinement. At T=550C, the total oxygen content, 6+x=6.7(1), whereas for T=725C, 6+x=6.4(1).

The structural results above the transition are illustrated in Fig. 4. By comparison with Fig. 1, the vibration amplitudes in this figure are much larger, reflecting the increase in vibration with higher temperature. However, this relative increase is not observed for the anions at O(1). The temperature-independent behavior at the O(1) sites clearly demonstrates that displacement of the anions around oxygen vacancies due to nonstoichiometric disorder are more important than the instrinsic disorder due to thermal effects. Again by comparison with Fig. 1, it is clear that an order-disorder process is the driving force for this orthorhombic to tetragonal transition. Below T_{ot}, only the O(1) sites are occupied and the O(5) sites are completely vacant. The anion vacancies form an ordered array on the O(5) sites and the remainder of the vacancies are on the O(1) sites. Above T_{ot}, the anion vacancies are disordered and their population among O(1) and O(5) sites are equal.

To more clearly illustrate the details of the oxygen rearrangements at T_{ot}, Fig. 5 shows the O(1) and O(5) site occupations as a function of temperature. For comparison, the results for the pure 123 compound[12] are also given in the figure. The orthorhombic-to-tetragonal transition is suppressed approximately 50C by the Sr-doping. As with the pure 123 material, at low temperature, the O(5) sites are completely vacant. The nonstoichiometry in both compounds inhibits full O(1) occupation.

CONCLUSIONS

High-temperature, in-situ neutron scattering experiments have been carried out on $YBa_{1.67}Sr_{0.33}Cu_3O_{6+x}$ in the temperature range $20<T<725C$. A second-order, orthorhombic-to-tetragonal phase transition at $T_{ot}=650C$ was observed. The order-disorder transition involves the rearrangement of anion vacancies on the O(1) and O(5) lattice sites. In the low-temperature phase, the crystal has orthorhombic symmetry, and for $T<<T_{ot}$, only O(1) lattices sites are occupied. Above the transition, the symmetry of the lattice is tetragonal. In this phase, the O(1) and O(5) lattice sites are equivalent and the anion vacancies are distributed randomly among these sites. The results of the analyses also show that the total vacancy concentration increases with increasing temperature. The roughly temperature-independent behavior of the vibration ellipsoids at the O(1) sites leads us to conclude that anion nonstoichiometric disorder is the dominant displacement effect at these

sites. The ability to observe these effects with high sensitivity is a direct consequence of the use of neutron scattering techniques.

To quantitatively describe the oxygen content and ordering in these materials, Shaked et. al.[18] have developed a statistical thermodynamic model based on the quasichemical approximation. Two important parameters in this model are the site binding energy for oxygen on the O(1) and O(5) lattice sites and a short-range order parameter defining the probability of near-neighbor (O(1) to O(5)) interactions. The solid lines in Fig.5 are the predictions of this model for the pure 123 material. Both isothermal and constant composition data were required to define the necessary parameters. Additional experiments are needed to apply this model to the Sr-doped material, $YBa_{2-y}Sr_yCu_3O_{6+x}$, in order to define the appropriate energy-related terms in the quasichemical approximation.

ACKNOWLEDGEMENTS

This work is supported by the U.S. Department of Energy, BES-Materials Sciences under contract W-31-109-ENG-38.

REFERENCES

1. J. D. Jorgensen, B. W. Veal, W. K. Kwok, G. W. Crabtree, A. Umezawa, L. J. Nowicki and A. P. Paulikas, "Structural and superconducting properties of orthorhombic and tetragonal $YBa_2Cu_3O_{7-x}$: the effect of oxygen stoichiometry", Phys. Rev. B36, 5731(1987).
2. J. D. Jorgensen, M. A. Beno, D. G. Hinks, L. Soderholm, K. J. Volin, R. L. Hitterman, J. D. Grace, I. K. Schuller, C. U. Segre, K. Zhang and M. S. Kleefisch, "Oxygen ordering and the orthorhombic-to-tetragonal phase transition in $YBa_2Cu_3O_{7-x}$", Phys. Rev. B36, 3608(1987).
3. P. P. Fritas and T. S. Plaskett, Phys. Rev. "High- Temperature order-disorder phase transition in the superconductor $YBa_2Cu_3O_{6+\delta}$ observed by electrical resistivity measurements", B36, 5723(1987).
4. Y. Kubo, Y. Nakabayashi, J. Tabuchi, T. Yoshitake, A. Ochi, K. Utsumi, H. Iagarashi and M. Yonezawa, "Determination of the orthorhombic-tetragonal $YBa_2Cu_3O_{7-\delta}$ phase boundary in the δ-T diagram", Japn. Jour. Appl. Phys. 26, L1888(1987).

5. K. Kishio, J. Shimoyana, T. Hasegawa, K. Kitazawa and K. Fueki, "Determination of Oxygen Nonstoichiometry in a high-Tc superconductor $YBa_2Cu_3O_{7-\delta}$", Japn. Jour. Appl. Phys. 26, L1228(1987).

6. E. D. Specht, C. J. Sparks, A. G. Dhere, J. Brynestad, O. B. Cavin, D. M. Kroeger, H. A. Oye and F. J. Seiler, "Effect of oxygen pressure on the orthorhombic-tetragonal transition in the high-temperature superconductor $YBa_2Cu_3O_x$", Phys. Rev. B37, 7426(1988).

7. W. R. McKinnon, M. L. Post, L. S. Selwyn, G. Pleizier, J. M. Tarascon, P. Barboux, L. H. Greene and W. Hull, "Oxygen intercalation in the perovskite superconductor $YBa_2Cu_3O_{6+x}$", Phys. Rev. B (submitted).

8. H. Bakker, D. O. Welch and O. W. Lazareth, Jr., "Theory of the orthorhombic-tetragonal phase transition in $YBa_2Cu_3O_{7-x}$", Solid State Commun. 64, 237(1987); H. Bakker, J. P. A. Westerveld, D. M. R. Lo Casio and D. O. Welch, "Theory of oxygen content, ordering and kinetics in 1,2,3 high-temperature superconductors", ibid (in press).

9. E. Salomons, N. Koeman, R. Brouwer, D. G. de Groot and R. Greissen, "Pressure-composition isotherms and lattice dilation of $YBa_2Cu_3O_x$", Solid State Commun. 64, 1141(1987).

10. L. T. Wille, A. Berera and D. de Fontaine, "Thermodynamics of oxygen ordering in $YBa_2Cu_3O_z$", Phys. Rev. Lett. 60, 1065(1988).

11. K. Nakamura and K. Ogawa, "Oxygen ordering in $YBa_2Cu_3O_{6+x}$", Japn. Jour. Appl. Phys. (submitted).

12. J. D. Jorgensen, H. Shaked, D. G. Hinks, B. Dabrowski, B. W. Veal, A. P. Paulikas, L. J. Nowicki, G. W. Crabtree, W. K. Kwok and L. H. Nunez, "Oxygen vacancy ordering and superconductivity in $YBa_2Cu_3O_{7-x}$", Physica C153-155, 578(1988).

13. M. A. Beno, L. Soderholm, D. W. Capone II, D. G. Hinks, J. D. Jorgensen, J. D. Grace, I. K. Schuller, C. U. Segre and K. Zhang, "Structure of the single-phase high-temperature superconductor $YBa_2Cu_3O_{7-\delta}$", Appl. Phys. Lett. 51, 57(1987).

14. B. W. Veal, W. K. Kwok, A. Umezawa, G. W. Crabtree, J. D. Jorgensen, J. W. Downey, L. J. Nowicki, A. W. Mitchell, A. P. Paulikas and C. H. Sowers, "Superconductivity in $YBa_{2-x}Sr_xCu_3O_{7-g}$d" Applied Physics Letters, 51, 279(1987).

15. J. Faber, Jr. and R. L. Hitterman, "Neutron powder diffraction at a pulsed neutron source: A study of resolution effects", Advances in X-Ray Analysis 29, Plenum Press, New York, 1985, pp119-130.

16. J. D. Jorgensen and J. Faber, Jr., "Electronically focussed powder diffractometers at IPNS-I", ICANS-VI Meeting, ANL, June 27-July2, 1982, ANL Report ANL-82-80(1983), pp105-114.
17. S. R. MacEwen, J. Faber, Jr. and A. P. L. Turner, "Overview No. 26. The use of time-of-flight neutron diffraction to study grain interaction stresses", Acta Met 31, 657(1983) contains a brief instrument description of the GPPD.
18. H. Shaked, J. D. Jorgensen, J. Faber, Jr., D. G. Hinks and B. Dabrowski, "A theory for oxygen content and ordering in $YBa_2Cu_3O_{6+x}$ in equilibrium with oxygen gas", submitted to Phys. Rev., September, 1988.

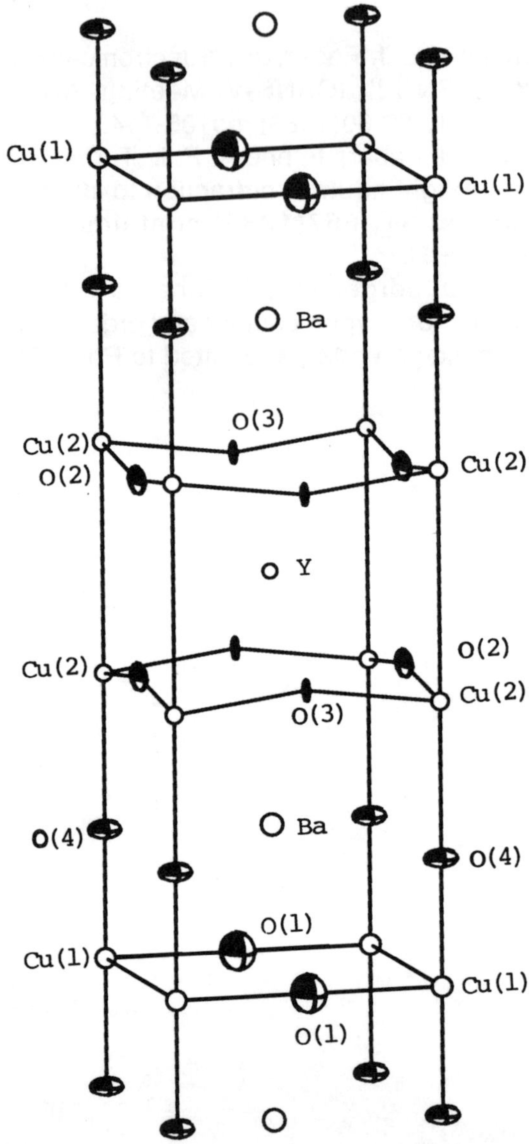

Fig.1. Room temperature, T=20C, orthorhombic (Pmmm) crystal structure of $YBa_{1.67}Sr_{0.33}Cu_3O_{6+x}$, with y=0.33 and x=0.95 determined from neutron powder diffraction experiments. The thermal vibration ellipsoids define contours with 50% probability that the atoms will be found inside these surfaces. The cations are defined by the open ellipsoids and the anions by the segmented ellipsoids.

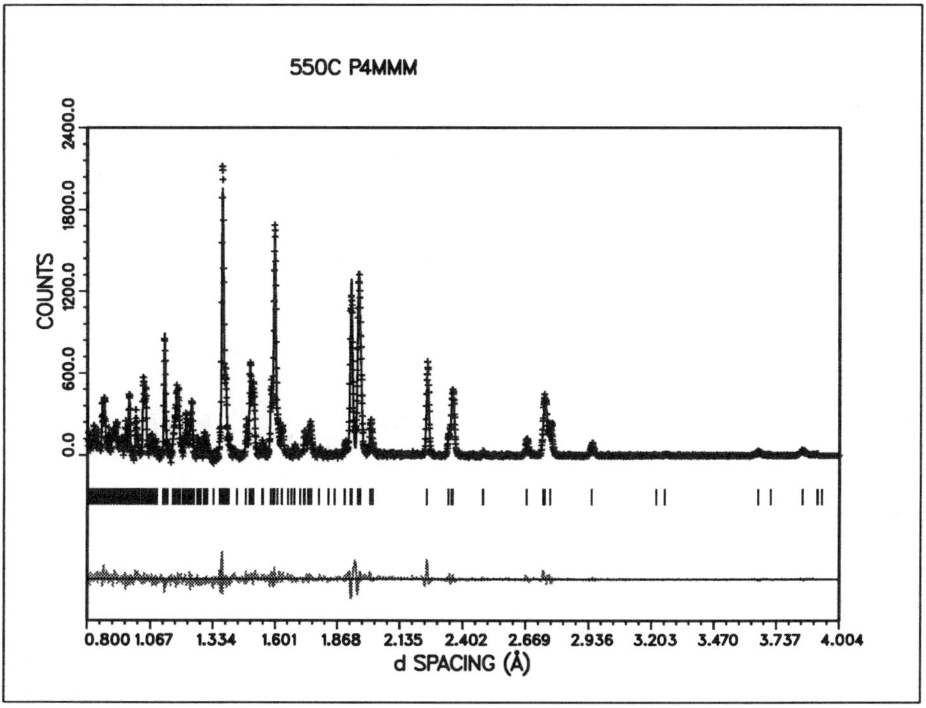

Fig.2. Results of Rietveld profile refinement on $Y(Ba_{1.67},Sr_{0.33})Cu_3O_{6+x}$ at T=550C and P_{O2}=1atm. The GPPD high-resolution powder diffractometer and a special restricted-angle, high-temperature furnace that provides controlled gaseous environments in the sample chamber were employed. The pluses are the experimental data points corrected for background, the solid line drawn through the points represents the least-squares best fit to the data, the tic marks define the positions of the Bragg peaks, and the lower portion of the figure illustrates the point-by-point differences between theory and experiment.

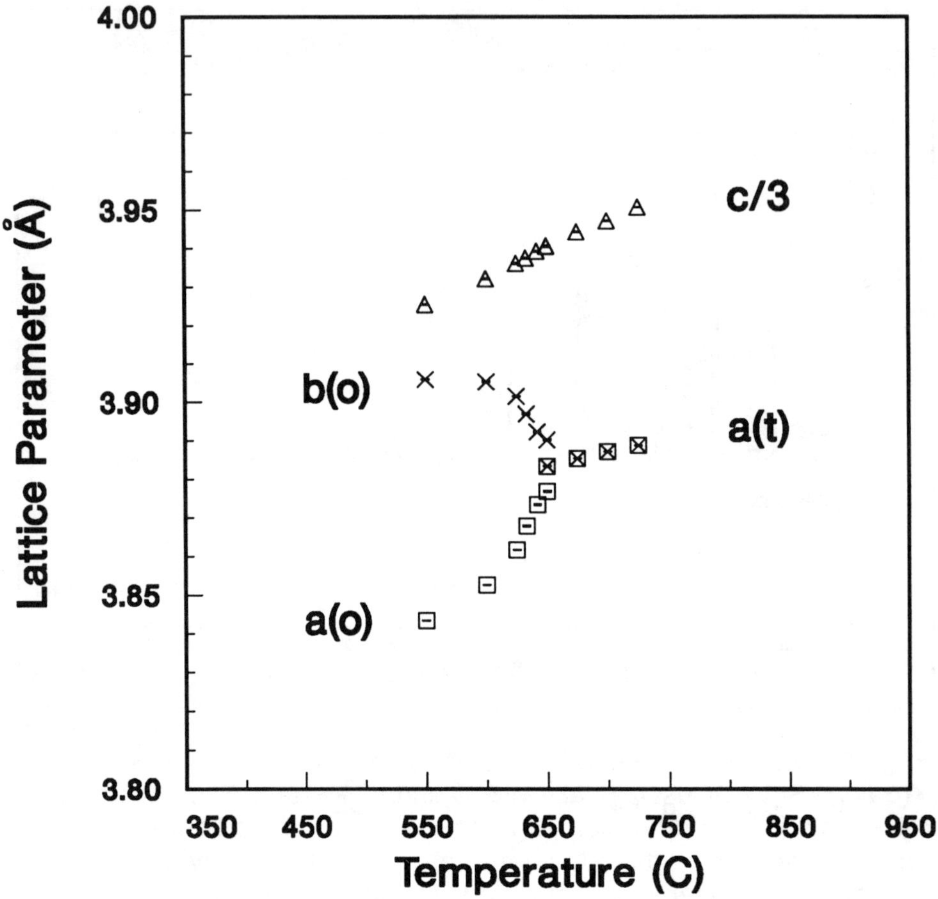

Fig. 3. Thermal expansion of $Y(Ba_{1.67},Sr_{0.33})Cu_3O_{6+x}$ in the temperature range 550<T<725C with P_{O2}=1atm. Note that the c-axis lattice parameter values have been scaled by a factor of 3.

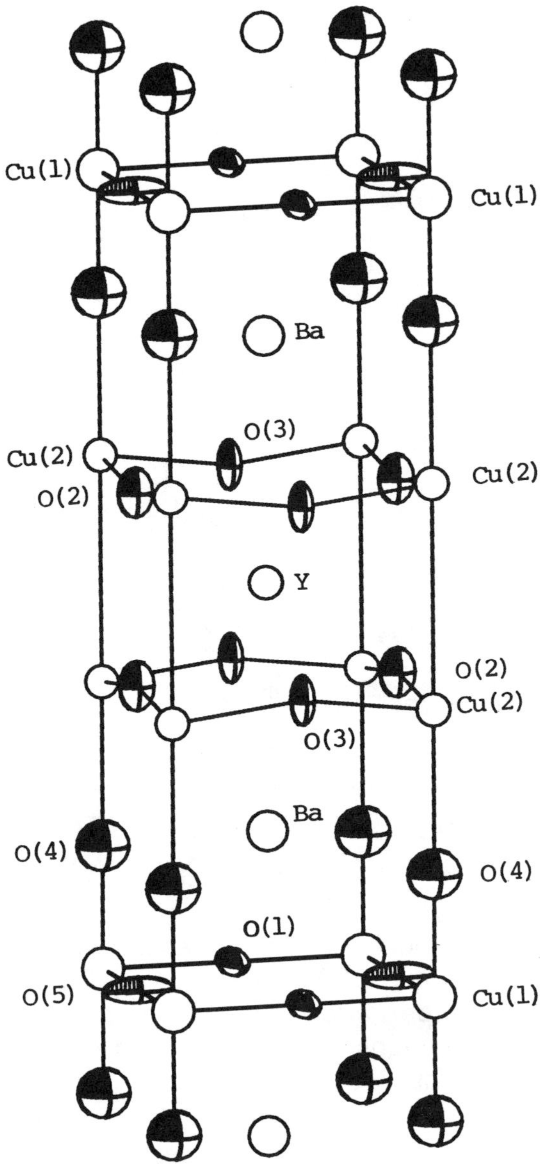

Fig. 4. High temperature, T=725C, tetragonal (P4/mmm) crystal structure of $Y(Ba_{1.67},Sr_{0.33})Cu_3O_{6+x}$, from high-temperature, in-situ measurements in an oxygen pressure, P_{O2}=1atm. In this structure, O(2) and O(3), and O(1) and O(5) atomic sites are equivalent.

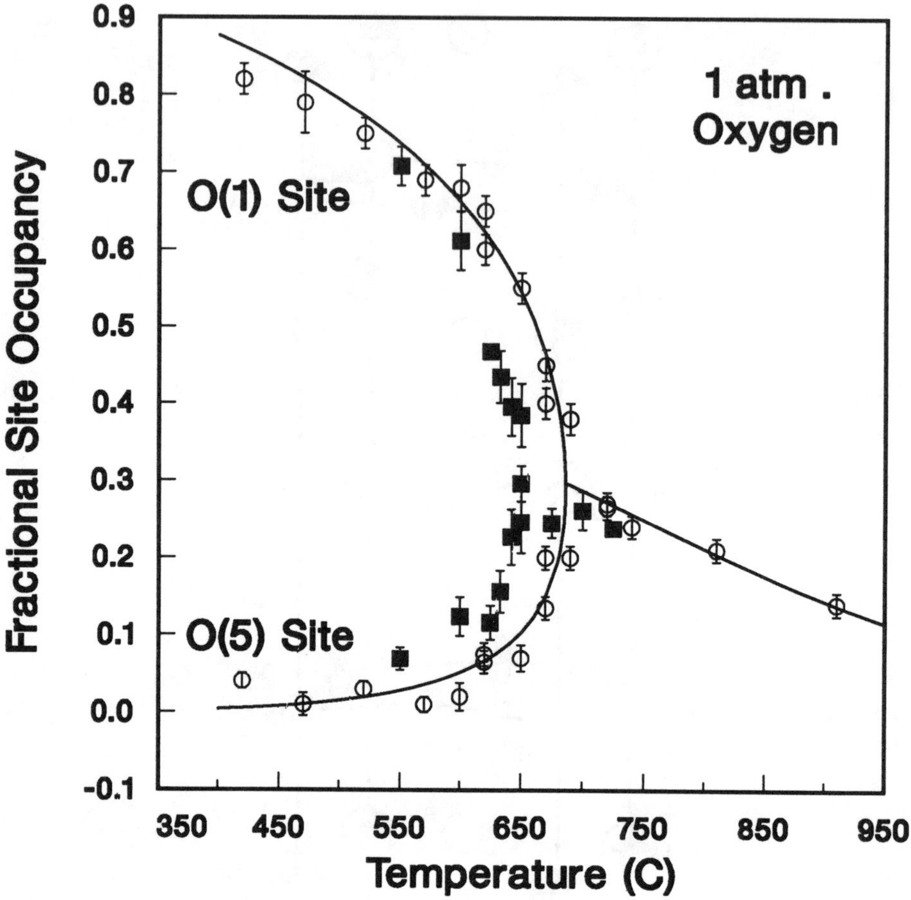

Fig. 5. Temperature dependence of the fractional site occupations for the anions on the O(1) and O(5) sites in $YBa_{1.67}Sr_{0.33}Cu_3O_{6+x}$ (the solid squares). The open circles are for the pure 123 compound, $YBa_2Cu_3O_{6+x}$ from Jorgensen et. al.[12] The solid lines correspond to the theoretical predictions by Shaked et. al.[18]

MICROTOMOGRAPHY OF SILICON NITRIDE / SILICON CARBIDE COMPOSITES

S. R. Stock, A. Guvenilir and T. L. Starr
Georgia Institute of Technology, Atlanta, GA 30332-0245, USA

J. C. Elliott and P. Anderson
London Hospital Medical College, London E1 2AD, UK

S. D. Dover
King's College, London WC2R 2LS, UK

D. K. Bowen
University of Warwick, Coventry CV4 7AL, UK

ABSTRACT

Microtomography nondestructively images the variation in density and composition in cross-sectional slices of a sample. A 0.010 mm diameter collimator and x-rays from a microfocus generator are used for microtomography of porosity in reaction bonded silicon nitride/silicon carbide fiber composites. The effect of vacuum degassing the aqueous fiber-powder blend is examined, and reconstructed slices from degassed and un-degassed samples are shown with two different resolutions (approximately 0.020 and 0.010 mm). Adjacent slices show the three-dimensional morphology of the pores. Application of microtomography to metals and metal matrix composites is also discussed.

INTRODUCTION

Recently advanced composites have received a tremendous amount of attention, and ceramic matrix composites (CMC) are filling important niches as structural materials. Increased toughness coupled with high strength is an important advantage of CMC compared to monolithic ceramics intended for advanced turbines and adiabatic heat engines, heat exchangers and recuperators.[1] The silicon nitride matrix/silicon carbide fiber system is of particular interest because commercial, continuous-polymer-derived silicon carbide fibers such as Nicalon provide good chemical compatibility and degradation resistance superior to carbon fibers in high temperature oxidizing environments.[2-6] The strong fibers act to prevent catastrophic brittle failure by dissipating energy during crack advance, improving fracture toughness and strength.[3-4] Complete densification of reaction sintered and of chemical vapor infiltrated specimens is frequently impossible, however, due to the enclosure of pores by growth of the matrix on surrounding fibers. The resulting lower density and large number of internal stress concentrators degrades the mechanical properties intrinsic to the materials system. If this porosity can not be eliminated during processing, most of the anticipated fracture toughness will be lost. Densification mechanisms must be understood thoroughly, therefore, if porosity is to be minimized.

The structural complexity of CMC such as silicon carbide/silicon nitride is accompanied by large sample-to-sample variability in microstructure. This inhomogeneity makes determination of the spatial variation of density imperative at each stage of the densification of each specimen. Characterization, therefore, must be nondestructive if incidental specimen differences are not to obscure fundamental mechanisms.

This paper reports microtomography of reaction-bonded silicon carbide fiber/silicon nitride matrix composites. Variations in porosity due to different processing conditions are studied using x-rays from a microfocus source and with emphasis on the three-dimensional morphology of pores. The minimum pixel dimensions approach 0.010 mm and pores with much smaller dimensions are readily detected. Prospects for higher resolution are also discussed.

X-RAY TOMOGRAPHY

A myriad of nondestructive evaluation techniques have been applied to processing defect characterization in composites. X-radiation offers significant advantages over other probes: large volumes of material (typically 100 mm^2 x 0.01-5 mm thickness) can be studied nondestructively with spatial resolution as low as 0.001 mm. Synchrotron radiation is generally required to obtain optimum sensitivity although, as outlined below, much can be done with laboratory microfocus sources.

High resolution radiography employing microfocus x-ray sources of synchrotron radiation are termed microradiography and can provide resolution better than 0.001 mm.[7] Determination of the three-dimensional distribution of features in a sample requires little more than recording stereo pairs -- providing that there is little overlap of images. One limitation of microradiography is geometrical invisibility: cracks produce virtually no contrast except when viewed from directions near to their planes. Another drawback is film saturation and its limited linearity of response.

Computed tomography and high resolution variants termed microtomography are closely related to radiography: the projected absorption in a thin slice is recorded for many different orientations of the object. The different projections are combined via a reconstruction algorithm to produce a two-dimensional map of x-ray absorption in the slice. The three-dimensional distribution of absorption in the sample is recovered by examining successive slices.

Spatial resolution has always been a limitation of conventional tomography apparatus. Medical units have resolution approaching 0.5 mm in most cases although specialized instruments have been reported with 0.05 mm resolution.[8,9] Industrial systems using fan or divergent x-ray beams from conventional or rotating anode generators and multiple detectors are designed for rapid inspection, and this coupled with the intrinsic dimness of conventional sources limits spatial resolution to no better than 0.025 mm.

The two approaches used for microtomography represent opposite extremes of complexity. Elliott and coworkers use pinhole collimation and a single detector to achieve high resolution.[10-12] Each view is produced by translating the specimen across the beam. This scheme is simple and has been used with both microfocus and synchrotron radiation. Parallel data collection is the basis of the more complex approach. Coupling a fine-grained fluorescent screen through a lens to a two-dimensional CCD (charge-coupled device) provides variable pixel sizes as small as 0.005 mm if synchrotron radiation is used.[13,14] Data collection is at least several orders of magnitude faster with the latter method, but until the CCD system can be modified for divergent microfocus radiation, the pinhole approach is the only available high resolution technique for the lengthy periods between synchrotron runs. It should be emphasized that both pinhole and CCD sampling produce excellent reconstructed slices.

MICROTOMOGRAPHY OF SILICON CARBIDE/SILICON NITRIDE COMPOSITES

The silicon carbide/silicon nitride composites were produced by reaction sintering fine silicon powder and Nicalon fibers.[6,15] The 0.015 mm diameter fibers were crushed to reduce their length to 0.5 mm prior to blending with an aqueous dispersion of silicon powder. Some of the blends were vacuum degassed to minimize gas incorporation. The composites were slip cast and reaction sintered below 1400°C. The final densities were between 0.8 and 0.9 of theoretical, and the volume fraction of fibers was about 0.20. The microscopy revealed little porosity and 0.3 mm diameter pores in the degassed and non-degassed samples, respectively.

Two 1x1x10 mm specimens were used for microtomography. Sample N-15-3 was from the batch produced with degassing, and N-1-4 was from the high porosity material. The data was obtained with the London Hospital Medical College translate-rotate apparatus (Fig. 1). Translational accuracy was better than 0.001 mm as was the wobble of the rotation axis.[11,13]

A Hilger and Watts Y33 microfocus generator was operated at 35 kV and 2 mA with a source size of 0.1x0.1 mm, a molybdenum target and a 0.075 mm zirconium filter. A 0.01 mm diameter platinum aperture was placed 30 mm from the x-ray source, and intensities were measured with a scintillation counter for 1 s per position. Approximately two hours were required to collect the data for the 64x64 pixel reconstructions and about eight hours for the 128x128 pixel reconstructions (shown in Fig. 2 and 3, respectively). X-ray output drift was checked every 10 min by measuring standard intensities. Slices were reconstructed with Fourier methods.[16]

Figure 2 shows relatively low resolution slices from the two samples (a. is from N-15-3 and b-f. are from N-1-4) obtained by measuring absorption at 64 points separated by 0.020 mm and with 60 projections. The slices of Fig. 2 are shown with 256 gray levels, and the lighter pixels represent lower absorption. The reconstruction is on a 64x64 pixel basis, each pixel is about 0.020 mm on a side and spatial resolution is, therefore, slightly greater than 0.020 mm. Little porosity is visible in N-15-3. The slices of N-1-4 are separated by 0.020 mm, and large, trapped pores and highly densified volumes are clearly visible. It is impossible, however, to discern the 0.015 mm fibers. The five adjacent slices reveal the complex geometry of the internal pores, and it would be very rewarding to examine a single specimen multiple times during densification to observe how the pores are sealed.

Figure 3 shows four adjacent slices of N-1-4 (0.010 mm apart) obtained by recording absorption at 128 points separated by 0.010 mm for each of 120 projections. The reconstructions are with 128x128 pixels (spatial resolution slightly greater than 0.010 mm) and with 256 gray levels. The pores are much more sharply delineated than in the reconstructions of Fig. 2. Variation in the "background" absorption indicates the inhomogeneous distribution of sub-pixel pores, and small clumps of very dark pixels show that there are also extremely dense regions in this sample.

Individual 0.015 mm fibers are not resolved, apparently because the pixel size is only slightly smaller than the fiber diameters and because the absorption of the fibers and matrix are very similar. It is possible that image processing combined with three-dimensional mapping using narrow subsets of the contrast range will allow the complex fiber structure to be resolved. The features revealed in this fashion can be mapped in three dimensions so that the resulting skeletal image emphasizes the fibers or the connected porosity.

Increased spatial resolution coupled with higher contrast and practical data collection rates for samples with cross-sectional areas larger than 1-2 mm^2 are important goals for microtomography. Contrast better than 5% with spatial resolution of 0.001 mm can be achieved, but considerable instrumentation development is necessary. The minimum meaningful specimen dimensions, however, are frequently larger than those which can be studied practically with a single pinhole. A simple combination of a row of pinholes and a position sensitive detector would be quite effective in enhancing data collection rates to the point where representative composite samples may be characterized during in situ, interrupted deformation tests.

A very important step in understanding densification is the microtomographic characterization of a sample from fiber/powder preform through the end of reaction sintering (or other processing steps). Mapping fiber positions prior to the formation of the final matrix phase will be critical in validating reconstructions of final composite and in identifying micrometer and smaller porosity which might otherwise be missed. Microtomographic characterization of a composite during interruptions in densification is already feasible, and it should be possible to use an in situ chamber to follow this process in real or near real time and without thermal cycling.

Metal matrix composites (MMC) can be studied with the same ease as CMC. Damage accumulation and cracking during mechanical testing are of considerable interest, and in situ deformation studies should be easier to implement than in situ processing characterization. The path of macroscopic cracks and the morphology of microcracks (e.g. matrix cracks, interface failure or fiber breakage) are areas in which microtomography would be invaluable.

Achievement of 0.001 mm spatial resolution over millimeter-wide dimensions will fill the gap between electron microscopy and statistical sampling techniques such as small angle scattering on the one hand and macroscopic measuring techniques on the other. The previously inaccessible size range is where microscopic mechanisms link with macroscopic behavior (and continuum models), and, as such, its study is critical for physically based modeling.

ACKNOWLEDGEMENTS

This research was partially supported by NSF through grants INT-8513629 and ENG/MSM-8614493, by the Medical Research

Council and the Special Trustees of The London Hospital Medical College and by the Science and Engineering Research Council.

REFERENCES

1. A. J. Caputo, D. P. Stinton, R. A. Lowden and T. M. Besman, "Fiber-Reinforced SiC Composites with Improved Mechanical Properties," Ceramic Bul. **66** (1987) 368.
2. T. I. Mah, M. G. Mendiratta, A. P. Katz and K. S. Madiyasni, "Recent Developments in Fiber-Reinforced High Temperature Ceramic Composites," Ceramic Bul. **66** (1987) 304.
3. S. T. Buljan, J. G. Galdoni and M. L. Huckabee, Si_3N_4 - SiC Composites," Ceramic Bul. **66** (1987) 347.
4. J. Homeny and W. L. Vaughn, "Whisker-Reinforced Ceramic Matrix Composites," MRS Bul. **XII** No. 11 (1987) 66.
5. R. Lundberg, L. Kahlman, R. Pompe, R. Carlsson and R. Warren, "SiC-Whisker-Reinforced Si_3N_4 Composites," Ceramic Bul. **66** (1987) 333.
6. T. L. Starr, J. N. Harris and D. L. Mohr, "Reaction Sintered Silicon Nitride Composites with Short Fiber Reinforcement," Ceramic Engineering and Science Proc. **8** No. 7-8 (1987) 985.
7. D. K. Bowen, "Microradiography and Absorption Microscopy," in: Characterization of Crystal Growth Defects by X-ray Methods (1980) 506.
8. F. H. Sequin, P. Burstein, P. J. Bjorkholm, F. Homburger and R. A. Adams, "X-ray Computed Tomography with 50-m Resolution," Appl. Optics **24** (1985) 4117.
9. R. A. Armistead, "CT: Quantitative 3-D Inspection," Advanced Materials Processes/Metals Progress, Mar. (1988) 42.
10. D. K. Bowen, J. C. Elliott, S. R. Stock and S. D. Dover, "X-ray Microtomography with Synchrotron Radiation," in: X-ray Imaging II **SPIE 691** (1986).
11. J. C. Elliott and S. D. Dover, "X-ray Microscopy Using Computerized Axial Tomography," J. Microscopy **138** (1985) 329.
12. J. C. Elliott, D. K. Bowen, S. D. Dover and S. T. Davies, "X-ray Microtomography of Biological Tissues Using Laboratory and Synchrotron Sources," Biological Trace Element Res. **13** (1987) 219.
13. J. H. Kinney, Q. C. Johnson, U. Bonse, M. D. Nichols, R. A.

Saroyan, R. Nusshardt, R. Pahl and J. M. Brase, "Three-Dimensional X-ray Computed Tomography in Materials Science," MRS Bul. **XIII** No. 1 (1988) 13.

14. B. P. Flannery, H. Deckman, W. Roberge and K. D'Amico, "Three-Dimensional X-ray Microtomography," Science **237** (1987) 1439.

15. T. L. Starr and J. N. Harris, "Development of Advanced Fiber Reinforced Ceramics," ORNL Report ORNL/Sub/83-43369/01.

16. S. D. Dover, J. C. Elliott, R. Boakes and D. K. Bowen, "Three-Dimensional X-ray Microscopy with Accurate Registration of Tomographic Sections," J. Microscopy (in press).

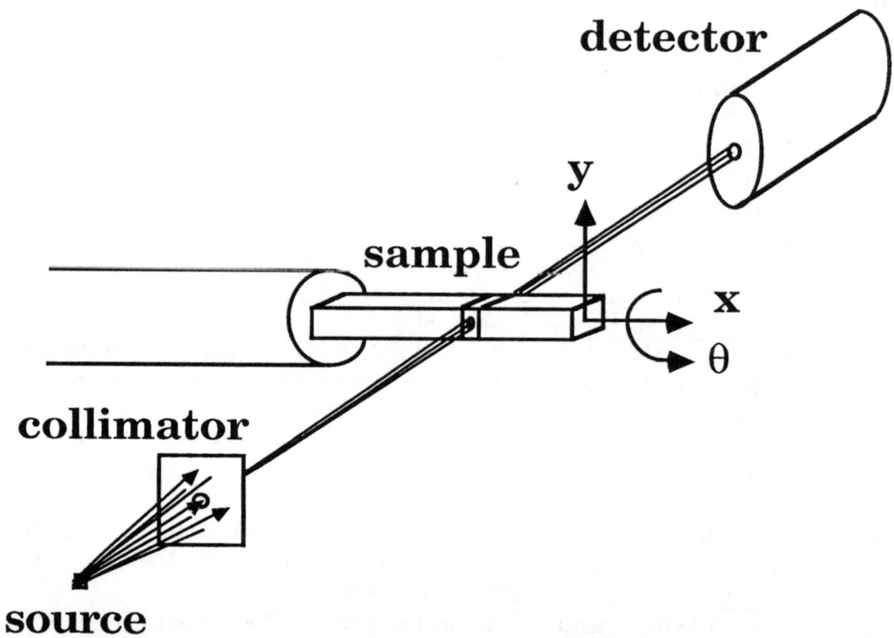

Figure 1. Schematic of the microtomography apparatus used to study the silicon carbide/silicon nitride composites. The sample is translated along the y-axis to record the absorption profile for each viewing angle θ. Adjacent slices are recorded by translation of the sample along x-axis.

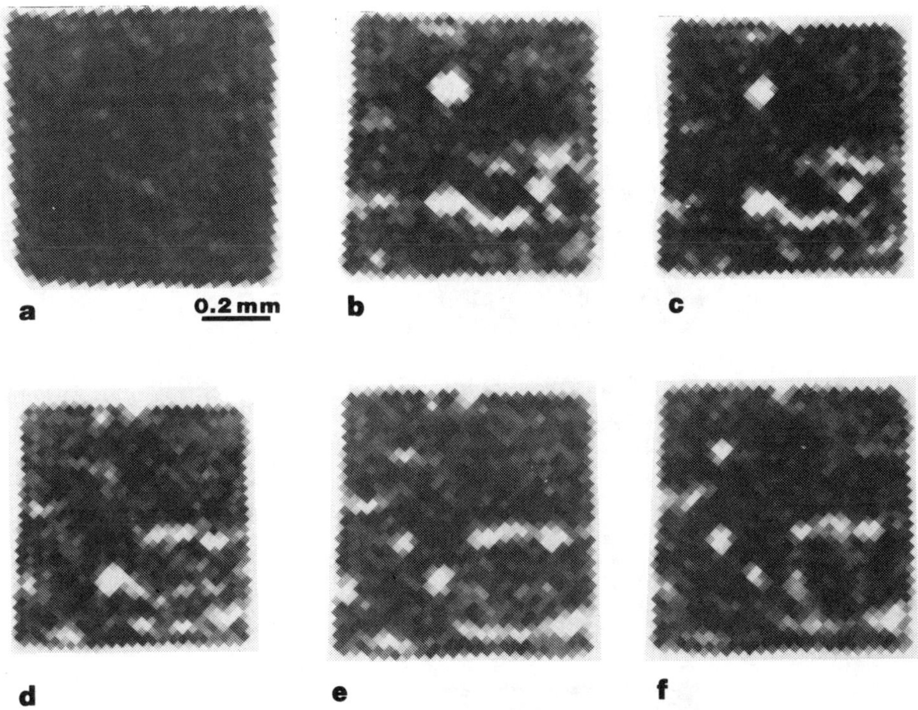

Figure 2. Slices of the nonporous (a) and porous (b-f) composites with the lighter pixels representing lower x-ray absorption. The slices from the porous sample are from adjacent volumes of material. The longest edge is 1.0 mm in these 64x64 pixel reconstructions.

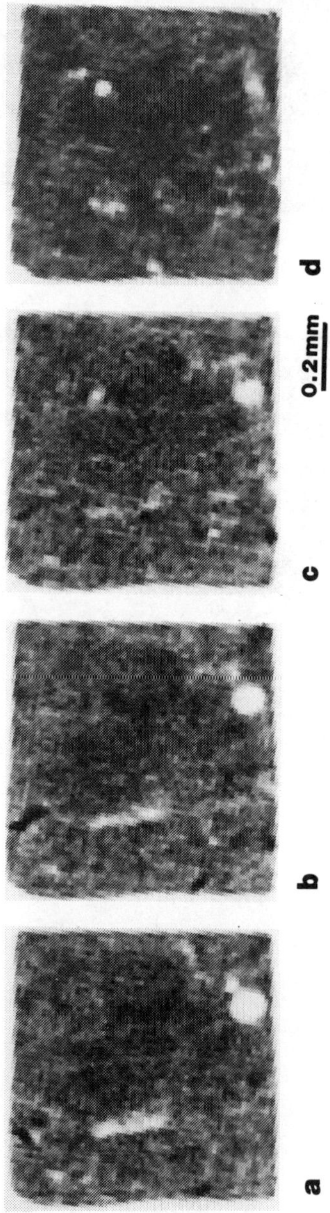

Figure 3. Four adjacent slices of the porous sample recorded with higher resolution than those shown in Fig. 2 and reconstructed with 128x128 pixels.

SMALL ANGLE NEUTRON SCATTERING CHARACTERIZATION OF MICROPOROUS SILICA

G. G. Long and S. T. Krueger
National Institute of Standards and Technology
Gaithersburg, MD 20899

and

R. A. Gerhardt
Center for Ceramics Research
Rutgers University
Piscataway, NJ 08855

ABSTRACT

The new technique of multiple small angle neutron scattering has been used to study the densification of microporous silica as a function of thermal processing. One of the advantages of small angle neutron scattering is that all of the scatterers, which in this case are pores within ceramic bodies, are measured whether they are open or closed, to yield a statistically significant determination of microstructure morphology. Furthermore, the neutron scattering techniques are nondestructive, and the time scale of data collection has been reduced, offering future possibilities for *in situ* measurements of microstructure evolution during processing.

INTRODUCTION

The sol-gel process can be used to prepare novel glasses, fibers, thin and thick films with chemical homogeneity and unique structural and electrical properties. Using the Shoup[1] technique, high purity and crack-free silica bodies with "made-to-order" pore size distributions can be fabricated. A major issue in the preparation of microporous silica bodies is the control of the pore sizes and the pore size distributions because it has been demonstrated that the amount and character of porosity bear a strong influence on the dielectric and other properties of the product material. The present research focuses on the characterization of microporous silica bodies by Small Angle Neutron Scattering (SANS), and by the new technique of Multiple Small Angle Neutron Scattering (MSANS)[2]. These are used to provide detailed information on pore morphology (pore sizes, pore fractions, and pore surface areas) as a function of thermal processing.

Multiple small angle neutron scattering is a technique first described by Berk and Hardman-Rhyne[3,4] for the investigation of the microstructure of samples that are thicker and less dilute than is usual for SANS investigations. The size regime includes microstructure between 0.08 and 10 micrometers, without increasing the instrumental resolution. Larger pore sizes in ceramics have traditionally been characterized by mercury penetration porosimetry[5]. The MSANS technique is, by comparison, a nondestructive technique that can be applied to the study of the microstructure evolution of ceramic materials as a function of processing parameters and even during thermal processing. Particle, or pore, sizes and size distributions are derived by analyzing the curvature of the scattering curve at its center when the data is dominated by multiple scattering.

NEUTRON SCATTERING THEORY

SANS diffraction, as it is usually applied, is sensitive to microstructure in the 1 nm to 0.1 μm size regime, where the scattering particles or pores are very dilute. As mentioned above, the sensitivity range of MSANS is to particle or pore radii between 0.08 and 10 μm, and it involves samples which are thicker and much less dilute.

The interaction of neutrons with matter is determined by the phase shift, ν, that a plane wave suffers in traversing a particle of radius R. $\nu = (4\pi/\lambda)\Delta nR$, where λ is the neutron wavelength, and $\Delta n = \Delta b\lambda^2/2\pi$, the index of refraction of the particle or void relative to the matrix. Δb is the relative scattering length density, or contrast, of the particle or void:

$$\Delta b = \Sigma_{cell}(b_i V_{cell}) - b_{matrix}.$$

Thus, the phase shift ν is equal to $2\Delta bR\lambda$, or directly dependent on the material contrast, on the dimensions of the particle or void, and on the neutron wavelength. If $\nu \ll 1$, the measurement is in the SANS diffraction regime, where the Born[6] approximation or the Rayleigh-Gans[7] theory hold. The scattering curve as a function of the scattering wavevector Q (where $Q = 2\pi\epsilon/\lambda$, and ϵ

is the scattering angle) is independent of neutron wavelength and depends only on the particle dimensions. When the scattering depends only on the particle size, an analysis of the curvature of the scattering curve near Q = 0 can be used to derive the Guinier radius[8] or the radius of gyration, R_g, of the scattering particle.

If $\nu \gg 1$, the experiment involves multiple refraction, where every particle scatters and geometrical optics applies. The scattering curve depends on the contrast index of refraction and therefore depends on neutron wavelength as λ^2.

When $\nu \approx 1$, multiple SANS diffraction, MSANS, applies[2,3] and the scattering curve broadens as a function of neutron wavelength. The predicted neutron intensity is approximately Gaussian and has a width approximately proportional to λ^2. Although the wavelength dependence appears the same for MSANS and multiple refraction, the underlying mechanism is different. In effect, when $\nu \approx 1$ the wavelength dependence of the radius of curvature, r_c, at Q = 0 of the scattering spectrum can be analyzed according to the formalism developed by Berk and Hardman-Rhyne[2,3] to derive the particle size $R_{eff}(0)$, where the zero indicates that Q = 0.

The assumptions made in the MSANS method are that the multiple scattering arises from a randomly distributed monodisperse population of spheres having no internal structure. Although calculations for polydisperse systems and experimental adjustments for internal structure can be made, these will not be discussed here. A measure of the amount of multiple scattering is given by the parameter $\bar{z}$ which is the sample thickness divided by the statistical mean free path length. $\bar{z}$ is related to the scattering radius of curvature *via* a universal relation[3]

$$r_c[QR] = 0.92\{\bar{z}\ \ln^{0.85}\bar{z}\}^{0.5} \quad \text{for } \bar{z} > 1.$$

For $\bar{z} > 5$, the agreement with the theory is better than 0.1%. The data for all values of thickness, scattering volume fraction, wavelength and phase shift lie on this curve. The analysis requires only the sample thickness, the wavelength dependence of the radius of curvature, the scattering volume fraction (which is a refinable parameter), and the scattering length density (which is 3.45 μÅ^{-2} in the present case) to derive $R_{eff}(0)$. In this work, the phase shifts were between 0.1 and 0.3, and $\bar{z}$ ranged from 10 to 70, or well within the MSANS formalism.

For large Q, the scattering is determined by the single scattering Porod law[8], which applies even when low Q is dominated by multiple scattering[3]. From the Porod scattering, the total scattering surface areas may be determined.

EXPERIMENT

1. Microporous Silica Samples

Samples were prepared by the Shoup[1] technique from 85 wt.% potassium silicate and 15 wt.% colloidal silica. Detailed information on sample preparation is given elsewhere[9]. The porosity of the nominally alkali-free (< 100 ppm Na or K) precursor body was between 81 and 86%, where the average pore radius was 0.10 μm as measured by mercury porosimetry. Since the calculated density was within 2% of that derived from simple geometric measurements, it was assumed that the pores in the precursor body were mostly open. Each of the samples was then subjected to a firing temperature above the onset of sintering, which occurs between 1150° and 1200°C depending on the average pore size of the dried precursors. The densities of the small angle neutron scattering samples were between 60 and 98% of Theoretical Density (TD). These densities are representative of the intermediate to final stages of thermal processing. The thicker samples with low density were originally intended for multiple scattering measurements and the thinner samples with higher density for SANS diffraction measurements. It is remarkable that a number of the thin samples were well within the range of applicability of MSANS, thanks to the availability of a cold neutron source. The sample thickness, densities, and percent TD are given in Table 1.

2. Neutron Scattering Experiment

The MSANS and SANS experiments were performed at the 20 MW research reactor operated by the National Institute of Standards and Technology (NIST, formerly the National Bureau of Standards) on its campus in Gaithersburg, Maryland. A cryogenic moderator has been installed in the reactor core, providing copious neutron fluxes in the wavelength region between 0.5 and 2.0 nm. The NIST SANS facility includes a helical-channel velocity selector for selecting the mean incident wavelength of the neutrons. A 12 mm collimating aperture follows the wavelength selector and an 8 mm aperture preceeds the sample position. Scattered neutrons are detected by a 64 x 64 cm^2 position sensitive counter divided into 128 columns and 128 rows. A detailed description of the NIST SANS facility is given in references 10 and 11.

MSANS measurements were carried out on each of the samples using at least five different neutron wavelengths between 0.6 and 2.0 nm. Scattering curves could be accumulated in as little as five minutes. The area detector displayed spectra which were circularly symmetric about the center of the incident beam, and thus could be circularly averaged to produce intensity *versus* scattering vector, Q, curves. These scattering curves displayed broadening as a function of incident neutron wavelength for all samples between 60% and 74% TD. Fig. 1 shows the circularly averaged scattering spectra as a function of wavelength for the 70% dense sample. At 95% TD and greater density, no broadening was detected, even for the thickest samples at the longest wavelengths. This indicates that the characteristically large scattering voids that were present in the intermediate stages of sintering have disappeared by the time the samples reach 95% TD. SANS diffraction measurements on the dense samples revealed pore populations with radii of gyration 33 nm and smaller.

After circular averaging, the I(Q) curves between 100% and 10% intensity were fitted with a Gaussian function and the radius of curvature of each broadened curve was extracted from the standard deviation, $r_c = \sqrt{1/2\sigma}$. The same fitting procedure was followed for the blank spectra (with no scatterer in place) to enable a correction for the instrument function. The corrected radii of curvature were then compared to the predictions of the theory to extract the effective radii of the scattering voids. A comparison of the derived radii of curvature with the theory is shown in Fig. 2, for $R_{eff}(0)$ = 0.224 μm and 0.30 void scattering density.

RESULTS AND DISCUSSION

MSANS[2,3] was used to measure the average pore sizes in microporous silica during the intermediate stages of thermal processing, where the radii could be expected to be in the 0.08 - 10 μm range. Total surface areas were estimated from the single scattering Porod curves, which were obtained by means of 0.6 nm neutrons. Pore populations in the late stages of sintering were studied by means of SANS diffraction measurements with 1.4 nm neutrons. The results are shown in Table 2.

The results for the effective pore radii as a function of sample density indicate that densification during the intermediate stages of sintering is accompanied by coarsening within the population of large pores. Although pore coarsening had previously been observed in the early stages of thermal processing, coarsening during the intermediate stages of processing was unexpected. The observed pore coarsening eventually levels off and by the time the material is 95% dense, there are no detectable pores with radii greater than 0.08 μm. SANS diffraction measurements on the 95% dense and on the 98% dense sample using 1.4 nm incident neutrons revealed the presence of populations of pores with radii of gyration 33 nm and 28 nm and smaller. The Porod data was used to derive values for the total surface scattering areas, and the results confirm that the total scattering area decreases monotonically as a function of thermal processing.

In a model proposed for this system, it has been suggested[9] that there are two distinct populations of pores in the microporous silica system. There is a population of "small" pores that form within loosely aggregated clusters (or domains) of closely packed fine silica particles. These pores within the domains are actually voids between touching particles. Then there is a population of "large" pores which are found between the domains. It is this latter population that dominates the MSANS data. The pores with smaller dimensions are less stable and will be eliminated earlier in the sintering process. Presumably, as the domains densify, the intradomain pores will coarsen. When the intradomain pores have disappeared, the interdomain pores can finally shrink. Previous estimates[9] of the pore dimensions using mercury penetration porosimetry suggested that intradomain densification was complete during the early stages of thermal processing, whereas this study indicates that intradomain densification is not complete until much later in the processing. Recently, the present results have been confirmed[12] up to 50% dense by electron microscopy. The discrepancy between the porosimetry results and the MSANS results may be attributed to the sealing-off of porosity in the intermediate stages of processing.

This study supports the model by Cao, Gerhardt and Wachtman[9] that microporous silica networks are composed of loosely aggregated clusters of particles separated from each other by a population of macropores. It demonstrates that the interdomain pores experience significant coarsening until the material is approximately 65% TD. The subsequent shrinkage of the interdomain pore population is very abrupt, and by the time the material is 95% TD the radius of gyration of the pores is of the order of 30 nm and less. The pore coarsening observed during sintering is explained by material transport from areas around interdomain pores to those around intradomain pores in order to minimize the total surface energy. Further details concerning the present work and statistical modeling[13] of the MSANS of this system are to be given in a subsequent publication[14].

ACKNOWLEDGEMENT

One of the authors (GGL) acknowledges with appreciation support for this work by a NIST Institute of Materials Science and Engineering Director's Innovation Award.

Table I. Microporous Silica Samples (15:85 Composition)

Sample No.	Thickness (mm)	Density (gm/cm^3)	%TD
1	4.872	1.30	60
2	1.420	1.42	65
4	1.230	1.56	70
5	1.221	1.63	74
6	4.855	2.09	95
7	1.130	2.10	95
9	1.300	2.15	98

Table II. Microstructure Results by means of MSANS and SANS Experiments.

Sample No.	Porosity	$R_{eff}(0)$ (μm)	$R_g(0)$ (nm)	Total surface area (cm^2/g)
1	0.40	0.185 ± 0.003	-	$1.5 \pm 0.3 \times 10^5$
2	0.35	0.243 ± 0.006	-	$7.0 \pm 0.8 \times 10^4$
4	0.30	0.224 ± 0.006	-	$7.5 \pm 3.5 \times 10^4$
5	0.26	0.167 ± 0.010	-	$6.9 \pm 0.9 \times 10^4$
6	0.05	<0.08	-	1900 ± 200
7	0.05	-	33 ± 4	-
9	0.02	<0.08	28 ± 3	800 ± 100

REFERENCES

1. R. D. Shoup, "Controlled Pore Silica Bodies Gelled from Silica Sol-Alkali Silicate Mixtures," Colloid and Interface Science 3, 63-69 (1976).

2. N. F. Berk and K. A. Hardman-Rhyne, "Analysis of SANS Data Dominated by Incoherent Multiple Scattering," J. Appl. Crystallogr. 21, (1988), in press.

3. N. F. Berk and K. A. Hardman-Rhyne, "Characterization of Alumina Powder Using Multiple Small Angle Neutron Scattering. I. Theory," J. Appl. Crystallogr. 18, 467-472 (1985).

4. K. A. Hardman-Rhyne and N. F. Berk, "Characterization of Alumina Powder Using Multiple Small Angle Neutron Scattering. II. Experiment," J. Appl. Crystallogr. 18, 473-479 (1985).

5. H. M. Rootare and C. F. Prenzlow, "Surface Areas from Mercury Porosimeter Measurements," J. Phys. Chem. 71, 2733-2736 (1967).

6. G. Porod, "The X-Ray Small-Angle Scattering of Close-Packed Colloid Systems I," Kolloid-Z. 124, 83-114 (1951); R. J. Weiss, "Small Angle Scattering of Neutrons," Phys. Rev. 83, 379-389, (1951).

7. J. W. Lord Rayleigh, "The Incidence of Light upon a Transparent Sphere of Dimensions Comparable with the Wavelength," Proc. Roy. Soc. A84, 25-46 (1910) and "Diffraction of Light by Spheres of Small Relative Index," A90, 219-225 (1914); R. Gans, "Diagrams of Radiation from Ultra-Microscopic Particles," Ann. Physik 76, 29-38 (1925).

8. G. Kostorz, Treatise on Materials Science and Technology, Vol. 15, (New York, Academic Press) 1979.

9. W. Cao, R. Gerhardt and J. B. Wachtman, Jr., "Processing and Sintering Characterization of Bulk Porous Silica," J. Am. Ceram. Soc. 71 (1988), in press.

10. C. J. Glinka, "The National Bureau of Standards Small-Angle Neutron Scattering Spectrometer," AIP Conference Proc. No. 89--Neutron Scattering, 395-397 (1981).

11. C. J. Glinka, J. M. Rowe and J. G. LaRock, "The Small-Angle Neutron Scattering Spectrometer at the National Bureau of Standards," J. Appl. Crystallogr. 19, 427-439, (1986).

12. H. Kerch and R. Gerhardt, "Ultramicrotomy of Porous Ceramics," in these Proceedings.

13. J. R. D. Copley, "The Significance of Multiple Scattering in the Interpretation of Small Angle Neutron Scattering Experiments," J. Appl. Crystallogr. 21, (1988), in press.

14. G. G. Long and S. T. Krueger, "Multiple Small Angle Neutron Scattering by Porous Silica." to be published.

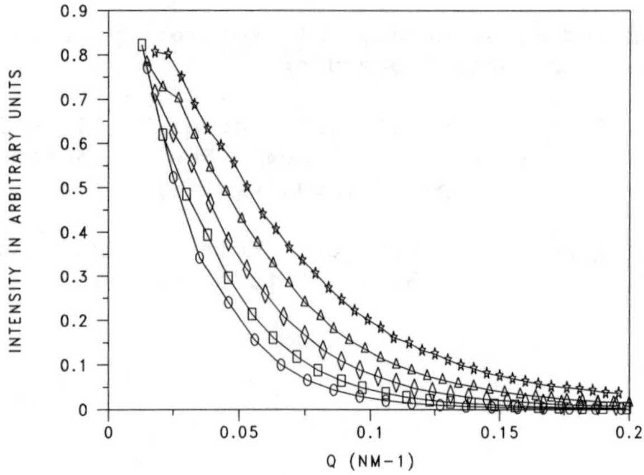

Fig. 1. Wavelength dependence of the MSANS spectra for Sample 4 (30% porosity). The circles represent $\lambda = 1.0$ nm data, the squares $\lambda = 1.2$ nm, the diamonds $\lambda = 1.4$ nm, the triangles $\lambda = 1.6$ nm, and the stars $\lambda = 1.8$ nm. The sample thickness was only slightly greater than 1.2 mm.

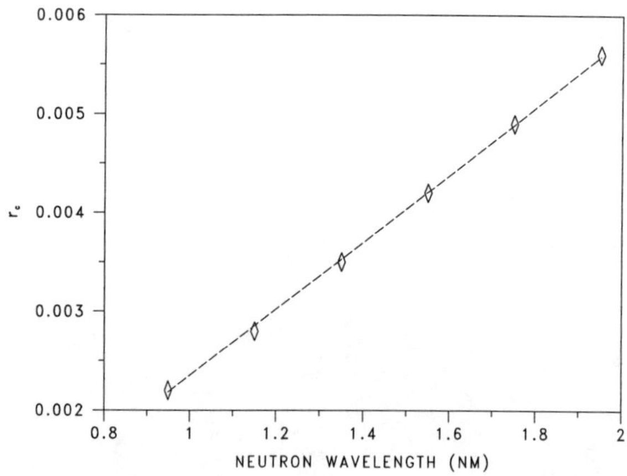

Fig. 2. Comparison of the fitted radii of curvature (diamonds) with the theoretical radii (dashed line) for $R_{eff}(0) = 0.224$ μm and 0.30 scattering void density.

ANALYSIS OF PACKED-PARTICLE REARRANGEMENT DURING COLLOIDAL CERAMIC PROCESSING: GAMMA-RAY DENSITOMETRY AND A REVIEW OF ALTERNATIVE METHODS

Christopher H. Schilling and William D. Samuels
Pacific Northwest Laboratory*
Richland, Washington 99352

Ilhan A. Aksay
Department of Materials Science and Engineering; and
Advanced Materials Technology Program,
The Washington Technology Center
University of Washington, Seattle, Washington 98195

SUMMARY

Electrostatic and steric repulsion forces that arise during colloidal processing of ceramics affect packed-particle arrangements in the powder consolidation stage and may influence microstructure development during sintering. A gamma-ray attenuation apparatus is used for nondestructive *in situ* measurement of local porosity and composition during drying, sintering, and a wide range of consolidation processes. A study is presented, which utilizes the apparatus to analyze solution chemistry effects on packed-particle rearrangement during sedimentation of colloidal α-Al_2O_3. Advantages of gamma-ray densitometry for the analysis of ceramic particle rearrangement are compared with those of alternative methods, such as rheology studies, consolidation kinetic modeling, permeability measurements, and various techniques for analyzing porous media. Gamma-ray densitometry may be used to augment these alternative techniques and assist in the development of more advanced rearrangement models.

---

* Operated by Battelle Memorial Institute for the U.S. Department of Energy under Contract DE-AC06-76RLO 1830.

INTRODUCTION

Recent studies have shown that physical properties of sintered ceramics may be significantly affected by packed-particle arrangements prior to sintering.[1-14] Achieving properties reliability in sintered ceramics requires that green ceramics be processed with uniform packing structures and a minimal size and frequency of packing-structure-related defects. Regulating such defects requires building a basic, scientific understanding of parameters which affect particle-particle rearrangement throughout successive processing stages (consolidation, drying, and sintering). In particular, effects of dispersant chemistry on ceramic-particle rearrangement are poorly understood, and improved analytical methods are required to evaluate the role of chemical reactions associated with powder surfaces, surface-adsorbed ions and polymers, and suspending media.

Nondestructive analyses of dynamic changes in particle-particle rearrangement are needed throughout successive processing stages to predict effects of pre-sintering parameters on final properties development. Density measurements based on gamma-ray attenuation are ideal for this purpose, as compositional evolution may be determined as a function of time, position, and processing history variables. A gamma-ray densitometer apparatus is described, which may be used for *in situ* analysis of local porosity and composition during a succession of various ceramic forming processes including slip casting, pressure filtration, sedimentation, dry powder pressing, drying, and sintering.[15,16] This paper describes the basic operation of the densitometer, demonstrates its sensitivity with a study of solution pH effects on alumina particle rearrangement, and finally, reviews advantages of the apparatus compared to alternative experimental methods.

APPARATUS

A collimated beam of high energy photons is passed through a sample, and changes in the intensity of transmitted radiation are measured to indicate changes in the average composition and porosity of the sample. Apparatus design and operating protocol are based on gamma-ray systems previously reported for analysis of soil composition, material thickness and composition, and slip casting kinetics.[17-27]

A radioactive isotope and radiation detector are mounted on opposite sides of the sample to be evaluated, and a positioning mechanism simultaneously moves the radioactive isotope and detector so that different elevations of the sample may be analyzed (Figure 1). At each elevation, the average composition and average porosity of the irradiated portion of the sample are determined by simultaneous solution of mass conservation and Beer-Lambert law equations:

Beer-Lambert Law

$$\frac{I}{I_o} = \exp - \sum_{i=1}^{n} \mu_i x_i$$

Mass Conservation

$$\sum_{i=1}^{n} x_i = L$$

I is the radiation intensity after attenuation by the sample, I_o is the radiation intensity without attenuation by the sample, and μ_i is the attenuation coefficient of the ith phase in the sample. μ_i is a function of the mass number of the ith phase and the wavelength of the radiation, which is in turn specific to a particular radioisotope. L is the irradiated length of sample, x_i is the concentration of the ith species, and n is the total number of phases in the sample.

Average concentrations of samples containing one or two phases may be determined by measuring attenuation from a single, monochromatic radioisotope. Sample compositions containing additional phases may be analyzed by simultaneous irradiation with two or more monochromatic radioisotopes and solving the above equations using wavelength-specific attenuation coefficients.[17-20] The apparatus is currently configured with 0.6 Curies of ^{137}Cesium, which is encapsulated in a shielded cask with shuttering mechanism and safety interlocks[a] to prevent radiation exposure to human operators. Dose rates to the samples can also be maintained well below threshold damage levels which are reported for various materials.[28-30] The radiation detector currently comprises a germanium diode;[b] NaI scintillation counters may also be used. Attenuation measurements are performed using a multichannel analyzer[c] by integration of energy-as-a-function-of-intensity data at the primary 662 keV (^{137}Cesium) peak.

---

a J. L. Shepherd & Assoc., Pacoima, CA; Model 28-6B irradiator.

b E. G. G. Ortec Corp., Oak Ridge, TN; high purity coaxial detector model GEM-23210.

c E. G. & G. Corp., Oak Ridge, TN; Model 6240B.

The densitometer is currently designed to accommodate a wide range of sample sizes and experimental configurations. Samples of up to 30 cm in size may be evaluated, and the beam may be collimated to less than 1 mm in size using various radiation shielding apparatus of relatively simple designs.[17-19] The elevating table may be positioned over a vertical distance of 22 cm in increments of 13 microns. The samples to be analyzed are mounted on a stationary pedestal, in order to minimize vibration transfer to samples. Various pieces of equipment may be attached to the top of the pedestal, such as sedimentation columns and filtration casting apparatus. A furnace may also be placed on the pedestal, so that the radiation beam passes through samples undergoing drying or sintering. The apparatus may also be configured to measure growth rates of powder compacts during sedimentation or filtration casting, as previously described by Miskin et al.[26,27]

SEDIMENTATION STUDY

Gamma-ray densitometry experiments were performed to evaluate effects of solution pH on particle-particle rearrangement during sedimentation of aqueous alumina suspensions. HCl and NH_4OH additions were used to establish a range of attractive- and repulsive double-layer forces, and striking variations in microstructure evolution were observed as a function of time and pH. Details of these studies were previously disclosed.[15-16]

As an example, Figure 2 illustrates the behavior of a flocculated suspension prepared at pH 8.6. The sediment accumulates by rapid settling and compaction of attached-particle networks, which form in suspension as a result of attractive double-layer forces. Rearrangement reached equilibrium after 90 minutes, resulting in a continuously decreasing packing efficiency between the lower and upper elevations of the sediment (approximately 28 to 6 vol %, respectively). The sediment appeared featureless to the naked eye; an erroneous assumption of position-invariant density would result if average density were calculated based on weight and apparent volume measurements. The sediment was subsequently dried, partially sintered, and vertical sections were cut and polished for examination by scanning electron microscopy. Pores with multiple hierarchies of size were observed, and a pore size gradient from the top to the bottom of the sample corresponded with the density profile in Figure 2.[16]

In comparison, sediments prepared from dispersed suspensions formed slowly and exhibited higher packing densities which were uniform with elevation.[16] Sediment densities of 46, 52, and 55 volume % powder were respectively produced by suspensions at pH 5, 7.3, and 2. During the early stages of consolidation, density gradients were exhibited by these samples, such that the upper elevations were less densely packed. Rearrangement to uniform densities ocurred during the later stages. Rearrangement rates were pH-dependent and may be attributed to differences in particle-particle repulsion forces based on double-layer theory and the restructuring of particle clusters.[31-33] SEM analysis was performed on sintered and polished samples, using aforementioned procedures. In comparison to the pH 8.6 sample, these samples exhibited significantly smaller pores with only 2 hierarchies of size. Packing structures were also uniform with elevation, which corresponded to the density profiles observed earlier by densitometry.

ALTERNATIVE ANALYTICAL METHODS

Advantages and disadvantages of various analytical methods are discussed with respect to evaluation of packed-particle rearrangement during colloidal ceramic processing. A primary advantage of gamma-ray densitometry, relative to alternative techniques, is that non-destructive, in-situ analysis of local porosity and composition may be performed throughout a succession of processing stages. Supplementary use of gamma-ray densitometry and these alternative techniques would be beneficial, as each method possesses certain advantages.

Porosimetry, Pycnometry, and Surface Area Analysis

Mercury porosimetry, gas pycnometry, and surface area analysis by adsorption techniques are powerful analytical tools for evaluating ceramic microstructures.[34] A major disadvantage of these approaches, relative to gamma-ray densitometry, is that real-time analyses of dynamical changes in packed-particle structures may not be performed. Additional drawbacks to these methods are as follows: 1) wet samples may not be analyzed because these techniques are applicable only to dried or sintered samples, 2) spatial gradations in porosity may not be evaluated within each sample, as uniform porosities are assumed, 3) small samples are used and precautions are required to ensure that these samples are representative of larger shapes, 4) porosimetry is destructive to samples, and 5) porosimetry and gas pycnometry may also be inappropriate for the special case of dried samples which are easily cracked or deformed.

Another concern with porosimetry pertains to possible errors in data interpretation. As an example, pore structures of colloidal ceramic compacts typically comprise networks of large pores, which are attached to one another by smaller entry pores. In such samples, mercury porosimetry measures the distribution of entry pore diameters,[34-35] and it may be possible to misinterpret microstructures based on porosimetry data alone.

Rheological Analyses

Rheological data may be correlated with various models to analyze colloidal structures in suspensions.[36-43] Additional research is needed to establish relationships between rheologically-deduced structures of colloidal suspensions and particle-particle rearrangement during consolidation of dense ceramic shapes. Rheological approaches are restricted in terms of analyzing rearrangement during consolidation, as experimental data are typically acquired from suspensions having time-invariant concentrations of particles. Gamma ray densitometry may be used to augment rheological techniques by providing analyses of real-time density changes associated with a suspension, cast layer, and suspension/cast layer interface.

Several authors have analyzed relationships between green microstructures and rheologically-deduced structures in colloidal suspensions.[6,11,44-45] Unfortunately, these attempts do not analyze the time- and position dependence of rearrangement during shape forming, mainly because of restraints associated with rheological models and the absence of available microstructural probes. As an example, Chou and Senna[44] correlated rheological behavior of alumina suspensions with porosimetry measurements obtained from slip cast bodies. Decreases in volume-averaged porosity and pore radius were exhibited in cast bodies as a result of greater packing efficiencies inside suspension agglomerates. The approach relied on use of the Thomas equation[40] to relate suspension viscosity to a bulkiness parameter, which is a ratio of the apparent volume fraction of agglomerates to the volume fraction of particles. This model does not reveal additional structural parameters such as agglomerate size.

In comparison, the elastic floc model[37-39] enables estimates of agglomerate size as a function of shear rate, based on an analysis of energy dissipation during the flow of flocculated suspensions. The model applies to suspensions containing monosize particles at relatively dilute concentrations of less than approximately 10 volume % powder. Additional research is needed to explore applications of this model to analysis of particle-particle rearrangement during ceramic shape forming.

As reviewed by Ramsay,[46] several authors have evaluated the viscoelastic behavior of ceramic suspensions using rheometry, although few attempts have analyzed particle-particle rearrangement. Recently, it was demonstrated that scaling theory may be used with rheometry to determine structural rearrangement parameters in colloidal systems.[41-42] The technique entails use of a borrowed approach from the analysis of network structures in polymer solutions. Scaling theory is based on the assumption that dense, ceramic suspensions comprise networks of monosize, fractal aggregates and that these aggregates are uniformly strained at low stresses in the linear viscoelastic region. Application of this approach to boehmite gels was successfully used to derive expressions for viscoelastic properties as functions of fractal aggregate size, which in turn is a function of particle concentration and fractal dimension. Structural information can also be obtained by measuring viscoelastic properties (storage and loss moduli) as a function of frequency. Additional research is needed to corroborate these data with other ceramic systems. Gamma ray densitometry may be used to augment this approach by revealing composition- and density changes as a function of time and position during cast-layer formation.

Another approach for analyzing colloidal structures in suspensions entails correlating rheological behavior with an equilibrium structures model based upon statistical mechanics.[9,32,33,47] Regimes of rheological behavior were evaluated in alumina suspensions as a function of solids content and particle-particle pair potential, and results were used to verify equilibrium phases predicted by this model. Specifically, transitions between Newtonian and Bingham-plastic flow behavior were associated with phase transitions among gas-like, liquid-like, and colloidal crystalline phases. Research by Aksay and colleagues suggests that similar phase transitions may be expected in the formation of ceramic casts.[2-4,9,14,32,33] Gamma ray densitometry may be used to augment the combined use of this model and rheological measurements by providing analyses of real-time density changes in consolidating casts.

Consolidation Kinetic Analyses

Micheals and Bolger[63] reported use of sedimentation kinetic measurements to correlate structural parameters of agglomerates in suspension with packed-particle arrangements in sediments. Their model assumes that flocs of dense-packed particles randomly associate to form weakly bonded aggregates, whose size and configuration are governed by the intensity of the local shear field. By the use of a force balance and Darcy's equation, the position of the sediment/supernate interface may be correlated with pore diameter, the ratio of floc volume to aggregate volume, and the ratio of floc volume to average particle concentration. The approach is limited to flocculated suspensions and does not reveal floc size or aggregate size. It also assumes that uniform packing structures are formed at the later stages of consolidation; as shown in Figure 2, this assumption may be inappropriate.

Davis and Russell[48] reported a kinetic model, which accounts for the nucleation and growth of colloidal crystals during sedimentation or filtration casting. The model may be used to determine the time- and position-dependence of cake height, average concentration of the cake and suspension, and the fraction of the cake which may be crystallized. Effects of various parameters may be analyzed, such as particle-particle pair potential, particle size, and suspension concentration.

Gamma-ray densitometry is incapable of analyzing some of the structural parameters revealed by the above models for sedimentation and colloidal crystal growth. Simultaneous use of densitometry with these models would be beneficial, however, as visual observations of sedimentation behavior do not reveal local variations in packing density.

Filtration kinetic measurements enable real-time analysis of average porosity and permeability during slip casting or pressure filtration.[2,4,13,14,49-54] The approach entails use of Darcy's equation to predict parabolic flow behavior as long as cast layers are deposited with uniform porosity and permeability. Deviations from parabolic flow kinetics are used to reveal packing-structure variations, which may be a result of various phenomena such as large-particle sedimentation and rheopexy effects. Numerical values of porosity and permeability may be calculated, both of which are assumed to be averaged across the volume of the sample. Average flow channel sizes may also be calculated by use of permeability models; these models assume highly simplified flow channel geometries which are generally atypical for ceramic casts.[34,55]

Several authors report use of permeability measurements to characterize packed-particle structures formed by slip casting or pressure filtration,[2,13,51,52,56,57] and it has been shown that permeabilities are affected by various parameters, such as electrophoretic mobility, applied pressure, and average concentration of particles in suspension. Assumptions of volume-averaged porosity and permeability are not always appropriate, based on reports of particle rearrangement which may occur as a result of the weight of the cast body and viscous drag forces exerted on pore walls.[54] These phenomena were modeled by Tiller and Tsai[54] using empirical relations for permeability and porosity as a function of position, local pressure, and empirical compressibility factors. These relations were used to derive volume-averaged porosity and permeability, which were both included in a modification of the Darcy equation.

Gamma-ray densitometry may be used to augment analysis of filtration kinetic measurements and/or permeability measurements by providing *in situ* measurement of density changes as a function of time and position. These data are also needed to assist the development of more advanced kinetic models for rearrangement. Such models would be useful to analyze effects of a wide range of processing parameters, such as applied pressure, mold geometry and wall friction, and various physiochemical aspects of the powder and suspending medium.

As an example, empirical studies of pressure filtration have shown that the average densities of filter cakes are proportional to the natural logarithm of the applied pressure, based on measurements of sample dimensions and weight after removing samples from a filtration apparatus.[13] A similar linear proportionality between average density and the natural logarithm of applied pressure is commonly observed in soil mechanics experiments used to analyze the mechanical behavior of granular ceramic media.[58] Additional research is needed in the areas of filtration and mechanical behavior of granular ceramic media in order to develop kinetic models describing rearrangement as a function of time, position, and applied pressure. Gamma ray densitometry may be used for this purpose, as such studies require real-time, non-destructive analysis of density changes without removing samples from the forming- or testing apparatus.

Another useful set of experiments entails evaluation of structural relaxation after shape formation.[d] Relaxation phenomena are not well understood, and research is needed to analyze the development of shrinkage-related microcracks, which may be retained during sintering and affect physical properties of the finished ceramic. In a recent study of pressure filtration, Lange[13] reported significant changes in relaxation behavior after the removal of applied pressure from alumina filter cakes. Relaxation varied depending on several parameters including mold shape, maximum applied pressure during filtration, electrophoretic mobility of the particles, and the addition of polymeric binders. Additional research is needed to evaluate relaxation kinetic models, which address the time- and position dependence of rearrange-

---

d Relaxation implies macroscopic shape changes in a green ceramic as a result of particle-particle rearrangement.

ment phenomena associated with a chosen system. Again, gamma-ray densitometry would be appropriate for this purpose.

Miscellaneous Analytical Methods

X-ray attenuation techniques have been used in a wide range of applications for measurements of density, porosity, and thickness.[25,59-63] X- and gamma-ray tomography are also valuable approaches for analyzing three-dimensional spatial arrangements of cracks and porosity in dried and sintered ceramic samples.[64-66] Additional research is needed to evaluate applications of these approaches for real-time, *in situ* analysis of particle-particle rearrangement phenomena during the shape forming of ceramics.

Gamma-rays, as opposed to x-rays, possess higher energies and therefore provide greater penetrating capabilities for packing-density analysis of ceramic materials. A primary advantage of gamma-ray densitometry is that large-scale ceramic samples above one to ten centimeters in size may be evaluated. Another advantage is that samples may be analyzed which are fully enclosed inside a thick-walled vessel, such as a furnace, uniaxial- or isostatic press, or a pressure filtration apparatus.

Optical microscopy techniques are advantageous from the standpoint of enabling real-time observation of colloidal structures in dilute suspensions.[67] However, these approaches are limited to the analysis of relatively small volumes of dilute suspensions; the behavior of colloidal structures in larger bulk volumes of more concentrated suspensions may be difficult to observe.

Small angle x-ray scattering (SAXS) and light scattering techniques may be used to evaluate structural parameters of colloidal suspensions, such as agglomerate size and fractal dimensions.[68-70] An advantage of x-ray scattering is that samples of relatively concentrated suspensions may be analyzed, whereas light scattering is usually limited to analysis of more dilute suspensions. Light scattering is applicable to analysis of heterogeneities of up to approximately 5 micrometers in size, whereas SAXS resolves smaller features between 1 Å and 1 micrometer in size. Small angle neutron scattering[71] (SANS) is also a valuable technique which may be used to analyze average porosity and average void diameter in both green and sintered ceramics. Additional research is needed to evaluate the application of SANS to studies of particle-particle rearrangement during shape-forming of ceramics.

While scanning laser acoustic microscopy (SLAM) and photoacoustic microscopy are useful techniques for the evaluation of sintered ceramics, it was recently reported that limitations exist for analysis of green ceramics.[60]

Difficulties with SLAM entailed breakage of fragile shapes, acoustic attenuation, introduction of artifacts, and limited detection of small defects. Similar problems were reported with photoacoustic microscopy, although additional research is needed to explore the applicability of this technique to green microstructures.

Stereological techniques are useful for direct, quantitative measurement of structural parameters of porous media.[35,60,72] Precautions are needed during sample preparation to alleviate formation of artifacts in colloidal ceramics. Several approaches may be used to preserve packed-particle structures of wet or dried ceramic samples, including freezing, freeze drying, polymer impregnation, and fixation.[60,73,74] Partial sintering has also been used to preserve packed-particle structures of alumina bodies during polishing.[2]

Molecular spectroscopy probes, such as nuclear magnetic resonance and laser raman spectroscopy, are useful for analyzing molecular interactions which may be responsible for particle-particle rearrangement during colloidal processing.[75,76] For example, particle-particle rearrangement may be influenced by variations in polymer- and solvent chemistry which affect polymer attachment to particle surfaces and the spatial arrangements that are assumed by the bound polymers. Additional research is needed to correlate molecular spectroscopy with analyses of microstructure evolution during successive processing stages. Again, gamma-ray densitometry would be useful to augment these studies.

ACKNOWLEDGMENTS

Partial support for the construction of the gamma-ray densitometer was provided by the Exploratory Research Program at the Pacific Northwest Laboratory. Experimental research was supported by the Materials Science Section of the Office of Basic Energy Sciences at the U.S. Department of Energy. The authors wish to thank G. J. Exarhos, R. W. Stephens, G. L. Graff, and J. J. Lannutti for technical assistance.

REFERENCES

1. I. A. Aksay, F. F. Lange, and B. I. Davis, "Uniformity of α-Al_2O_3-ZrO_2 Composites by Colloidal Filtration," *J. Am. Cer. Soc.*, **66**, C190 (1983).

2. C. H. Schilling, "Microstructure Development by Colloidal Filtration," M.Sc. Thesis, University of California at Los Angeles, Los Angeles, California, (1983).
3. I. A. Aksay, "Microstructure Control Through Colloidal Consolidation," in ***Advances in Ceramics, Volume 9, Forming of Ceramics***, edited by J. A. Mangels and G. L. Messing, American Ceramic Society, Columbus, Ohio, (1984), pp. 94-104.
4. I. A. Aksay and C. H. Schilling, "Colloidal Filtration Route to Uniform Microstructures," in ***Ultrastructure Processing of Ceramics, Glasses, and Composites***, edited by L. L. Hench and D. R. Ulrich, John Wiley and Sons, New York, (1984), pp. 439-447.
5. F. F. Lange, "Sinterability of Agglomerated Powders," *J. Am. Ceram. Soc.*, **67**, 83 (1984).
6. M. D. Sacks and T. Y. Tseng, "Preparation of SiO_2 Glass from Model Powder Compacts: I, Formation and Characterization of Powders, Suspensions, and Green Compacts," *J. Am. Ceram. Soc.*, **67**, 526 (1984).
7. C. Han, "Sintering of Bimodal Compact," M. Sc. Thesis, University of California at Los Angeles, Los Angeles, California, (1985).
8. C. Han, I. A. Aksay, and O. J. Whittemore, "Characterization of Microstructural Evolution by Mercury Porosimetry," in ***Advances in Material Characterization***, edited by R. L. Snyder, R. A. Condrate, Sr., and P. F. Johnson, Plenum Press, (1985), pp. 339-47.
9. I. A. Aksay and R. Kikuchi, "Structures of Colloidal Solids," in ***Science of Ceramic Chemical Processing***, edited by L. L. Hench and D. R. Ulrich, Wiley Interscience, New York, (1986), pp. 513-21.
10. F. F. Lange, B. I. Davis, and E. Wright, "Processing-Related Fracture Origins: IV, Elimination of Voids Produced by Organic Inclusions," *J. Am. Ceram. Soc.*, **69**, 66 (1986).
11. M. D. Sacks, "Rheological Science in Ceramic Processing," in ***Science of Ceramic Chemical Processing***, edited by L. L. Hench and D. R. Ulrich, Wiley Interscience, New York, (1986), pp. 522-38.
12. I. A. Aksay, W. Y. Shih, and M. Sarikaya, "Colloidal Processing of Ceramics with Ultrafine Particles," in ***Proc. Third Int. Conf. Ultrastructure Processing of Ceramics, Glasses, and Composites***, San Diego, California, February 23-27 (1987).
13. F. F. Lange and K. T. Miller, "Pressure Filtration: Consolidation Kinetics and Mechanics," *Am. Ceram. Soc. Bull.*, **66**, 1498 (1987).
14. C. H. Schilling and I. A. Aksay, "Slip Casting of Advanced Ceramics and Composites," in ***Transactions of the Canadian University-Industry Council on Advanced Ceramics Third Workshop***, edited by P. S. Nicholson, February 24-25, Montreal, Quebec (1987).

15. C. H. Schilling, G. L. Graff, W. D. Samuels, and I. A. Aksay, "Gamma Ray Densitometry: Non-Destructive Analysis of Density Evolution During Ceramic Powder Processing," in ***Proceedings of Twenty-Third University Conference on Ceramic Science, Atomic and Molecular Processing of Electronic and Ceramic Materials: Preparation, Characterization, and Properties***, Materials Research Society, August 30-September 2, Seattle, Washington, pp. 239-251 (1987).
16. C. H. Schilling, I. A. Aksay, "Gamma-Ray Attenuation Analysis of Packing Structure Evolution during Powder Consolidation," in ***Ceramic Transactions, Vol. I, Ceramic Powder Science II***, edited by G. L. Messing, E. R. Fuller, Jr., and H. Hausner, American Ceramic Society, Westerville, OH, pp. 800-808 (1988).
17. L. Stroosnijder, and J. G. De Swart. "Column Scanning with Simultaneous Use of ^{241}Am and ^{137}Cs Gamma Radiation," *Soil Science*, **118**, pp. 61-69, (1974).
18. B. J. Bridge, and N. Collis-George. "A Dual Source Gamma Ray Traversing Mechanism Suitable for the Non-Destructive Simultaneous Measurement of Bulk Density and Water Content in Columns of Swelling Soil," *Aust. J. Soil. Res.*, **11**, 83-92 (1973).
19. D. L. Nofziger, and D. Swartzendruber. "Material Content of Binary Physical Mixtures as Measured with a Dual-Energy Beam of Gamma-Rays," *J. Appl. Phys.*, **45** [12], 5443-5449 (1974).
20. R. J. Reginato, "Swelling and Shrinking of Soil In Situ as Determined by a Dual Energy Gamma-Ray Technique," Ph.D. Thesis, University of California at Riverside (1973).
21. J. K. Henshall, and D. J. Campbell. "The Calibration of a High Resolution Gamma-Ray Transmission System for Measuring Soil Bulk Density and an Assessment of its Field Performance," *J. Soil Sci.*, **34**, pp. 453-463 (1983).
22. J. W. Hitchon, "Use of a Radiometric Technique to Investigate the Settling of Aqueous Thoria Slurries," United Kingdom Atomic Energy Research Establishment Report CE/M 188, 1956.
23. H. A. Kearsey, and L. E. Gill. "A Study of the Sedimentation of Flocculated Thoria Slurries using a Gamma-Ray Technique," *Trans. Instn. Chem. Engrs.*, **41**, pp. 296-306 (1963).
24. S. A. Colgate, "Gamma-Ray Absorption Measurements," *Phys. Rev.*, **87** [4], pp. 592-601 (1952).
25. H. R. Rhodes, "Some Uses of Radioisotopes in Continuous Measurement," *Research*, **14**, pp. 147-153 (1961).

26. C. F. Cooper, and S. F. A. Miskin. "The Slip Casting Kinetics of Alumina by Gamma-Ray Absorption," *Proc. Brit. Ceram. Soc.*, No. 3 (Fabrication Science), pp. 135-154 (1965).
27. R. F. Deacon, and S. F. A. Miskin. "Gamma-Ray Absorption for the Continuous Observation of Slip Casting Kinetics," *Trans. Brit. Ceram. Soc.*, **63**, 473-486 (1964).
28. A. Charlesby, ***Atomic Radiation and Polymers***, Pergamon Press (1960).
29. M. Dole, ***The Radiation Chemistry of Macromolecules, Vol. 1 & 2***, Academic Press (1973).
30. M. B. Bruce and M. V. Davis, ***Radiation Effects on Organic Materials in Nuclear Plants***, Electrical Power Research Institute Report EPRI NP-2129, Project 1707-3, Final Report November 1981, Electrical Power Research Institute, Palo Alto, CA.
31. J. Th. H. Overbeek, "Recent Developments in the Understanding of Colloidal Stability," *J. Colloid and Interface Sci.*, **58**, 408 (1977).
32. W. Y. Shih, I. A. Aksay, and R. Kikuchi, "Phase Diagrams of Charged Colloidal Particles," *J. Chem. Phys.*, **86**, 5127 (1987).
33. W. Y. Shih, I. A. Aksay, and R. Kikuchi, "Reversible-Growth Model: Cluster-Cluster Aggregation with Finite Binding Energies," *Phys. Rev. A*, **36**, 5015 (1987).
34. F. A. L. Dullien, ***Porous Media: Fluid Transport and Pore Structure***, Academic Press, New York (1979).
35. F. A. L. Dullien, and P. N. Mehta, "Particle Size and Pore (Void) Size Distribution Determination by Photomicrographic Methods," *Powder Technol.*, **5**, 179 (1972).
36. A. S. Michaels, and J. C. Bolger. "The Plastic Flow Behavior of Flocculated Kaolin Suspensions," *Ind. Eng. Chem. Fund.*, **1**, 153-162 (1962).
37. B. A. Firth, and R. J. Hunter, "Flow Properties of Coagulated Colloidal Suspensions I. Energy Dissipation in the Flow Units," *J. Colloid Interface Sci.*, **57** [2], 248-256 (1976).
38. B. A. Firth, "Flow Properties of Coagulated Colloidal Suspensions II. Experimental Properties of the Flow Curve Parameters," *J. Colloid Interface Sci.*, **57** [2], 257-265 (1976).
39. R. J. Hunter and J. Frayne, "Flow Behavior of Coagulated Colloidal Sols V. Dynamics of Floc Growth under Shear," *J. Colloid Interface Sci.*, **76** [1], 107-115 (1980).
40. D. G. Thomas, "Transport Characteristics of Suspension VIII: A Note on the Viscosity of Newtonian Suspensions of Uniform Spherical Particles," *J. Colloid Interface Sci.*, **20**, 267-277 (1965).

41. S. I. Kim, "Modification of Hierarchical Structure and Rheological Behavior of Aqueous Ceramic Particulate Systems," Ph.D. Thesis, University of Washington, Seattle, WA (1989).
42. W. H. Shih, W. Y. Shih, S. I. Kim, and I. A. Aksay, "Scaling of the Elastic and Yield Behavior of Colloidal Gels," to be published.
43. H. D. Weymann, M. C. Chuang, and R. A. Ross, "Structure of Thixotropic Suspensions in Shear Flow: I. Mechanical Properties," *The Physics of Fluids*, **16** [6], 775-783 (1973).
44. C. C. Chou, and M. Senna, "Correlation Between Rheological Behavior of Aqueous Suspensions of Al_2O_3 and Properties of Cast Bodies: Effects of Dispersant and Ultrafine Powders," *Amer. Ceram. Soc. Bull.*, **66** [7], 1129-1133 (1987).
45. E. Jones, P. D. A. Mills, R. F. Stewart, and D. Sutton, "Structure Formation in Binary Dispersions," in ***Advances in Ceramics, Vol. 21: Ceramic Powder Science***, The American Ceramic Society, Inc., pp. 587-600 (1987).
46. J. D. F. Ramsay, "Colloidal Properties of Synthetic Hectorite Clay Dispersions I. Rheology," *J. Colloid Interface Sci.*, **109** [2], 441-447 (1986).
47. M. Yasrebi, "Kinetics of Flocculation in Alpha Alumina Colloidal Suspensions," M. Sc. Thesis, University of California at Los Angeles (1983).
48. K. E. Davis and W. B. Russel, "A Model of Crystal Growth in the Sedimentation and Ultrafiltration of Colloidal Hard Spheres," in ***Advances in Ceramics, Vol. 21: Ceramic Powder Science***, The American Ceramic Society, Inc., pp. 573-585 (1987).
49. D. S. Adcock and I. C. McDowall, "The Mechanism of Filter Pressing and Slip Casting," *J. Amer. Ceram. Soc.*, **40** [10], 355 (1957).
50. P. H. Dal and W. Deen, "Die Scherbenbildung Reim Keramischen Giessverfahren," in ***Proceedings of the Sixth International Ceramic Congress, Weisbaden***, Deut. Keram. Gessell, pp. 219-242 (1958).
51. K. Bridger, M. Tadros, W. Leu, and F. Tiller, "Filtration Behavior of Suspensions of Uniform Polystyrene Particles in Aqueous Media," *Separation Sci. Tech.*, **18** [12-13], 1417-1438 (1983).
52. T. J. Fennelly and J. S. Reed, "Mechanics of Pressure Slip Casting," *J. Amer. Ceram. Soc.*, **55** [5], 264-268 (1972).
53. I. A. Aksay and C. H. Schilling, "Mechanics of Colloidal Filtration," in ***Advances in Ceramics, Volume 9***, edited by J. A. Mangels and G. L. Messing, American Ceramic Society, Inc., Columbus, OH, pp. 85-93 (1984).
54. F. M. Tiller, and C. D. Tsai, "Theory of Filtration of Ceramics: I, Slip Casting," *J. Amer. Ceram. Soc.*, **69** [12], 882-887 (1986).
55. A. E. Scheidegger, ***The Physics of Flow Through Porous Media***, Third Edition, University of Toronto Press, Toronto, (1974).

56. T. J. Fennelly and J. S. Reed, "Compression Permeability of Al_2O_3 Cakes Formed by Pressure Slip Casting," *J. Amer. Ceram. Soc.*, **55** [8], 381-383 (1972).
57. S. C. Danforth and M. Velasquez, "Permeability of Filter Pressed Monsized SiO_2," ***Mat. Res. Soc. Symp. Proc.*** Vol. 24, Elsevier Science Publishing Co., pp. 239-243 (1984).
58. M. A. Janney and G. Y. Onoda, "Particulate Mechanics of Highly Loaded Ceramic Systems," in ***Advances in Ceramics, Vol. 21, Ceramic Powder Science***, Editors G. L. Messing, K. S. Mazdiyasni, J. W. McCauley, and R. A. Haber, American Ceramic Society, (1987), pp. 615-626.
59. G. Y. Baaklini and K. J. Roth, "Probability of Detection of Internal Voids in Structural Ceramics Using Microfocus Radiography," *J. Mater. Res.*, **3** [3], 457-467 (1986).
60. M. D. Weeks and J. W. Laughner, "Characterization of Unfired Ceramic Microstructures," ***Advances in Ceramics, Vol. 21: Ceramic Powder Science***, The American Ceramic Society, Inc., pp. 793-800, (1987).
61. A. M. Gaudin, and M. C. Fuerstenau. "The Transviewer-X Rays to Measure Suspended Solids Concentration," *Eng. and Mining. J.*, **159** [9], 110-112 (1958).
62. A. M. Gaudin, and M. C. Fuerstenau. "On the Mechanism of Thickening," Preprint, Int. Minin Proc. Congr. London, April 1960, pp. 115-127.
63. A. S. Michaels, and M. C. Bolger. "Settling Rates and Sediment Volumes of Flocculated Kaolin Suspensions," *Ind. Eng. Chem. Fund.*, **1**, 24-33 (1962).
64. M. Srivinasan, W. P. Binnie, R. A. Youngman, M. Sherman, and W. Friedman, "Characterization of SiC Powder and SiC Green Bodies," in ***Ceramic Transactions, Vol. I, Ceramic Powder Science II***, edited by G. L. Messing, E. R. Fuller, Jr., and H. Hausner, American Ceramic Society, Westerville, OH, pp. 792-799 (1988).
65. W. A. Ellingson, R. A. Roberts, P. S. Wong, M. W. Vannier, T. Hentea, R. J. Kriz, and J. J. Vaitekunas, "Nondestructive Testing for Structural Ceramics," DOE Advanced Research and Technology Development Fossil Energy Materials Program Semiannual Report for the Period Ending September 30, 1987, ORNL/FMP/2, pp. 27-58.
66. T. Taylor, W. A. Ellingson, and W. D. Koenigsberg, "Evaluation of Engineering Ceramics by Gamma-Ray Computed Tomography," *Ceram. Eng. Sci. Proc.*, **7** [7-8], 772-783 (1986).
67. A. Kose, M. Ozaki, K. Takano, Y. Kobayashi, and S. Hachisu, "Direct Observation of Ordered Latex Suspension by Metallurgical Microscope," *J. Colloid. Interface Sci.*, **44**, 330 (1973).
68. D. W. Schaefer and K. D. Keefer, "Structure of Random Porous Materials: Silica Aerogel," *Phys. Rev. Lett.*, **56** [20], 2199-2202 (1986).

69. D. W. Schaefer, K. D. Keefer, R. A. Shelleman, and J. E. Martin, "Fractal Growth and Structure in Ceramic Science," in ***Advances in Ceramics, Vol. 21, Ceramic Powder Science***, Editors G. L. Messing, K. S. Mazdiyasni, J. W. McCawley, and R. A. Haber, American Ceramic Society, (1987), pp. 561-566.
70. G. Dietler, C. Aubert, D. S. Cannell, and P. Wiltzius, "Gelation of Colloidal Silica," *Phys. Rev. Lett.*, **57** [24], 3117-3120 (1986).
71. K. A. Hardman-Rhyne, and N. F. Berk, "Small-Angle Neutron Scattering from Porosity in Sintered Alumina," *J. Amer. Ceram. Soc.*, **69** [11], C285-C287 (1986).
72. Underwood, E. E., ***Quantitative Stereology***, Addison-Wesley, Reading, MA, (1970).
73. D. G. Pickles, and E. Lilley, "Ultramicrotoming of Ceramic Powders for Electron Microscopy," *J. Amer. Ceram. Soc.*, **68** [9], C222-C223 (1985).
74. M. T. Postek, J. S. Howard, A. H. Johnson, and K. L. McMichael, ***Scanning Electron Microscopy: A Student's Handbook***, Ladd Research Industries, Inc., pp. 115-240 (1980).
75. G. J. Exarhos, K. F. Ferris, D. M. Friederich, and W. D. Samuels, "Molecular Spectroscopic Characterization of Binding Interactions in Phosphazene-Stabilized Alumina Dispersions," in ***Proceedings of the 23rd Conference in Ceramic Science, Atomic and Molecular Processing of Electronic and Ceramic Materials: Preparation, Characterization, and Properties***, Editors I. A. Aksay, G. L. McVay, T. G. Stoebe, and J. F. Wager, Materials Research Society, August 30-Sept. 2, Seattle, Washington, pp. 127-145 (1987).
76. G. J. Exarhos, K. F. Ferris, D. M. Friedrich, and W. D. Samuels, "^{31}P NMR Studies of Aqueous Colloidal Dispersions Stablized by Polyphosphazene Adsorption." *J. Amer. Ceram. Soc.* 71, [9] pp. C406-C407, (1988).

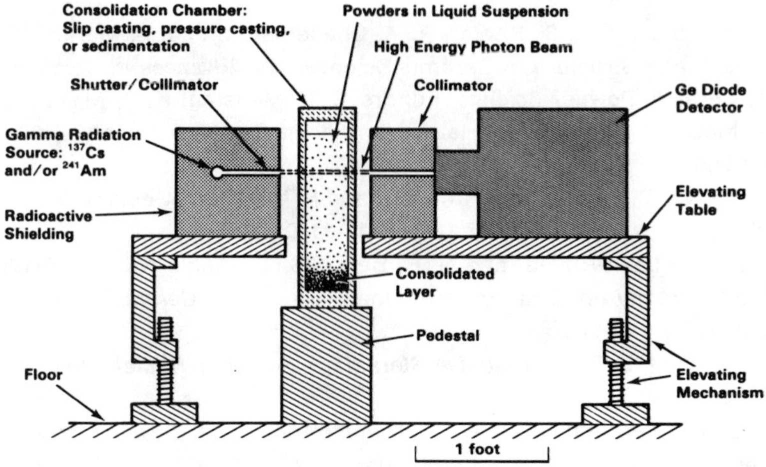

Figure 1. Schematic illustration of gamma-ray densitometer. A collimated beam of gamma radiation is directed through a ceramic as it is processed. An elevating table varies the beam position, and a germanium diode measures beam attenuation. Multiple phases in ceramic composites may be analyzed by multiple radioactive isotopes.

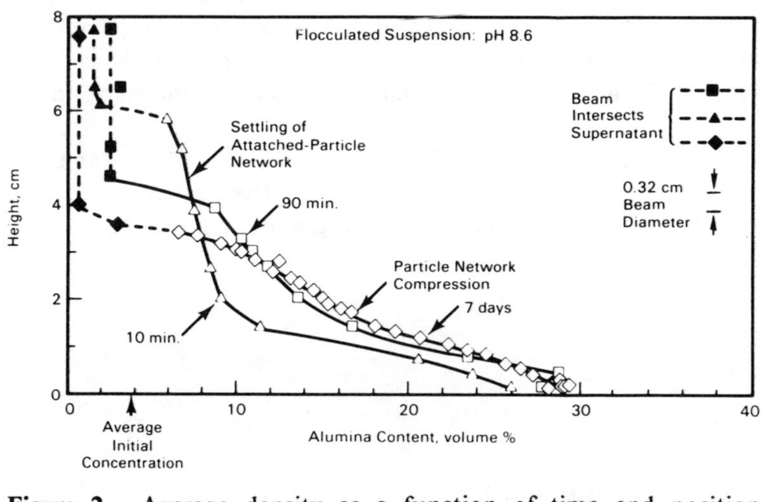

Figure 2. Average density as a function of time and position during sedimentation of flocculated alumina suspension at pH 8.6. The sediment is rapidly formed by settling and compaction of attached-particle networks, resulting in gradient packing densities with elevation. Attenuation was measured using a 3.1 mm diameter beam.

CHARACTERIZATION OF CERAMIC SUPERCONDUCTORS BY MAGNETICALLY MODULATED MICROWAVE REFLECTION (MMR)

N. S. Dalal, T. A. Mahl, P. K. Kahol, and J. P. DeLooze,
The Department of Chemistry, West Virginia University,
Morgantown, WV 26506-6045

J. S. Moodera, National Magnet Laboratory, Massachusetts Institute of Technology, Cambridge, Massachusetts 02139

A. S. Bhalla, Materials Research Laboratory, The Pennsylvania State University, University Park, Pennsylvania 16802

ABSTRACT

A microwave technique is described which provides a novel, contactless procedure for the characterization of high T_c ceramic superconductors. The method utilizes a standard electron paramagnetic resonance (EPR) spectrometer. The output signal is proportional to changes in the magnetically-modulated microwave reflection, hence the term MMR. The MMR methodology appears to have the potential for becoming a quick, highly-sensitive, and non-abrasive technique for determining the T_c and the 'relative' superconducting quality of a large number of samples. The technique is illustrated by its application to variously prepared $YBa_2Cu_3O_{7-x}$ samples.

INTRODUCTION

This presentation outlines what we believe is a highly-sensitive, quick, and reliable procedure for detecting superconducting properties of the recently discovered, high T_c, ceramic superconductors. This procedure is based on the fact that the surface impedance of these high T_c superconductors starts to decrease abruptly at the temperature of the onset of superconductivity ($\approx T_c$) in these materials. Perhaps an even more significant property of these materials is that their microwave impedance is a sensitive function of an externally applied magnetic field, especially at temperatures close to but below the T_c. It is also found that this high sensitivity of the microwave impedance is preserved when the externally applied magnetic field is modulated at audio frequencies up to 100 kHz. This makes it possible to take advantage of the high sensitivity inherent in the phase-sensitive detection and narrow band amplification techniques. Apart from some minor differences, the method was developed more-or-less simultaneously by Rao and coworkers [1], Moorjani *et al.* [2], Symko *et al.* [3], in our own laboratory [4], Müller and coworkers [5], Khachaturyan *et al.* [6], Rittori *et al.* [7], Rakvin and Dulcic [8,9], and Sastry [10]. Here we summarize some details of the methodology as developed in our laboratory, with illustrations of some typical results.

EXPERIMENTAL ARRANGEMENT

Figure 1 shows a schematic diagram of the experimental arrangement. The main part of the apparatus is an electron paramagnetic resonance (EPR) spectrometer incorporating a high Q, resonant microwave cavity. In our device, we employ a Bruker ER-200D, X-Band ($\approx$ 9.5 GHz) EPR spectrometer using a TE_{102} microwave cavity with Q $\approx$ 5,000. As is the case with most EPR spectrometers, the Bruker ER-200D model employs 100kHz magnetic field modulation and lock-in detection. The only significant modification needed is a pair of Helmholtz coils capable of generation a steady magnetic field of 0-50 Gauss magnitude in such a way as to nullify the 20-30 Gauss remnant field of the electromagnet of the EPR spectrometer.

The microwave reflectivity measurements are made by placing a few milligrams of the ceramic (powder or pellet form) in a 3-4 mm diameter quartz tube. This tube is positioned at the center of the cavity such that the sample is subjected to the maximum of the microwave magnetic field in the cavity (Fig 1(b)). The sample temperature is varied by a digital temperature controller and the microwave output is monitored as the power reflected from the cavity. As the sample becomes superconducting, its microwave impedance decreases, thereby resulting in a concomitant increase in the amount of the microwave power reflected from the cavity which constitutes the MMR signal. Thus, for a given sample, a plot of the MMR signal versus temperature would yield direct information on the evolution of superconductivity in the sample. We stress that this interpretation differs from that offered by Moorjani *et al.* [2] who ascribe the signal to a change in the T_c due to the 100 kHz modulating field

RESULTS AND DISCUSSION

Figure 2 shows some typical MMR signals from a 5 mg $YBa_2Cu_3O_{7-x}$ sample in the vicinity of T_c (≈ 91.5 K). It is clearly seen that the signal increases rapidly below T_c and can thus be used as a probe of the phase transition (≈ T_c). Figure 3 shows a plot of the signal amplitude as a function of temperature. The temperature at which the MMR signal starts to increase abruptly (≈ 91.5 K) corresponds to the T_c of this material, as measured independently via standard four probe resistivity measurements [4]. Fig. 4 illustrates the resolution of the methodology for measuring the T_c. The plot in Fig. 4 was obtained from a sample tube in which we deliberately placed two different preparations of YBaCuO, with their T_c's estimated to differ by about 1 K. The two steps observed at 91 and 92 K correspond to the T_c's of the two preparations, pointing to the surprisingly high resolution of the methodology for T_c determination. The results clearly demonstrate that the method provides a quick, sensitive and non-abrasive procedure for determining the T_c without any need for attaching contacts or metal deposition. This method should be valuable for examining effects of process variables on the change in the superconducting behavior of a given sample, since the same sample can be investigated by the MMR method at successive stages.

Finally, we mention that the MMR method can also be used for estimating the superconducting grain size via the change in the width of the MMR signal [4]. The same procedure also leads to an estimate of the lower critical field, H_{c1}, for these samples as demonstrated elsewhere [4,5,7].

REFERENCES

[1] S.V. Bhat, P Ganguly, T.V. Ramakrishna, and C.N.R. Rao, "Absorption of electromagnetic radiation by superconducting $YBa_2Cu_3O_7$: an oxygen-induced phenomenon," *J.Phys. C.* 20, L559 (1987).

[2] K. Moorjani *et al.*, "Superconductivity in bulk and thin films of $La_{1.85}Sr_{0.15}CuO_{4-\delta}$ and $Ba_2YCu_3O_{7-\delta}$," *Phys. Rev. B* 36, 8309 (1987).

[3] R. Durny *et al.*, "Microwave absorption in the superconducting and normal phases of Y-Ba-Cu-O," *Phys Rev. B* 36, 2361 (1987).

[4] J. Stankowski, P.K. Kahol, N.S. Dalal, and J.S. Moodera, "Possible Josephson oscillations spectra and electron paramagnetic resonance of Cu^{2+} in Y-Ba-Cu-O," *Phys. Rev. B* 36, 7126 (1987).

[5] K.W. Blazey *et al.*, "Low-field microwave absorption in the superconducting copper oxides," *Phys. Rev. B* 36, 7241 (1987).

[6] K. Khachaturyan, E.R. Weber, P. Tejedor, A.M. Stacey and A.M. Portis, "Microwave observation of magnetic field penetration of high-T_c superconducting oxides," *Phys. Rev. B* 36, 8309 (1987).

[7] C. Rettori, D. Davidov, I. Belaish, and I. Felner, "Magnetism and critical fields in the high-T_c superconductors $YBa_2Cu_3O_{7-x}S_x$ (x=0,1): An ESR study," *Phys. Rev. B* 36, 4028 (1987).

[8] M. Peric, B. Rakvin, M. Prester, N. Brnicevic, "Size of Josephson junctions in Ba-Y-Cu-O compounds," *Phys. Rev. B* 37, 522 (1988).

[9] A. Dulcic, B. Leontic, M. Peric, and B. Rakvin, "Microwave Study of Josephson Junctions in Gd-Ba-Cu-O Compounds," *Europhys. Letters* 4, 1403 (1987).

[10] M.D. Sastry, A.G.I. Dalvi, Y. Babu, R.M. Kadam, Y. Yakhmi and R.M. Iyer, "Possible role of Cu^{2+}- Cu^{4+} pairs in the superconductivity of $YBa_2Cu_3O_{7-x}$ from electron spin resonance observations *Nature* (London) 330, 49 (1987).

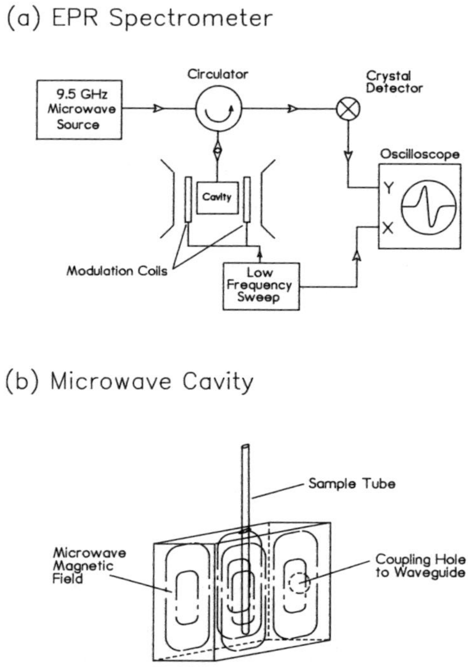

Fig. 1 Block diagram of the MMR apparatus.

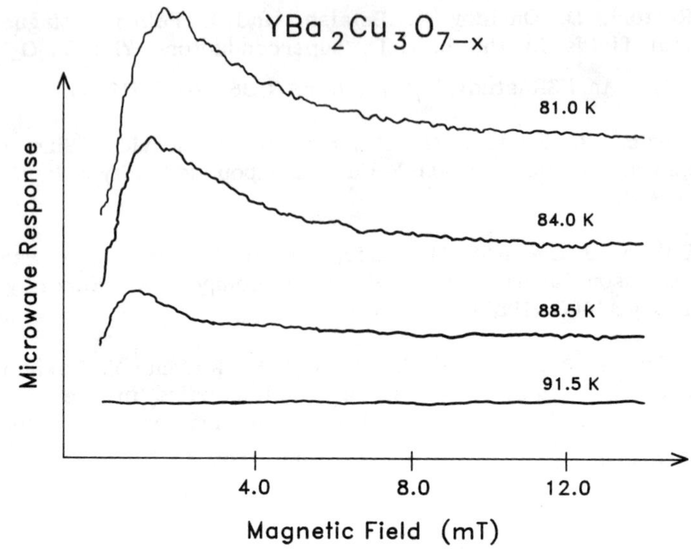

Fig. 2 Typical MMR peaks for a YBaCuO sample, T ≈ 91 K.

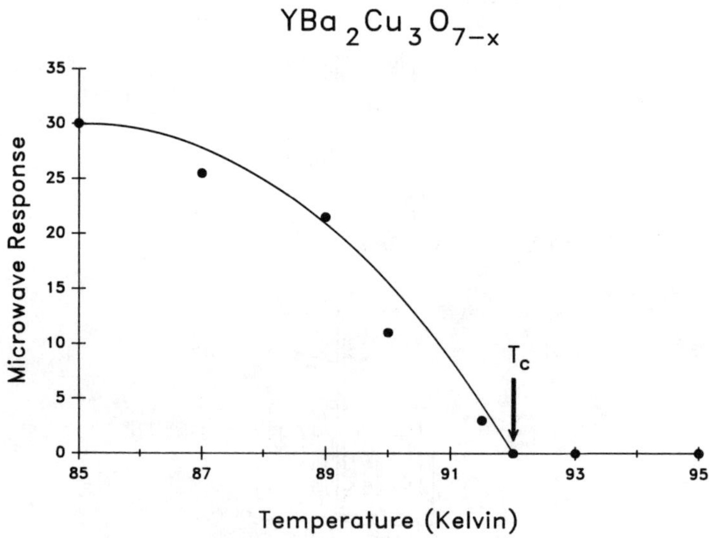

Fig. 3 Temperature dependence of the MMR signals for the YBaCuO sample used for the plots in Fig. 2.

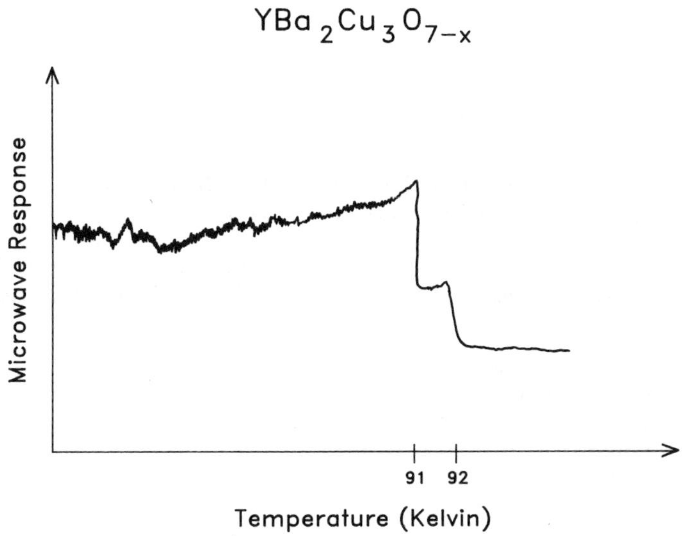

Fig. 4 MMR plots from a mixture of two YBaCuO samples with T_c's of 91 and 92 K, demonstrating the high resolution of the methodology for measuring T_c's of mixtures.

SYNTHESIS OF TOBERMORITE: NMR SPECTROSCOPY AND ANALYTICAL ELECTRON MICROSCOPY

Takeshi Mitsuda*, Hideo Toraya*, Yoshihiko Okada** and Masao Shimoda**

* Ceramics Research Laboratory
Nagoya Institute of Technology
Asahigaoka, Tajimi, Gifu-ken, 507 Japan
** Onoda Cement Co.
Toyosu, Tokyo, 135 Japan

ABSTRACT

Mixtures of CaO and silicic acid with or without γ-Al_2O_3, and CaO and quartz were hydrothermally treated in suspensions at 180°C. Reaction always proceeded through the formation of calcium silicate hydrate gels (C-S-H) to 1.1 nm tobermorite. For preparations using quartz, the C-S-H that were initially formed show a wide range of Ca/Si ratios following 0.5 h of processing. This ratio decreased as a function of processing time. With silicic acid, the initial reaction to give C-S-H having a Ca/Si ratio of 0.8 is very fast, but the subsequent crystallization to tobermorite is slower than with quartz. However, addition of Al markedly accelerates the conversion of C-S-H into Al-substituted tobermorite. ^{29}Si NMR spectroscopy results show that double chains are predominant in the tobermorites and single chains in the C-S-H.

INTRODUCTION

1.1 nm tobermorite, $Ca_5Si_6O_{18}H_2.4H_2O$ is an important constituent of some kinds of autoclaved calcium silicate products, such as aerated concretes and thermal insulation. In general, the formation of 1.1 nm tobermorite depends on the nature of the starting materials, particularly of silica. The sequence of hydrothermal reactions and associated kinetics have been studied in stirred suspensions for mixtures of CaO and various forms of silica with or without Al (1,2,3,4).

In this paper, the forming process of the 1.1 nm tobermorite is evaluated using analytical transmission electron microscopy and ^{29}Si NMR spectroscopy.

EXPERIMENTAL PROCEDURE

Starting Materials and Reaction Conditions

The starting materials were mixtures of CaO with reagent grade silicic acid (with or without added γ-Al_2O_3) and CaO with quartz. The CaO was made from reagent grade $CaCO_3$ at 1050°C. The γ-Al_2O_3 was Gamma Micropolish having a grain size below 0.05 µm. The quartz was 99.9% SiO_2 and exhibited particle size below 10 µm. The mixtures had Ca/Si ratios of 0.8; when γ-Al_2O_3 was added, the Ca/(Al+Si) ratio was 0.8 and the Al/(Al+Si) ratio was 0.10. All reactions were done at 180°C under saturated steam pressures. Those in stirred suspensions were carried out using a water/solids weight ratio of 20 and a stirring rate of 200 r.p.m. The products were filtered, washed, and dried in a vacuum at 80°C. All products were examined by X-ray diffractometry, electron microscopy and ^{29}Si NMR spectroscopy.

Examination of Products

NMR Spectroscopy

The ^{29}Si chemical shifts were recorded with respect to tetramethlysilane (TMS) on a JEOL GX-270 solid state high-resolution NMR spectrometer at 53.54 MHz using magic-angle sample spinning. As reference standards for silicate anion structure of calcium silicate hydrates, xonotlite (double chain), hillebrandite (single chain) and gyrolite (layer) were prepared; and natural tobermorite, specimen 11, described earlier (5) was used. The total measuring time was 20-30 h for each sample.

Analytical Electron Microscopy

All products were examined using an analytical transmission electron microscope (JEOL, JEM-2000 FX) fitted with an energy-dispersive X-ray detector (TN-5500). The samples were dispersed in isopropyl alcohol and were then settled on an electron-microscope grid. As standards for determining the correction factor, xonotlite and kaolinite were used for Ca/Si and Al/Si respectively. In the electron microscope, the samples consisted mainly of crumpled foils or fibres for C-S-H and of lath crystals for tobermorite; they were thin, and less than 20 nm thick.

RESULTS AND DISCUSSION

Reaction Sequence

Fig. 1 shows typical X-ray powder patterns of hydrothermally formed products using different mixtures at increasing processing times.

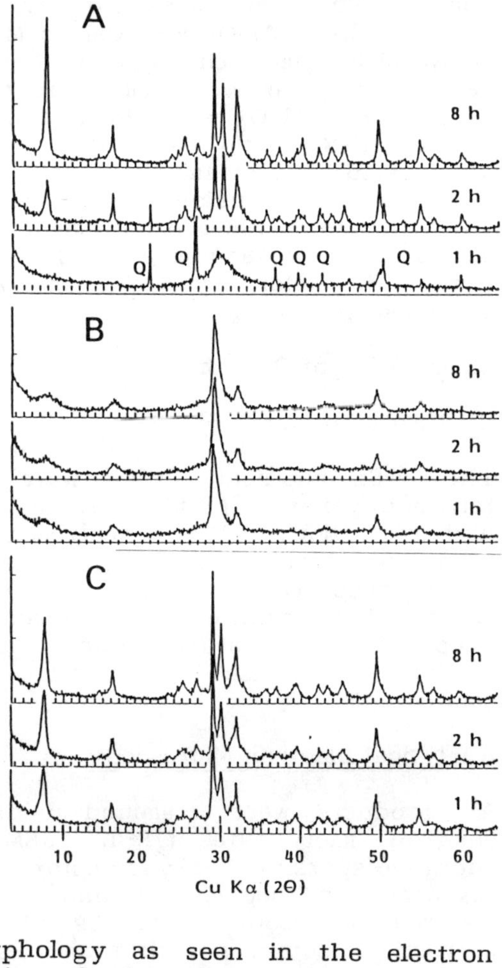

Fig. 1. X-ray diffraction patterns of products using (A) quartz + CaO, (B) silicic acid + CaO, and (C) silicic acid + CaO + γ-Al_2O_3.
Q = unreacted quartz.

Fig. 2 illustrates the morphology as seen in the electron microscope during the formation of tobermorite. The runs using the mixtures of CaO and silicic acid gave C-S-H with crumpled foils (A in Fig. 2) or fibrous aggregates (B), which did not

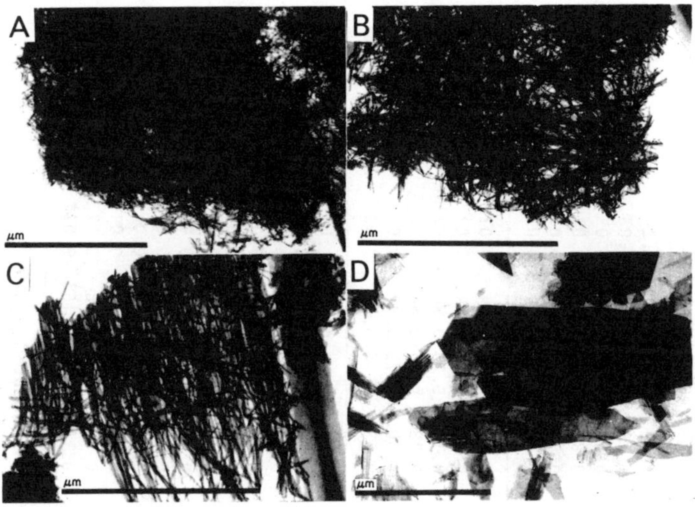

Fig. 2. TEM micrographs of typical morphology showing (A) crumpled foils, (B) aggregates of crumpled fibres, (C) groups of orientated fibres, and (D) well-formed plates of 1.1 nm tobermorite.

convert into highly crystalline tobermorite (plate or lath in D) even after 16 h. However, addition of Al markedly accelerated the conversion of C-S-H into Al-substituted tobermorite. On the other hand, the mixtures of CaO and quartz gave initially C-S-H with unreacted quartz, which subsequently reacted to give a highly crystalline tobermorite with increase in reaction time. These results agree with those reported previously by Mitsuda et al. (1,3).

Variations in Composition

Table 1 summarizes the results of analytical electron microscopy. There are significant differences in composition during formation of tobermorite by using mixtures of CaO and quartz. The mean Ca/Si ratio for the C-S-H aggregates prepared for 0.5 h was 1.69; this value agrees with the mean upper limit of C-S-H in previous characterization studies of hydrothermally formed C-S-H (6). With increase of reaction time, the C-S-H formed initially consisted of crumpled foils (A in Fig. 2) and, reacted further to give mixtures of crumpled foils and fibre aggregates (B) having the mean Ca/Si ratio of 1.13. The material finally crystallized to tobermorite (lath or plate in shape) at a ratio below 0.9.

Table 1. Results of Hydrothermal Runs and Analytical Electron Microscopy

Starting materials	Curing Time (h)	Products	Thermal behaviour	Morphology	No. of analyses	Ca/(Al+Si), Al/(Al+Si)* Mean	Range	Std. dev.
	0.5	C-S-H, Q	w	A	60	1.69	1.35-2.07	0.20
CaO	1.0	C-S-H, Q	w	A,B	59	1.13	1.04-1.17	0.03
+	2.0	Tob	N	C,D	70	0.89	0.83-1.01	0.03
Quartz	4.0	Tob	N	D	60	0.80	0.75-0.86	0.02
	8.0	Tob	M	D	26	0.79	0.75-0.84	0.02
	16.0	Tob	M	D	28	0.78	0.75-0.84	0.02
	0.5	C-S-H	w	A	51	0.81	0.77-0.84	0.02
CaO	1.0	C-S-H	w	A	56	0.81	0.77-0.84	0.02
+	2.0	C-S-H	w	A,B	75	0.79	0.75-0.84	0.02
Silicic	4.0	C-S-H	w	A,B	70	0.78	0.75-0.84	0.02
Acid	8.0	C-S-H	w	B	58	0.80	0.75-0.87	0.03
	16.0	C-S-H	w	B	60	0.82	0.75-0.89	0.03
	0.5	Tob	N	C,D	79	0.82	0.72-0.99	0.04
						0.10*	0.02-0.18*	0.03*
CaO	1.0	Tob	N	D	39	0.81	0.73-0.92	0.05
+						0.11*	0.08-0.14*	0.01*
Silicic	2.0	Tob	N	D	40	0.81	0.76-0.91	0.04
Acid						0.10*	0.06-0.13*	0.02*
+	4.0	Tob	N	D	40	0.81	0.73-0.89	0.04
γ-Al_2O_3						0.11*	0.08-0.12*	0.01*
	8.0	Tob	N	D	40	0.81	0.74-0.89	0.04
						0.11*	0.08-0.14*	0.01*
	16.0	Tob	N	D	40	0.80	0.71-0.88	0.02
						0.11*	0.08-0.15*	0.02*

Products; Q = unreacted quartz, C-S-H = poorly crystallized calcium silicate hydrate, Tob = 1.1 nm tobermorite.
Thermal behaviour: N = normal type, M = mixed type, w = basal spacing weakened.
Morphology: see text.

With the mixtures of CaO and silicic acid, the initial reaction to give C-S-H was very fast; the mean Ca/Si ratio was 0.81 after only 0.5 h, almost the same value as in starting mixtures. However, the subsequent crystallization to tobermorite was slower than with quartz; the fiber aggregates were retained even after a 16 h treatment. On the other hand, the addition of Al gave poorly crystallized Al-substituted tobermorite after only a 0.5 h treatment. For all Al-tobermorites, the mean range of Ca/(Si+Al) and Al/(Al+Si) ratios were 0.80-0.82 and 0.10-0.11 respectively, almost nearly the same as those of bulk mixtures.

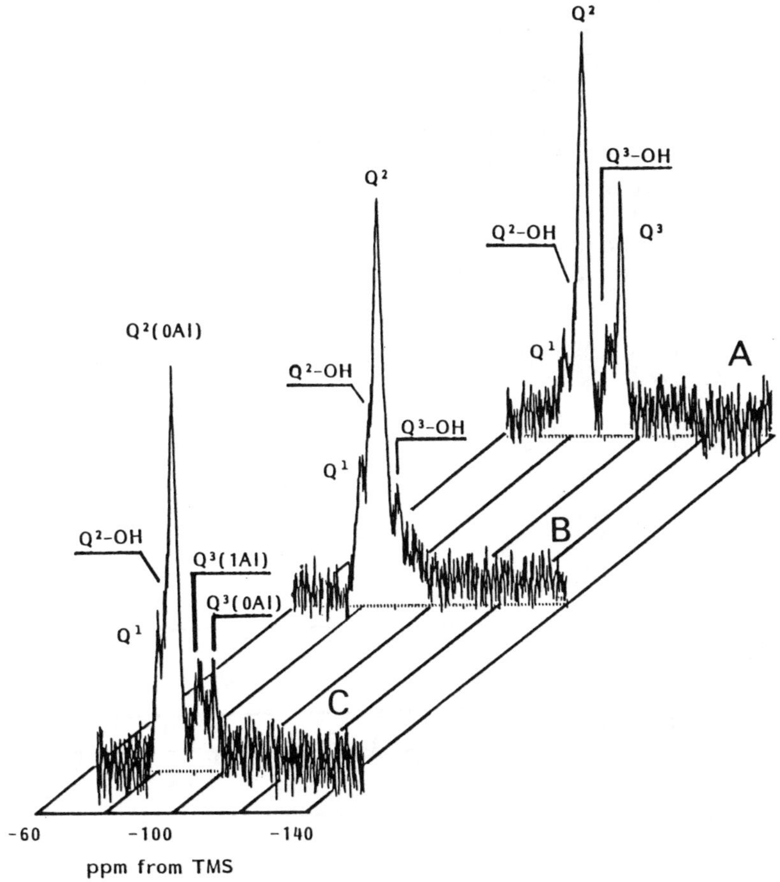

Fig. 3. ^{29}Si NMR spectra of 1.1 nm tobermorites (A,C) and poorly crystallized calcium silicate hydrate(B). Products were prepared at 180°C for 16 h using (A) quartz + CaO, (B) silicic acid + CaO, and (C) silicic acid + CaO + γ-Al_2O_3.

Silicate Anion Structure

Fig. 3 shows typical NMR spectra for tobermorite (A), C-S-H (B), and Al-substituted tobermorite (C), all of which were reacted for 16 h. The results indicate that the tobermorites prepared with or without Al exhibit the presence of chain middle groups Q^2 and branching sites Q^3 at -85 and -91 to -95 ppm, respectively; these shifts are very similar to those of synthetic xonotlite and natural tobermorite (anomalous type in thermal behaviour) showing Q^2 at -86 and -89, and Q^3 at -97 and -95 to -99 ppm, respectively. On the other hand, the C-S-H prepared from silicic acid without Al gives a Q^2 peak with very weak Q^3 feature. Fig. 4 shows the plots of the relative intensities of the Q^2 and Q^3 peaks in products at various times of hydrothermal treatment. The results indicate that highly crystalline tobermorite prepared from quartz gave a Q^2/Q^3 ratio of 1.6 after 8 h and 1.5 after 16 h. For the C-S-H using silicic acid without Al, the content of Q^3 begins to increase with time, but the Q^2/Q^3 ratios are higher than with quartz (5.0 after 4 h and 4.8 after 16 h). Al-substituted tobermorites prepared with added Al gave Q^2/Q^3 ratios between the C-S-H and the Al-free tobermorites.

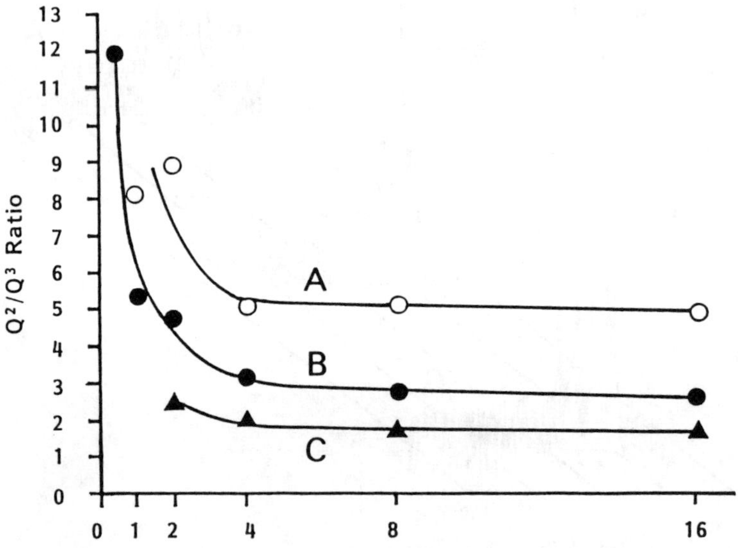

Fig. 4. Intensity ratio of Q^2/Q^3 in products prepared using (A) quartz + CaO, (B) silicic acid + CaO, and (C) silicic acid + CaO + γ-Al_2O_3.

In this paper, we confirmed the NMR results of synthetic tobermorites reported by Wieker et al. (7) and Komarneni et al. (8). Wieker et al. first showed that 1.1 nm tobermorite made from CaO and quartz at 140°C consists of double chain silicate anions, and exhibits an increasing Q^2/Q^3 ratio with increase of Ca/Si ratio of the starting mixtures. These results suggest that 1.1 nm tobermorite prepared with or without Al above 140°C gives mainly double-chain silicate anions. Hydrothermally formed C-S-H, prior to subsequent crystallization to tobermorite, yields small amounts of bridging sites between the chains, or mixtures of single and double chains.

ACKNOWLEDGEMENTS

We thank the Japanese Government for a grant-in aid for scientific research to T. Mitsuda and H. Toraya. We also thank the Cement Association of Japan for a research fellowship to H. Toraya.

REFERENCES

(1). T. Mitsuda and H.F.W. Taylor, "Influence of Aluminium on the Formation of Calcium Silicate Hydrate Gels into 11 Å tobermorite at 90° C and 120° C," Cem. Concr. Res., 5(3) 203-210 (1975).

(2). S.A.S. El-Hemaly, T. Mitsuda and H.F.W. Taylor, "Synthesis of Normal and Anomalous Tobermorites," Cem. Concr. Res., 7(4) 429-438 (1977).

(3). C.F. Chan, M. Sakiyama and T. Mitsuda, "Kinetics of the CaO - Quartz - H_2O Reaction at 120° to 180°C in Suspensions," Cem. Concr. Res., 8(1) 1-6 (1978).

(4). C.F. Chan and T. Mitsuda, "Formation of 11 Å Tobermorite from Mixtures of Lime and Colloidal Silica with Quartz," Cem. Concr. Res., 8(1) 1-6 (1978).

(5). T. Mitsuda, and H.F.W. Taylor, "Normal and Anomalous Tobermorites," Mineral. Mag., 42 229-235 (1978).

(6). T. Mitsuda, S. Kobayakawa and H. Toraya, "Characterization of Hydrothermally Formed C-S-H," 8th Int. Cong. on Chemistry of Cement, Vol. 3 173-178 (1986).

(7). W. Wieker, A.-R. Grimmer, A. Winkler, M. Mägi, M. Tarmak and E. Lippmaa, "Solid-State High-Resolution ^{29}Si NMR of Synthetic 14 Å, 11 Å and 9 Å Tobermorites," Cem. Concr. Res., 12(3) 333-339 (1982).

(8). S. Komarneni, R. Roy, D.M. Roy, C.A. Fyfe, G.J. Kennedy, A.A. Bothner-By, J. Dadok and A.S. Chesnick," ^{27}Al and ^{29}Si Magic Angle Spinning Nuclear Magnetic Resonance Spectroscopy of Al-Substituted Tobermorites," J. Mat. Sci., 20 4209-4214 (1985).

MEASUREMENT OF THE THERMAL DIFFUSIVITY OF TRANSLUCENT ALUMINUM NITRIDE

R. C. Enck, R. D. Harris, and R. A. Youngman
BP Research International
Research Center Warrensville
Cleveland, OH 44128

ABSTRACT

The laser flash method is the preferred technique for determining the thermal diffusivity for thin, high thermal conductivity materials. However, the optical translucency of AlN has created serious measurement problems for this technique and has caused substantial controversy in the interpretation of the experimental results for this material. We have developed an approach which eliminates these measurement ambiguities and eliminates the requirements for opaque surface coating and measurement correction factors by taking advantage of the strong AlN vibrational absorption in the spectral region near 14 µm and by using shorter laser pulses. This technique utilizes a short-pulse-length TEA CO_2 laser, operating between 9 and 11 µm, and a HgCdTe detector, operating between 11 and 14 µm. We routinely obtain values that are repeatable to ± 1%, are accurate to ± 3 - 5%, and are independent of sample thickness. A similar system to the one described here is being used by our Carborundum division for QA in its AlN substrate production facility.

INTRODUCTION

Aluminum nitride, because of its combination of very high intrinsic thermal conductivity and excellent electrical properties, is a promising new material for integrated circuit substrates and packages. Since the thermal conductivity of AlN is a strong function of a number of process and starting material parameters, it is critical that an accurate, rapid, non-destructive thermal conductivity measurement technique be available for process development studies and for production monitoring and quality assurance.

The technique of choice for thin, high thermal conductivity samples is the laser flash thermal diffusivity measurement.[1,2] This involves uniform pulse heating of the front surface of a disk shaped sample with a short, strongly absorbed laser pulse and monitoring the resultant time evolution of the temperature of the rear surface, usually with an infrared detector. Sample thermal diffusivity (α) is determined by fitting the shape of this temperature transient with a simple one-dimensional heat flow model.

The expression derived from this model[1] is shown in Eq. (1):

$$\frac{T(t)-T(0)}{T_m-T(0)} = 1+2\sum_{n=1}^{\infty} (-1)^n \exp(-n^2 \pi^2 \alpha\ t/L^2), \qquad (1)$$

where t is the time after the laser pulse, T(t) is the time dependent sample rear surface temperature, T(0) is the initial rear surface temperature before the laser pulse, T_m is the maximum rear surface temperature after the laser pulse, and L is the sample thickness.

Since the theory contains only the single adjustable parameter α, this fitting is usually done by setting the theoretical expression for $t_{1/2}$, the time after the laser pulse at which the rear surface temperature reaches 0.5 of its maximum value, equal to its experimental value. The resulting expression[1] for α is shown in Eq. (2):

$$\alpha = 0.1388 \frac{L^2}{t_{1/2}}. \qquad (2)$$

Thermal conductivity (λ) is obtained by multiplying α by the sample density and heat capacity (c_p). Accuracy of this fit is sometimes checked by comparing α derived from fits at different times (e.g. $t_{1/4}$, $t_{3/4}$) or from a non-linear least squares fit to some portion of the experimental curve.

There have been many reports of efforts to use the conventional laser flash thermal diffusivity measurement to determine thermal conductivity of AlN substrates.[3 - 8] These have met with limited success, particularly for high thermal conductivity materials, with investigators generally reporting very strong apparent sample thickness dependence of thermal diffusivity.[3 - 5] Dependence on laser pulse energy is also observed and has been proposed as one cause of the thickness dependence.[8] These effects have been variously ascribed to use of a laser pulse length comparable to or greater than $t_{1/2}$,[3] the transparency of the sample at the laser and detector wavelengths, operating the infrared detector in its non-linear response region,[8] having inadequate bandwidth in the signal amplification system,[7] and variation of sample thermal properties with laser pulse induced temperature changes.[8] These all violate the assumptions on which the theoretical analysis of the experiment depends and all probably make some contribution in most reported studies.

The problems caused by non-linear infrared detector operation and temperature dependence of α and c_p are eliminated by limiting laser pulse energy to a regime where these parameters are independent of pulse energy. Similarly, adequate amplifier bandwidth is demonstrated by lack of dependence of thermal diffusivity on sample thickness. Lack of thickness dependence of thermal diffusivity also demonstrates that laser pulse width is short enough that it has negligible effect on measurement of thermal diffusivity. Generally however, the conventional laser pulse width (typically 0.7 ms to 1 ms) is comparable to or greater than $t_{1/2}$ for a typical AlN substrate ($t_{1/2}$ = 0.65 ms for λ = 200 W/m•K and thickness = 0.635 mm (0.025")), the measured α is strongly thickness dependent, and very large correction factors (factors of 3 to 5 or greater) are applied to convert measured values to what are presumed to be the correct values of α. (Short pulse measurements at 1.06 μm have been shown to eliminate some of these problems.[6]) Finally, transparency of the sample to the usual laser wavelengths used (1.06 μm, 0.694 μm) and to the wavelength range of the usual infrared detectors used (InSb, < 5.5 μm) can also cause an apparent thickness dependence in α. Opaque coatings are applied to both sides of the sample to prevent this effect. The preferred coating is a thin (~ 0.1 μm) evaporated metal layer, frequently gold, followed by a thin sputtered or sprayed carbon/graphite layer. These layers are not thought to have a substantial effect on α, but no systematic study of this has been reported. The coatings must become a factor for thin enough samples and will cause an additional thickness dependence of α.

Given the multitude of possible sources of error inherent in the conventional implementation of the laser flash thermal diffusivity measurement and the fact that the correction factors applied can be large compared to the measured values, it seems that the accuracy of the technique is problematical, particularly for commercially important geometries and materials (thickness ≤ 0.635 mm, λ > 200 W/m•K). Multilayer sputter or vacuum coating is also very labor intensive and thus not suitable for large scale routine testing during process development or pilot plant studies. We will describe a laser flash thermal diffusivity measurement system which rapidly provides accurate results in real time and requires <u>no</u> coatings, <u>no</u> corrections to measured values, and <u>no</u> additional sample preparation.

EXPERIMENTAL

A schematic diagram of the experimental apparatus is shown in Fig. 1. The pulse width of the atmospheric pressure tunable TEA CO_2 laser (Model TEA-103-2, Lumonics, Inc.) is about 0.5 μs with maximum repetition rate of about 0.5 Hz. Maximum multimode tunable pulse energy that can be deposited on the front surface of the sample is about 10 joules in a 3 cm square beam. Tuning range is about 9.2 μm to 11 μm. All measurements were done using 11.79 μm (10P38 line) unless otherwise noted. The strong AlN vibrational absorption due to the Al-N "stretch" mode occurs between about 10 μm and 25 μm. We measure the absorption coefficient at 10.79 μm to be about 675 cm^{-1}, so that about 63% of the incident pulse

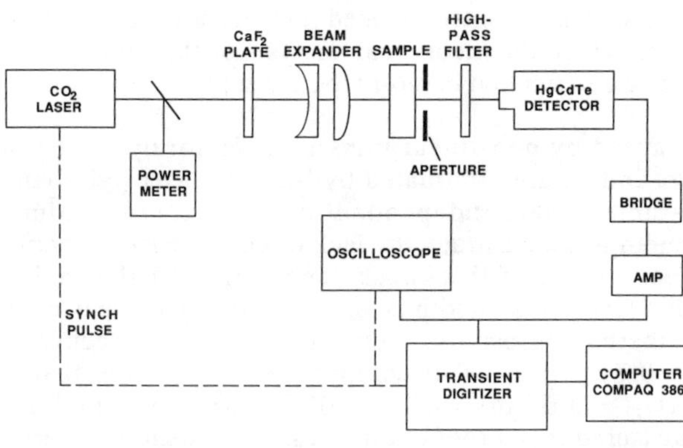

FIG. 1. Schematic diagram of the thermal diffusivity measurement apparatus.

is absorbed in the first 15 μm of the sample. CaF_2 plates of various thicknesses are used as filters to adjust pulse energy on the sample.

The sample is supported on a rubber plate with an aperture (≤ 11 mm) smaller than the sample size. For thick (> 1 mm) samples larger than 2.5 cm diameter, the 3 cm square beam is expanded with a 2X beam expander. (Thin 2" square, or larger, samples need no beam expansion since transverse heat flow is negligible during the measurement time.) Black body radiation emitted from throughout the sample (including the initially hot front surface) emerges from the sample back surface since AlN is transparent below about 10 μm (except for some relatively narrow multiphonon peaks near 8 μm). A long pass filter (Model LP-100, Balzers Optical Corp.) blocks the weakly absorbed light below 10.5 μm allowing us to sense only the radiation from (and thus the temperature of) the sample rear surface. The infrared detector (Model HCI-100, Infrared Associates) is a 2.5 mm square HgCdTe photoconductive device operating at 77 K and sensitive at wavelengths less than about 14.5 μm. The detector has an impedance of about 30 ohms and is operated in a low noise bridge circuit. Detector output goes into a low noise dc differential amplifier (Model 675, Perry Amplifier) with dc to 500 kHz bandwidth.

The electronics are dc coupled throughout. The Ithaco 1201 voltage amplifier has a maximum bandwidth of dc to 400 kHz. The LeCroy 6810 Waveform Digitizer has dc to 5 MHz bandwidth, 5 megasample/sec maximum sampling rate (0.2 μsec/point), and 12 bit resolution. The digitizer is triggered by a synch pulse from the laser power supply transmitted through a fiber optic link to eliminate noise coupling. A Compaq 386 computer running LeCroy CATALYST software in a DESQVIEW (Quarterdeck Office Systems) window acquires the waveform data (1024 points before and 3072 points after the laser pulse), averages a number of waveforms if necessary, reformats the data for use by the analysis program and saves both versions to disk. Sample information can be added from the keyboard. DESQVIEW then passes control to ASYST (ASYST Software Technologies), the analysis program, which reads the data from disk, calculates thermal diffusivity by two different methods, first matching theory and experiment at $t_{1/4}$, $t_{1/2}$, $t_{3/4}$ and then using a nonlinear least squares fit of the experimental waveform to the first four terms of the exponential series expression derived theoretically. It then displays the results, saves the results to disk, and provides a graphical comparison of theory and experiment if required. After incrementing the file number, control is passed back to CATALYST and acquisition of a new waveform is enabled. This entire process can be carried out under software control and takes less than 30 sec if the least squares fit is calculated and about 15 sec if it is not, exclusive of any manual data entry. Software revisions are in progress which will allow acquisition as well as analysis to be carried out in ASYST. This will reduce these times to 50% or less of the above values and allow entry of sample or test information from disk files or other programs. After a measurement session the data is merged into a database and is available for further analysis.

RESULTS

Figure 2 compares a theoretical and experimental thermal diffusivity transient for an AlN sample about 13 mm square and 0.607 mm thick. The experimental curve is the average of about 20 transients, each with a laser energy density of 0.3 J/cm^2 and

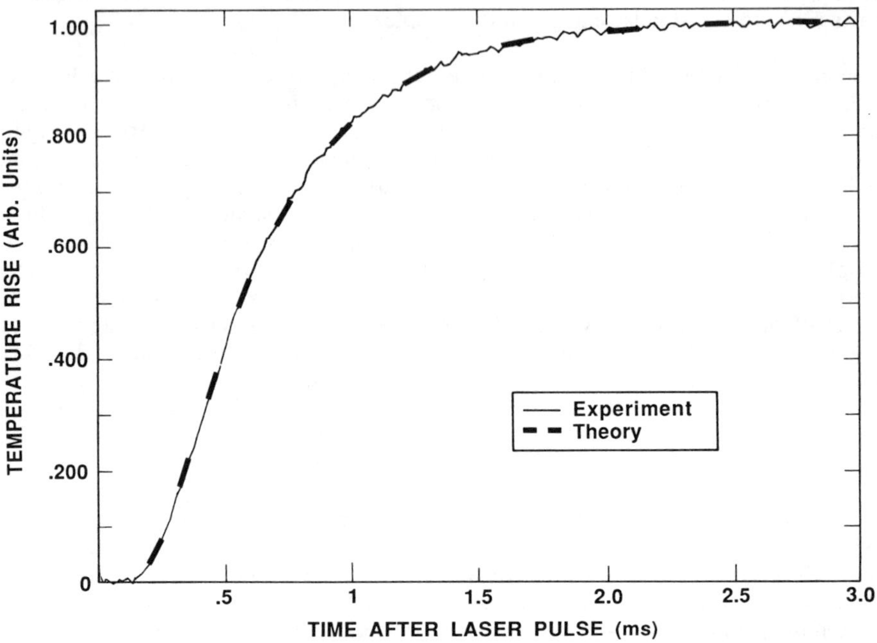

FIG. 2. Comparision of experimental and theoretical rear surface temperature transients.

separated by 10 seconds, long enough to eliminate effects of pulse repetition rate. The theoretical curve was calculated from Eq. (1) using terms through n = 4. Use of 4 terms of the series allows at least 96% of the waveform to be used for the least squares fit. Inclusion of higher order terms would be required to extend the fit substantially closer to the origin. The agreement is excellent and gives values for α and λ of 0.895 cm^2/s and 209.2 W/m•K, respectively, for this sample (using 0.172 cal/gK for AlN heat capacity). Thermal conductivity values derived from $t_{1/4}$, $t_{1/2}$, $t_{3/4}$ were 209.5, 211.9 and 212.5 W/m•K, respectively. For this measurement, the results from single point fits tend to be slightly larger (~ 1%) than from the least squares fit. This is because this measurement only lasts 5 or 6 half rise times so that the T_m used by the single point fits is about 1% below the true infinite time value, causing α to be 1% high. This effect disappears when longer measurement times are used and is avoided in the least squares fit. A second independent measurement gave 209.4 W/m•K using the least squares fit even though there was a spread of 10% among the three single point fits due to increased noise. This is representative of the improved precision obtained by fitting the entire curve.

The maximum rear surface temperature rise, T_m, has not been directly measured. However, for a 0.440 mm thick AlN sample and a laser pulse of 1.0 J/cm^2, the maximum possible T_m, if all the incident light is absorbed, is calculated to be 9 K.

Figure 3 shows α for this sample as a function of incident pulse energy density. α is constant within $\pm$ 1% below 0.5 J/cm^2 and is only down by about 3% at 1 J/cm^2. This indicates that there are no significant effects of sample temperature rise or non-linearity of temperature measurement in this range. All measurements reported in this study keep T_m near or below the value produced by 0.1 J/cm^2 in Fig. 3.

In order to test the absolute accuracy of our system for a wide range of potential applications, three thermal conductivity standards were used: NBS graphite (RM 8425, sample 023C; λ = 88.8 W/m•K, α = 0.736 cm^2/s, diameter (d) = 1.27 cm), NBS electrolytic iron (RM 8421; λ = 76.8 W/m•K, α = 0.218 cm^2/s, d = 3.17 cm), and commercial high purity copper (99.999%, Aesar; λ = 398 W/m•K,[9] α = 1.16 cm^2/s, d = 1.9 cm). Each standard was cut into a number of samples with a range of thicknesses. Thermal diffusivities were calculated using NBS or published values of density (confirmed by our own measurements) and values of heat capacity from the JANAF Thermodynamic Tables (Third Edition). An expanded beam was used for the iron measurements. Very thin colloidal graphite spray coatings (dgf-123, Miracle Power Products) were applied to both sides of the iron and copper samples to reduce their reflectivity. The results of this study for samples with thickness ranging from 0.192 mm to 4 mm and $t_{1/2}$ ranging from 0.14 ms to 100 ms are shown in Fig. 4. The solid lines show literature or NBS values for α. Each point is the average of at least 3 (average 4.4) measurements with each measurement an average of 10 waveforms. Generally 2 intensities and/or 2 sampling rates are included in each point. The maximum total spread in α within a point is $\pm$ 1% (less than the size of the symbols in the figure) except for the thickest samples (4 mm) and the 2 mm copper sample where it is less than $\pm$ 2%. All of the copper and iron measurements for samples less than 2 mm thick are within $\pm$ 3% of the literature values. All measurements, including those on graphite, are within $\pm$ 5% of the recommended values. Rapid baseline drift (on a 0.5 second time scale) in this dc coupled system is the apparent cause of the increased scatter for thick samples. The greater scatter observed for graphite may be due to material inhomogeneity or sample preparation technique and is being studied. Preliminary results indicate that at least $\pm$ 5% accuracy can be obtained using single transients by increasing incident laser energy (see Fig. 3) and using the non-linear least squares fit for analysis.

The lack of dependence of α on sample thickness demonstrated in Fig. 4 demonstrates that none of the system limitations discussed earlier are present over a

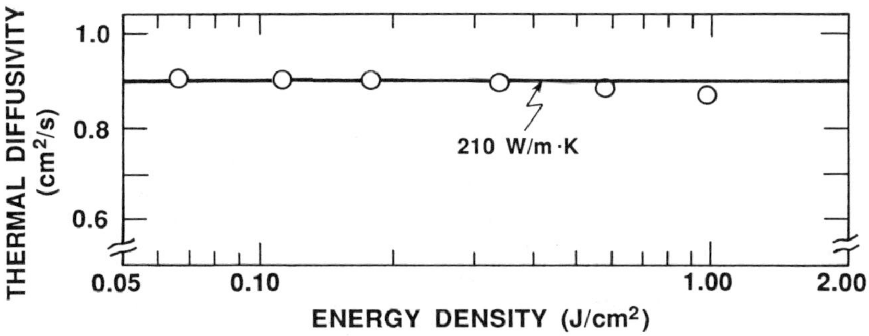

FIG. 3. Apparent dependance of thermal diffusivity on excitation laser pulse energy.

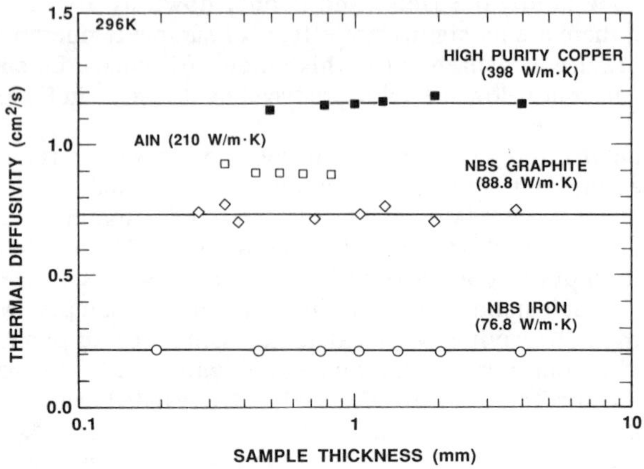

FIG. 4. Dependance of thermal diffusivity on sample thickness for AlN and three standard materials.

very wide range of sample parameters including all that are likely to be important for substrate applications. In fact there is no indication of a problem at the shortest $t_{1/2}$ (140 μs) so that thinner and/or higher thermal conductivity materials can be accommodated.

The remaining unanswered question concerns the effect of the optical transmission of AlN. Even with the relatively strong absorption at 10.79 μm, this effect must become important below some thickness. The reported strong thickness dependence occurs for samples thinner than about 3 mm. Shown in Fig. 4 is the thickness dependence of α for AlN from 0.341 mm to 0.822 mm. (Contrary to the case of the standard samples, this data results from progressive grinding of the same sample.) The lack of dependence of α on thickness except for a 3% rise for the thinnest sample shows that transparency is not a problem for thicknesses of current interest. This is confirmed by our preliminary observation that no substantial difference in α is seen for 0.635 mm AlN samples when switching between 10.79 μm and 10.59 μm excitation wavelengths, although there is a substantial difference in absorption coefficient.

CONCLUSIONS

We have described the construction, operation and testing of a rapid, automated, accurate laser flash thermal diffusivity measurement system optimized for thin, high thermal conductivity materials, particularly AlN. Thermal diffusivity of AlN substrates can be rapidly measured with no additional sample preparation, no opaque coatings, and using no correction factors. For thin samples (≤ 1 mm, thicker with an expanded beam) arbitrarily large plates can be tested. This is done either in real time or batch mode with absolute accuracy of ± 3 - 5%. (Our repeatability of ± 1% suggests that accuracy can be improved.) This has been accomplished by using a laser pulse (0.5 μs) short compared to rear surface temperature risetimes and laser

(10.79 μm) and detector (> 10.5 μm) wavelengths which are strongly absorbed by AlN. The system is directly applicable to other similar materials (e.g. Al_2O_3, SiC, BeO), although the degree of optical absorption at the laser wavelength must be individually determined.

This array of capabilities makes this system ideal both for process research where we provide rapid turnaround for many experimental samples with minimum manpower investment, and for production development and quality assurance use where we have a second system, which, when completely automated, will provide very high volume capability.

REFERENCES

1. W. J. Parker, R. J. Jenkins, C. P. Butler and G. L. Abbott, "Flash Method of Determining Thermal Diffusivity, Heat Capacity and Thermal Conductivity," J. Appl. Phys., 32(9) 1679-84 (1961).

2. R. E. Taylor and K. D. Maglic, "Pulse Method for Thermal Diffusivity Measurement", Compendium of Thermophysical Property Measurement Methods 1, Edited by K. D. Maglic, A. Cezairliyan, and V. E. Peletsky, Plenum Press, New York (1984).

3. Technical Bulletin, "Measurement of Thermal Conductivity of SHAPAL AlN Substrate," Tokuyama Soda Co. Ltd., Tokyo, Japan Feb(1987).

4. Nobuyuki Kuramoto, Hitofumi Taniguchi and Isao Asa, "Sintering and Properties of High Purity AlN Ceramics", presented at International Symposium on Ceramic Substrates and Packages, Oct. 19-21 (1988), Denver.

5. J. L. Hastert, "A Simple Technique for Thermal Diffusivity Measurements," presented at International Symposium on Ceramic Substrates and Packages, Oct. 19-21 (1988), Denver.

6. A. K. Knudsen, C. A. Langhoff, and S. H. Delzer, "Apparatus for the Measurement of Thermal Conductivities via the Laser-Flash Technique" presented at Annual Meeting of American Ceramic Society, April 26-30 (1987) Pittsburgh, PA.

7. A. J. Whittaker, H. M. Zhang, D. G. Onn, "Thermal Conductivity of Ceramic Fibers", presented at Annual Meeting of American Ceramic Society, May 1-5 (1988) Cincinnati, OH.

8. D. P. H. Hasselman and G. A. Merkel, "On the Specimen Size Effect of the Thermal Diffusivity/Conductivity of Aluminum Nitride", Presented at Pacific Coast Regional Meeting of American Ceramic Society, Oct. 23-26 (1988), San Francisco, CA. To be published.

9. Y. S. Touloukian, R. W. Powell, C. Y. Ho, and P. G. Klemens, Thermophysical Properties of Matter, Vol. 1, IFI/Plenum, New York (1970).

Section III. Electronic Techniques

DLTS INVESTIGATION OF ELECTRONIC CERAMICS

James F. Cordaro

James C. Simpson and Rodson A. Winston
New York State College of Ceramics
at Alfred University
Alfred, New York 14802

ABSTRACT

Deep level transient spectroscopy (DLTS) has become an important analytical technique in the semiconductor industry. This technique has a large potential to assist in the development of a fundamental understanding of grain boundary controlled electronic ceramic devices such as varistors and thermistors, yet has been used infrequently in ceramics. Applications of DLTS to the characterization of electronic defects in the depletion regions of grain boundaries in polycrystalline semiconducting ceramics and to grain boundary interface states are discussed. Considered in this paper are methods which utilize lock-in amplifier, dual-gated boxcar averager, Fourier analysis, and modulating functions. Experimental problems encountered such as nonexponential capacitance transients and distributions of defect energies are also addressed.

I. INTRODUCTION

Since its development approximately 14 years ago, deep level transient spectroscopy (DLTS) has been used extensively in the semiconductor field to characterize electronic defects in Si, Ge, and the III-V compound semiconductors.[1] The utilization of DLTS in the ceramics field to characterize defects in oxide compound semiconductors however, has not been so extensive and limited mainly to ZnO for varistors.[2]

Deep levels present in conventional p-n junction semiconductor devices are often detrimental to device performance. In contrast, ceramic devices such as positive temperature coefficient of resistance (PTCR) thermistors and varistors operate solely due to the presence of deep levels, both interface states at grain boundaries and bulk states within the grains. It thus appears appropriate to utilize techniques such as DLTS and admittance spectroscopy[3] in order to determine the electronic signatures and origins of deep levels in such devices. This can help to remove much of the empiricism now used in improving device performance.

There are several problems encountered in working with oxide compound semiconductors which often make the interpretation of data difficult. Low purity by comparison with Si, Ge, and GaAs, large nonstoichiometry, sensitivity to oxygen, and polarization from ion drift are often present. In addition, other problems, although not specific to ceramics, are usually encountered such as nonexponential capacitance transients, combined signals from both interface states and bulk states in polycrystalline materials, and distributions in the energies of deep levels. These problems should be addressed during interpretation of data from oxide semiconductors and extra caution needs to be exercised during experiments.

II. EXPERIMENTAL METHODS

DLTS measurements start with the repetitive filling and then emptying of traps in the depletion region of a semiconductor, while the junction is monitored by a fast capacitance meter. Capacitance transients are produced by the relaxation of the junction as trapped carriers are released from deep levels within the depletion region. The capacitance transients, F(t), are usually assumed to have the general form of eq. (1).

$$F(t) = A \exp(-et) + B \qquad (1)$$

e is the trap emission rate, A is the initial change in capacitance, B is a baseline capacitance and t is time. The filling of empty traps is normally accomplished by coupling a fast voltage pulse into the bias applied to the junction. Two methods which are used are a pulse transformer[4] and an opto-isolator pulse switching circuit.[5]

Experimental methods differ primarily in the analysis of capacitance transient signals. The analysis of these transients is aimed at finding the emission rates of the traps as a function of temperature. Two general categories of methods are used; methods which are based upon a linear filtering operation of F(t) and methods which are used to calculate the emission rates of the traps directly from the transients. The methods which utilize a linear filtering operation include lock-in amplifier[6], dual-gated boxcar averager[1], and correlation spectrometer[7] whereas Fourier analysis and modulating functions[8] calculate the emission rates directly.

Of the first group of methods, the DLTS signal is derived basically from a linear filtering operation of the capacitance transient, F(t). This operation can be described by eq. (2).

$$s(t_p) = \int_{t_p} F(t)w(t)dt \qquad (2)$$

t_p is the period of the transient and w(t) is a weighting function. Lock-in amplifier, dual-gated boxcar averager, and correlator methods all have different weighting functions which result in different signal to noise ratios. The actual DLTS signal, $S(t_p)$, is the time average of $s(t_p)$ over period t_p, i.e. $S(t_p) = s(t_p)/t_p$. In the specific case of the lock-in amplifier, eq. (2) is modified slightly and the output is given by eq. (3).

$$S(t_p) = \frac{1}{t_p} \int_{t_p} f(t)w(t)dt \qquad (3)$$

In eq. (3), f(t) is the fundamental Fourier component of the capacitance transient, F(t).

The above methods utilizing a linear filtering operation are all real-time methods in that the DLTS signal is continuously recorded as the temperature of the specimen is changed, generally from 77 K to 300 K. The temperature of the peak DLTS signal is then noted. Several scans over temperature must be made in order to obtain trap characteristics; in the cases of the lock-in amplifier and correlation spectrometer by using different t_p, and in the case of the dual-gated boxcar averager by using different gate widths and times. There is some advantage however, to recording the transients directly using a digital computer for later analysis. In this way, complete information on trap energy and capture cross section can be obtained from only one scan over temperature. The operation in eq. (2) is carried out numerically, using weighting functions which, for example, simulate a dual-gated boxcar averager.

Fourier analysis and modulating functions can be used to calculate time constants for each recorded transient as a function of temperature. Consider Fourier analysis as an example. If $f(\omega)$ is the Fourier transform of F(t) in eq. (1), as defined by (4)

$$f(\omega) = \int_0^\infty F(t) \exp(i\omega t)\, dt \qquad (4)$$

then the real and imaginary components of $f(\omega)$ are given by eqs. (5) and (6).

$$\mathrm{Re}\,[f(\omega)] = \frac{Ae}{e^2 + \omega^2}\,[\,1 - \exp(-et_p)] \qquad (5)$$

$$\mathrm{Im}\,[f(\omega)] = \frac{A\omega}{e^2 + \omega^2}\,[\,1 - \exp(-et_p)] \qquad (6)$$

From (5) and (6), the emission rate, e, for each transient is obtained by the following expression.

$$e = \frac{\omega \, \mathrm{Re}\,[f(\omega)]}{\mathrm{Im}\,[f(\omega)]} \qquad (7)$$

In practice, Re [f(ω)] and Im [f(ω)] in eq. (7) are carried out numerically using a discrete Fourier transform. Modulating functions[8], by comparison with the above Fourier analysis, has an advantage in being able to fit multiple exponentials to the transients and hence separate traps with similar emission rates.

When the emission rates have been obtained as a function of temperature, the trap energies are calculated by constructing an Arrhenius plot, as $\ln(e_n/T^2)$ versus $1/T$, according to the relationship[9] in eq. (8), as given for an electron trap.

$$e_n = \frac{s_n v_{th} N_c}{g} \exp(-E/kT) \qquad (8)$$

e_n is the electron emission rate, N_c the conduction band effective density of states, v_{th} the thermal velocity, s_n the capture cross section, E the trap energy, and g the spin degeneracy. It is assumed here that the capture cross sections are independent of temperature.

III. RESULTS ON POLYCRYSTALLINE ZINC OXIDE

Figures 1(a) and 1(b) show a double Schottky barrier at a grain boundary in a semiconducting ceramic with no bias and with an applied bias V_a, respectively. The depletion regions are typically 0.1 μm wide in ZnO varistors. Interface states are shown as several levels, each of which has a distribution in energy. The bulk states are deep levels which reside in the depletion regions and into the interior of the grains. The densities of these bulk states are not likely to be uniform with distance from the grain boundary. In this work, the DLTS signals, as for example those shown in Figure 2, originate mainly from the bulk states.

Figure 2 is a scan of a commercial GE V100ZA4 ZnO varistor material. The material was ground to 0.43 mm thickness and polished to a 15 μm surface finish. Aluminum ohmic contacts were vacuum evaporated onto two opposing surfaces of the specimen. The data were taken with no bias applied to the specimen. A majority carrier voltage filling pulse of 20 V and 50 μs duration was applied at a frequency of 20 Hz using a pulse transformer. The transients, each consisting of 400 data points, were recorded digitally and analyzed according to eq. (2) by using a weighting function which simulates that of a lock-in amplifier. The starting time of the weighting function was 0.001 s while the ending time was varied from 0.007 s to 0.01, 0.02, 0.03, and 0.045 s. This resulted in calculated emission rates of 2.89, 2.13, 1.15, 0.787, and 0.534 (x 10^2) sec^{-1}. One peak indicating an electron trap is clearly shown.

The results of a similar analysis of polycrystalline ZnO are shown in Figure 3. In this case, reagent grade, undoped ZnO was pressed into a pellet and then sintered at a temperature of 1083 C in air for a period of 12 hr. Since the double Schottky barriers are not developed in this material, the material was ground, polished, and fabricated into a metal-semiconductor Schottky diode. A silver Schottky contact was applied to one face of the material, while an ohmic aluminum contact was applied to the other face. This resulted in current rectifying characteristics. A 3 V reverse bias was applied to the diode. Majority carrier voltage filling pulses of 3.0 V and 50 μs duration were applied at a frequency of 5 Hz using an opto-isolator pulse circuit. The transients were collected digitally and analyzed in a similar fashion to that of the commercial varistor. The calculated emission rates for the curves in Figure 3 are 33.3, 21.8, 16.2, and 12.5 sec^{-1}. Three electronic defects, E1, E2, and E3 present in the depletion region of the Ag/ZnO contact are clearly visible, although the peak associated with E2 is very weak.

An Arrhenius plot of the calculated emission rates versus peak temperatures is given in Figure 4. Trap E1 in the pure ZnO has a calculated energy of 0.15 eV and a capture cross section of 7.8 x 10^{-18} cm^2. It is likely that this is the same defect which is found in other commercial varistors using admittance spectroscopy[10].

Although the emission rates for trap E2 in pure ZnO have a scatter due to the very small peak encountered, these rates follow exactly the rates found for the trap in the commercial varistor. The energies and capture cross sections of the two defects match, indicating that exactly the same defect is responsible for both peaks. Using a composite of the information from the two materials, the energy is calculated to be 0.32 eV and the capture cross section is 2.7 x 10^{-15} cm^2. This is the same defect[10] observed previously in varistors and in single crystal ZnO[11]. The defects are very likely to be defects native to ZnO, and therefore due to the first or second ionizations of either zinc interstitials or oxygen vacancies.

The interpretation of defect E3 in Figure 3 is not straightforward. The peak found is extremely broad. The emission rates are plotted in Figure 5 which yields not a straight line, but rather a curve. The calculation of a trap depth from this is thus highly questionable. There are several possible reasons why the data does not yield a straight line as in traps E1 and E2. One explanation is a capture cross section which could be dependent upon temperature. In this case the emission rate from this trap will not strictly follow a simple Arrhenius behavior. Other explanations include a distribution of trap energies, contributions from interface states, and instrumental error in the capacitance meter due to the impedance of the junction.

Trap densities in the materials can be obtained from calculations of the amplitudes of the capacitance transients[4].

$$N_T \cong 2\,(N_D - N_A)\,\frac{\Delta C}{C} \qquad (9)$$

N_T is the trap density, N_D and N_A are the donor and acceptor densities respectively, ΔC is the capacitance change due to filling the traps, and C is the capacitance due to the depletion regions. For the commercial varistor, N_T from the scan in Figure 2 is estimated at 10^{13} cm^{-3} using an estimated donor density of 10^{17} cm^{-3}.

IV. NONEXPONENTIAL TRANSIENTS

Some difficulties arise in the interpretation of capacitance transients from materials such as ZnO varistors. One of the problems is that the capacitance transients can be nonexponential. All DLTS studies of ZnO varistors to date are based upon the assumption that the capacitance transients follow simple exponential behavior as presented in eq. (1). This assumption is valid only to a first approximation. When the transients are nonexponential, and analyzed by methods such as lock-in amplifier, dual-gated boxcar averager, and the Fourier transform of eq. (1), the reported values of trap energy and capture cross section are only as accurate as the original assumption. Figure 5 shows a such a nonexponential transient obtained from a Schottky contact on single crystal ZnO. The calculated linc is from the Fourier analysis method. The transients obtained on commercial varistors show similar behavior. The same problem in the frequency domain, i.e. using admittance or impedance, is referred to as non-Debye behavior.

There can be several reasons for such nonexponential behavior. The presence of more than one deep level, states having a distribution of energies, and a broad Debye tail[12] to the edge of the depletion region are possible causes. Distinguishing between such possible causes remains a problem. Modulating functions is good as a first approach to determine if there is more than one trap present having very similar emission rates. If the time constants obtained are well-behaved on an Arrhenius plot, the presence of more than one deep level becomes clear. If the time constants are not well-behaved, other causes must be sought. Fitting two time constants by modulating functions to the transients for single crystal ZnO as in Figure 6 produces one group of time constants with an energy of 0.30 ± 0.02 eV and another group of time constants which are not well-behaved. This suggests that the nonexponential behavior in this case is not due to the presence of more than one deep level.

V. DISTRIBUTION OF DEFECT ENERGIES

The presence of a deep level having a distribution of energies is possible both for bulk states and interface states. In solids such as ZnO and TiO_2, the deep levels are also likely to be the native donors where the native donors can have two possible ionizations and reside more than 0.1 eV from the band edge. Since these materials are wide band gap semiconductors, defect densities close to the free electron densities might then be expected. When the solids are very nonstoichiometric, some broadening of the energy level can occur. As suggested previously, a broadening of energy levels may lead to nonexponential transients. Such a broadening might be detected by the method of Hjalmarson, Samara, and Farmer[13] in which a plateau in the suggested DLTS spectrum might be observed.

Interface states reside at or near (within ~1 nm) of the grain boundary interface and originate from local nonstoichiometry (i.e., uncompensated bonds) or impurity atoms which have segregated to the interface. The interface states may or may not exist in discrete levels. In ZnO varistors, studies by admittance spectroscopy and other techniques[14] suggest that several interface levels may exist, each with a distribution in energy. Detection of interface states[15] in polycrystalline semiconductors is generally through the application of a bias across the double Schottky barrier and the use of majority carrier voltage filling pulses. Interface states in a pure, conventional, semiconductor are readily detectable by this method. The distribution of the interface density of states with respect to energy is obtained by systematically varying the bias on the junction in several experiments. Increasing the bias forces the Fermi level at the interface toward shallower energies as illustrated in Figure 1(b). In this way, the Fermi level can be made to sweep through the interface states.

In an electronic ceramic with many impurities (a commercial varistor contains on the order of 10 chemical components) and several deep bulk levels, some portion of the signal will originate from bulk states in the depletion regions. Thus, separation of the signals which originate from the interface states, from those which originate from the bulk states, becomes important. One method for performing this with certainty is to apply a metal Schottky contact to one grain on a chemically cleaned surface of the same material, in order to avoid signals from the double Schottky barriers.

The capacitance transients arising from a distribution in trap energy at an interface, as pictured in Figure 1(b), might be expected to have an integral form similar to that given by eq. (10).

$$F(t) = \int_E A(E) \exp(-e(E)t)\, dt + B \qquad (10)$$

In (10), F(t) is the transient, A is the amplitude, B is a baseline capacitance, E is trap energy, and t is time. The amplitude A and the emission rate e are both functions of energy. In this case, obtaining the interface density of states above the zero bias Fermi level possibly requires systematic application of bias and a solution to the signals in the form of eq. (10). Finally, provided the electronic signatures of the interface states can be obtained experimentally using DLTS, the identification of the origin of the states is difficult without complementary calculations of the energy levels associated with dangling bonds[16] and impurities.

VI. CONCLUSION

Capacitance transient spectroscopy techniques should be applicable to a wide variety of semiconducting ceramic materials including ZnO, TiO_2 and other titanates, and semiconducting metal oxides in general which exhibit normal band-type conduction. Problems can arise however, such as those due to the typically low purities found and deep native donors and acceptors. Investigations of deep levels, particularly interface states can help to remove some of the empiricism now present in electronic ceramic components.

ACKNOWLEDGEMENT

This research was supported by NSF grant DMR-8709582.

REFERENCES

1. D. V. Lang, "Deep-Level Transient Spectroscopy: A New Method to Characterize Traps in Semiconductors," J. Appl. Phys., 45 [7] 3023-3032 (1974).

2. N. Shohata, T. Matsumura, and T. Ohno, "DLTS Measurement on Non-Ohmic Zinc Oxide Ceramic Varistor," Jpn. J. Appl. Phys., 19 [9] 1793-1794 (1980); T. Arakawa, H. Kurachi, M. Tsujita, and J. Shiokawa, "Deep-Level Transient Spectroscopy of Rare-Earth Zinc Oxide Varistors," J. Mater. Sci. Lett., 4, 1442-1444 (1985).

3. D. L. Losee, "Admittance Spectroscopy of Impurity Levels in Schottky Barriers," J. Appl. Phys., 46 [5] 2204-2214 (1975).

4. G. L. Miller, D. V. Lang, and L. C. Kimmerling, "Capacitance Transient Spectroscopy," pp. 377-476 in Ann. Rev. Mater. Sci., Vol. 7. Edited by R. A. Huggins, R. H. Bube, and R. W. Roberts, Annual Reviews, Palo Alto, CA, 1977.

5. D. Smith, M.S. Thesis, "An Automated Pulser for the Cornell DLTS System," Cornell University (1983); W. Schaff, private communication.

6. J. T. Schott, et al., "Transient Capacitance Measurement of Deep Defect Levels in GaAs and Si," Nat. Tech. Information Serv., U.S. Dept. Commerce, AD-A022 652 (1976); D. S. Day, M. Y. Tsai, B. G. Streetman, and D. V. Lang, "Deep-Level-Transient Spectroscopy: System Effects and Data Analysis," J. Appl. Phys., 50 [8] 5093-5098 (1979).

7. G. L. Miller, J. V. Ramirez, and D. A. H. Robinson, "A Correlation Method for Semiconductor Transient Signal Measurements," J. Appl. Phys., 46 [6] 2638-2644 (1975).

8. C. M. Ransom, T. I. Chappell, J. L. Freeouf and P. D. Kirchner, "Modulating Functions Waveform Analysis of Multi-Exponential Transients for Deep-Level Transient Spectroscopy," IBM Research Report, RC 11825 (Log #52955) Yorktown Heights, NY (1986).

9. W. Shockley and W. T. Read, Jr., "Statistics of the Recombinations of Holes and Electrons," Phys. Rev., 87 [5] 835-842 (1952).

10. J. F. Cordaro, Y. Shim, and J. E. May, "Bulk Electron Traps in Zinc Oxide Varistors," J. Appl. Phys., 60 [12] 4186-4190 (1986).

11. J. C. Simpson and J. F. Cordaro, "Characterization of Deep Levels in Zinc Oxide," J. Appl. Phys., 63 [5] 1781-1783 (1988).

12. J. M. Noras, "Transient Distortion and DLTS," Solid St. Comm., 39, 1225-1227 (1981).

13. H. P. Hjalmarson, G. A. Samara, and J. W. Farmer, "An Improved Deep Level Transient Spectroscopy Method," J. Appl. Phys., 63 [5] 1801-1804 (1988).

14. F. Greuter, G. Blatter, M. Rossinelli, and F. Schmuckle, "Bulk and Grain Boundary Defects in Polycrystalline ZnO," pp. 235-240 in Defects in Semiconductors. Edited by H. J. von Bardeleben, Trans Tech Pub., Switzerland (1986); J. P. Gambino, et al., "Grain Boundary Electronic States in Some Simple ZnO Varistors," J. Appl. Phys., 61 [7] 2571-2574 (1987).

15. A. Broniatowski, "Electronic States at Grain Boundaries in Semiconductors," pp. 95-117 in Polycrystalline Semiconductors, Edited by G. Harbeke, Springer-Verlag, Berlin (1985).

16. M. H. Sukkar, H. L. Tuller, and K. H. Johnson, "Electronic Structure of Grain Boundaries and Interfaces in Polycrystalline Zinc Oxide," pp. 141-145 in Grain Boundaries in Semiconductors, Edited by H. J. Leamy, G. E. Pike, and C. H. Seager, North Holland, New York, 1982.

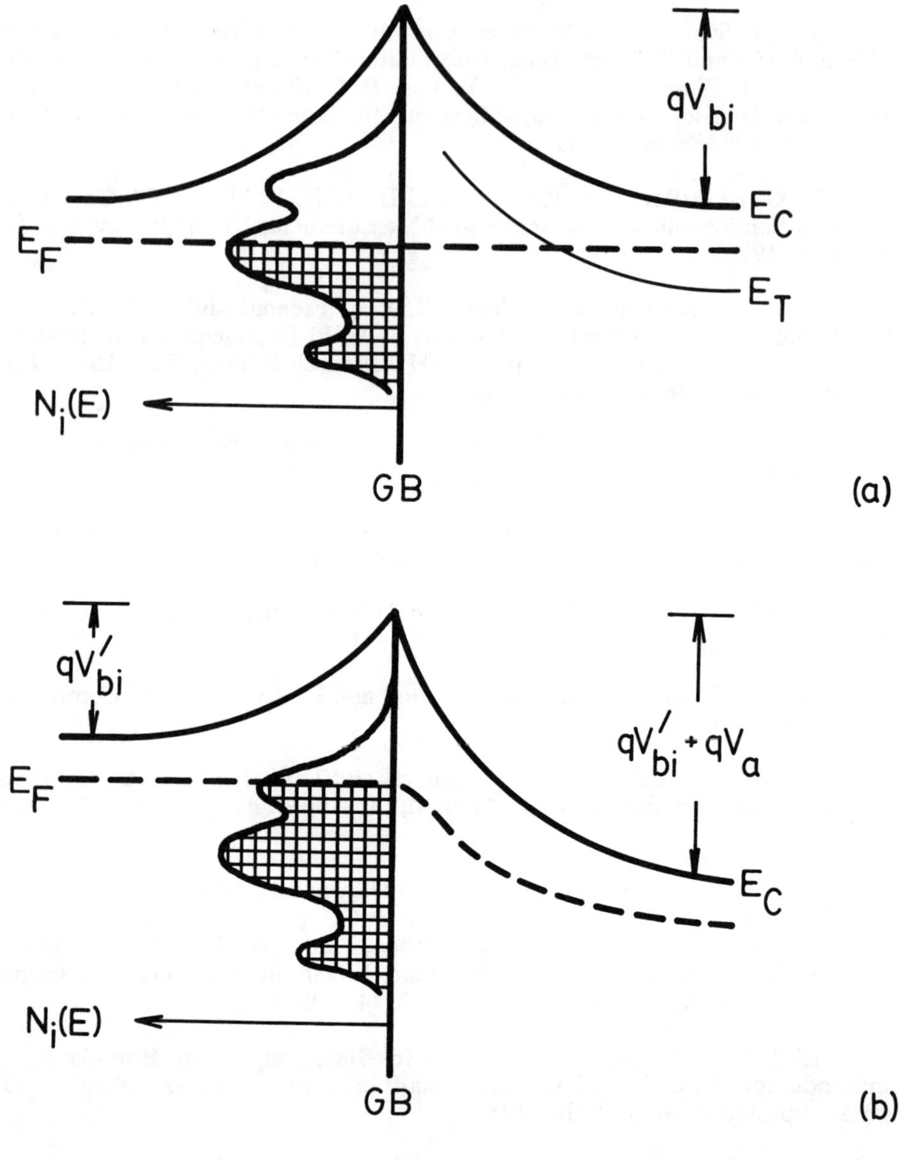

1. (a) Ideal double schottky barrier at grain boundary showing both interface states and bulk states under condition of zero bias. $N_i(E)$ is the interface density of states.

(b) Double Schottky barrier with applied bias.

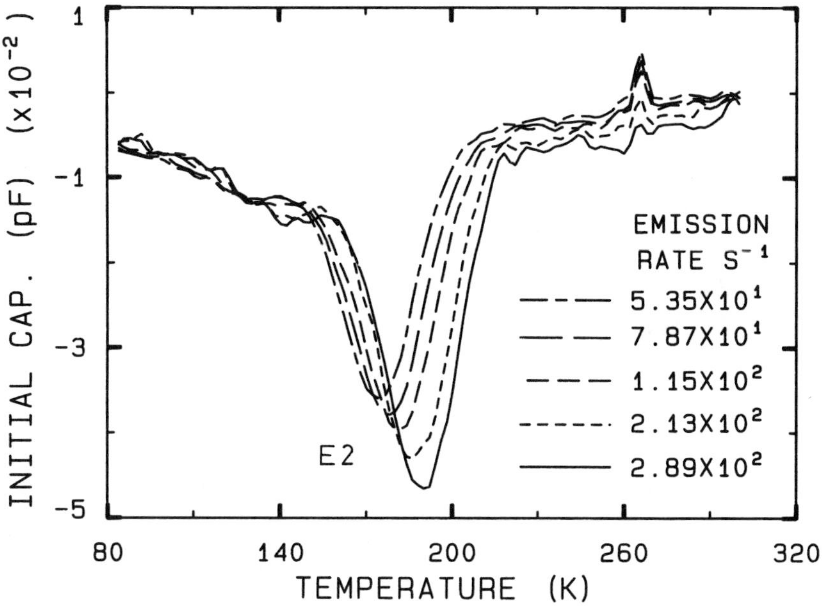

2. DLTS scan of commercial GE V100ZA4 varistor material using digitally collected data and a weighting function which simulates that of a lock-in amplifier.

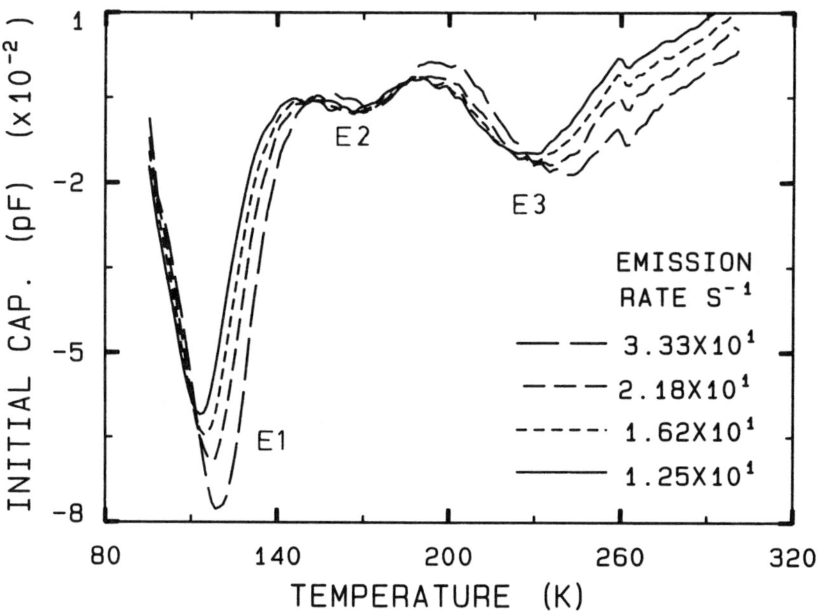

3. DLTS scan of reagent grade, polycrystalline ZnO using a silver Schottky contact. The data were collected digitally and a weighting function which simulates that of a lock-in amplifier was used.

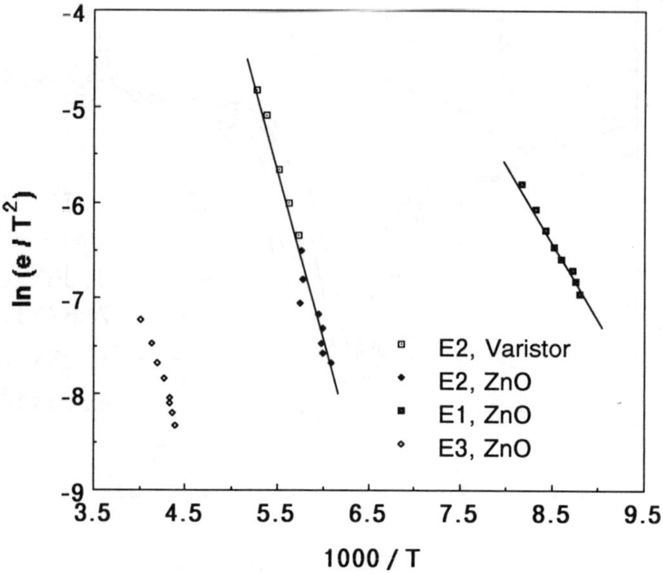

4. Arrhenius plot of the emission rates of traps found in the commercial varistor material and in ZnO.

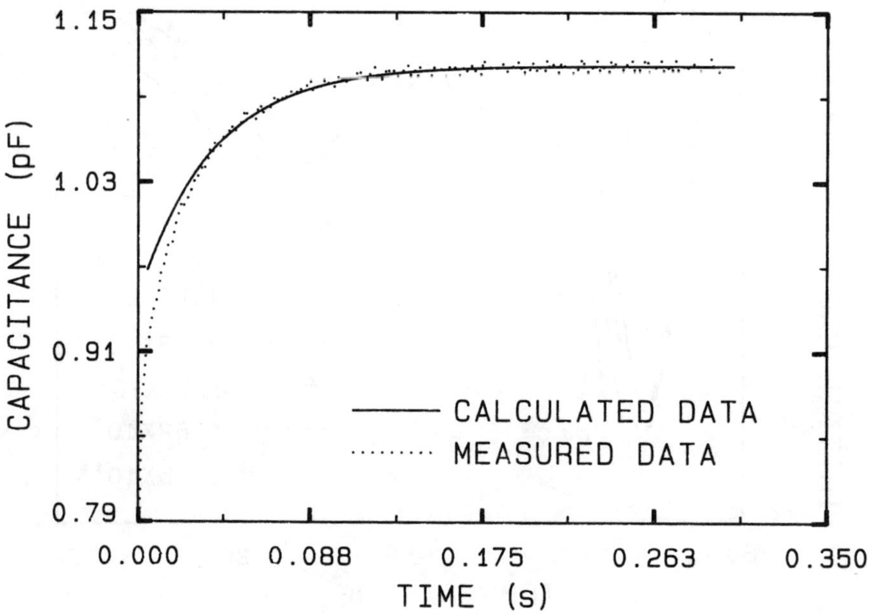

5. Capacitance transient from single crystal ZnO and calculated fit to the transient using a single time constant. The fit was made using the Fourier analysis.

INVESTIGATION OF LOW TEMPERATURE DEGRADATION OF Y-TZP BY COMPLEX IMPEDANCE SPECTROSCOPY

E. J. Kennedy and I. K. Lloyd
University of Pittsburgh, Materials Science and Engineering
Pittsburgh, PA 15261

ABSTRACT

Complex impedance spectroscopy (CIS) was used to examine the aging of yttria stabilized tetragonal zirconia at 300°C. CIS was found to be more sensitive to the aging process than x-ray diffraction. The minimum critical grain size for aging was less than 0.37μm; 0.3μm samples showed no aging effects after 360 hours while 0.37 and 0.40μm grain size samples showed aging after 24 hours. The aging observed appeared to be a surface effect and it was evident in the grain boundary activation energy for resistance. While aging occurred in the 0.37 and 0.40μm grain size samples, it was not sufficient to cause microcracking which could be observed in the SEM.

INTRODUCTION

The tetragonal polymorph of zirconium oxide, normally stable only in the temperature range 1200°C to 2350°C, can be stabilized at room temperature by microstructure control and the addition of controlled quantities of dopant.[1] The resulting material is potentially a dense, fine-grained, 100% tetragonal zirconia, with mechanical and electrical properties similar to that of a partially stabilized zirconia (PSZ), but with enhanced strength and toughness.[2] When aged in the temperature range 200°C to 300°C for a period of time, however, the material undergoes a partial transformation to the monoclinic structure, resulting in a degradation in strength and toughness.[3] This behavior has been examined using x-ray diffraction and strength and toughness measurements,[4, 5] but not using conductivity measurements. The aging is associated with the transformation of the tetragonal zirconia to monoclinic zirconia and is often accompanied by microcracking. As in the tetragonal to monoclinic transformation upon cooling from above 1000°C, there appears to be a critical grain size (~0.3μm to 0.5μm) below which the tetragonal phase is stable. The present study was designed to examine and understand the aging process using the impedance spectroscopy technique pioneered by Bauerle[6] to follow the changes in the electrical behavior of the grain boundaries and the bulk.

Since conductivity is usually very sensitive to changes in defect state (point defect concentrations, microcracking, ...) it was expected to be more sensitive to the early stages of aging than x-ray diffraction (which can only measure phase changes and may have a detection limit as large as five volume percent). The mechanical behavior of Y-TZP has been shown to be very sensitive to aging; however, it requires large numbers of samples and it is destructive, making sequential aging experiments more difficult. Conductivity measurements are non-destructive and are well suited to sequential aging experiments.

COMPLEX IMPEDANCE SPECTROSCOPY

CIS is a spectroscopy technique which uses AC impedance measurements as a function of frequency to resolve several polarizations in the same material. A voltage

$$V = V_0 e^{j\omega t}$$

is applied to the cell, and the resulting current

$$I = I_0 e^{j(\omega t + \phi)}$$

is measured.[7] The impedance is calculated by

$$Z = \frac{V}{I} = Z_0 e^{-j\phi}$$

or, equivelently,

$$Z = R + jX$$

where R is the resistance (real component) and X is the reactance (imaginary component).

For analysis, the resulting data is plotted in the complex impedance plane, obtaining a plot like figure 1. Each semi circular arc corresponds to a unique polarization that may be modeled in the simplest form by a resistor and capacitor in parallel. The impedance diagram of such a model is a semicircle with diameter d equal to the resistance R, whose center is on the real axis. In a real material, there are several such polarizations, resulting in several semi-circles. In t-ZrO_2 there are three semi-circles that may be seen, resulting from the electrode, grain boundary, and bulk grain polarizations. Kleitz *et al.*[7] reported that the high frequency polarization corresponds to the bulk grains, the intermediate frequency polarization to the grain boundaries, and the low frequency polarization to the electodes. Bauerle[6] showed that the high frequency arc corresponded to the bulk impedance by varying the aspect ratio of the sample. He also showed that the lowest frequency feature corresponded to the electrode effect by varying electrode size.[6] The middle range frequency arc has been assumed to be the grain boundary effect although specific proof is not readily available. Ongoing studies[8] examining conductivity (using CIS) in Y-TZP as a function of grain size are expected to confirm this assumption. It is generally difficult to see all three semi-circles at any given temperature. In figure 1, the semi-circles for the bulk grains and the grain boundaries are visible, while the electrode polarization was below the minimum frequency range of our equipment.

EXPERIMENTAL PROCEDURE

Rod shaped samples 25mm long and 4.5mm in diameter and 19mm long and 3.5mm in diameter were prepared from Toyo Soda 3 mol% Y_2O_3-ZrO_2 (TZ-3Y) powder. The samples were isostatically pressed at 340 MPa (50 kpsi) without binder and then sintered at 1400°C for 2 hours in air. The sintered samples were >95% tetragonal ZrO_2 and were 99% dense with an average grain diameter of about 0.3μm. Some samples were annealed at 1400°C for an additional 3 or 8 hours to achieve a grain sizes of 0.37μm and 0.40μm. Platinum bands were painted on the ends of the samples, and platinum electrodes were wrapped around the bands. The samples were characterized using XRD (x-ray diffraction) and complex impedance spectroscopy (CIS) before aging and after each aging interval. The microstructure was examined before aging and at the end of the aging process.

Some samples from each group were aged in dry air. The 0.3μm samples were also aged in air saturated with water vapor at room temperature and in 0.01% O_2 in argon at 300°C to determine if atmosphere effects (especially water vapor) accelerated the aging process. The 0.3μm grain samples were aged for 360 hours total (528 hours for wet air) and were examined after 72, 168, 360, and 528 hours. The 0.37μm samples were aged for 24 hours and examined before and after aging. The 0.40μm samples were aged for 48 hours and examined before aging and after 24 and 48 hours.

The samples were characterized using XRD after each aging interval to determine the change in the monoclinic phase content. The relative height of the $(11\bar{1})$ monoclinic peak compared to that of the (111) tetragonal peak was used to estimate the percent monoclinic zirconia. Electrical properties were also characterized after each aging interval using complex impedance spectroscopy. (Note: see Bauerle[6] or Gur[9] for a more general treatment of complex impedance spectroscopy.) The conductivity was measured using a four terminal pair configuration between 5Hz and 4MHz with a HP-4192A impedance analyzer at temperatures between 400°C and 1100°C in air. Figure 1 shows a typical "Cole-Cole" type plot of the data. The data was analyzed by computer using an iterative fit to two semicircles[8] to determine the grain boundary resistivity (diameter of the grain boundary semicircle), the grain resistivity (the diameter of the grain semicircle), the angle of depression of the semicircular arcs, and the relaxation time of the arcs. The angle of depression was used as a measure of non-ideality. The resistivities were used to calculate the activation energies. The relaxation times were used to calculate an equivalent capacitance for each arc. For more details see reference [8].

The microstructure was characterized using as annealed polished surfaces or polished sections which were thermally etched at 1300°C for 30 minutes. The grain size was determined by lineal analysis. Since the grains were equiaxed, the correction for spherical grains was used. Density was determined by the water immersion technique. The microstructure was only characterized before aging and after 360 hours for the 0.3μm grain size samples, and before aging and after 24 and 48 hours for the 0.37μm and 0.40μm samples, respectively.

RESULTS AND DISCUSSION

This study was designed to determine if complex impedance spectroscopy could be used to monitor and understand the degradation that usually accompanies low temperature (~100-400°C) aging in tetragonal zirconia.[3, 4, 5] Aging affects the flexural strength of Y-TZP[3] and is a concern in structural applications. Aging would also be important in electrical applications since microcracking would affect both the conductivity and integrity of the material and the conductivity of monoclinic zirconia is not the same as that of tetragonal zirconia. While the aging process is known to be related to the transformation of the Y-TZP to the stable monoclinic phase,[3, 4, 5] it is not yet well understood. Our matrix of experiments examined the role of the grain size, grain boundaries, surfaces, and atmosphere in the aging process.

Complex impedance spectroscopy measures conductivity as a function of frequency. When the relaxation frequencies for the grain and grain boundary contributions are well separated, a Cole-Cole plot like figure 1 is obtained. (Electrode interface effects may also be seen at low frequencies near DC.) The resistivity of both the grains and grain boundaries can be obtained from the radius of their semicircular arcs. (See reference [8] for details.) The depression of the origin of the arcs below the imaginary axis is a measure of the non-ideality of the system. When the impedance measurements are made as a function of temperature as well as frequency, the activation energy for grain and grain boundary conductivity can also be determined. Since complex impedance spectroscopy will show mechanism changes through activation energy and non-ideality changes, it can be used to monitor reactions occurring in the grains or grain boundaries which will affect transport behavior.

Samples with three different grain sizes, 0.30, 0.37, and 0.40μm were examined. The 0.3μm grain size samples were chosen since they were expected to be relatively resistant to aging based on the literature,[3, 4] and would thus allow an examination of the early stages of aging. The larger grain size samples were expected to age much more rapidly and were included to show overall aging behavior. Samples with two different diameters and thus two different relative surface areas were investigated to help distinguish between surface and grain boundary effects during aging.

The aging mechanism is not yet well understood. Aging is known to be related to the tetragonal → monoclinic transformation but the transformation process is not known. Since water has been implicated as an aging accelerator,[10, 11] and since the transformation appears to start on the sample surface,[4, 12] some investigators[11] have postulated that the transformation is enhanced by the formation of $Y(OH)_3$ on the surface. This type of theory does not necessarily fit the minimum critical grain size observed for transformation, since smaller grain size samples would have higher surface areas to form hydroxide and thus accelerate the reaction. Several investigators[3, 5, 10] have postulated a nucleated martensitic type reaction. This type of theory fits the minimum critical grain size data since below the critical size the strain energy for transformation exceeds the chemical driving force. The nucleation idea helps to explain possible water vapor effects[10] as well as the dependance of the monoclinic phase content on aging time.[3, 5, 10] Our high water vapor pressure experiments were included to help determine if water was a strong enough accelerator to decrease the critical grain size and if so, to help distinguish the mechanism.

The 0.3μm grain size samples showed no signs of aging after at least 360 hours at 300°C in dry air, wet air, or 0.01% O_2 in argon. The microstructure was unchanged by aging. The percent monoclinic zirconia did not change after aging; XRD indicated it remained constant at less than five volume percent. The conductivity was not affected by 360 hours of aging. The activation energies for grain boundary conductivity and bulk conductivity were not a function of aging. (See table 1 and figure 2a.) If aging had introduced monoclinic zirconia at the grain boundaries, the grain boundary activation energy would be expected to change. Microcracking would also be expected to affect the grain boundary activation energy as well as increasing the grain boundary resistance. The angle of depression and relaxation frequency for grain boundary and grain conductivity were also unaffected by aging in the 0.30μm samples. This behavior fits well with the idea of a nucleated martensitic type transformation reaction. It also indicates that the critical grain size for aging may be greater than the 0.3μm which would be expected from Watanabe's strength and toughness results[3]. Water vapor was not a sufficiently strong aging accelerator to cause the 0.3μm samples to age.

The larger grain size samples (0.37 and 0.4μm) were intermediate in grain size between the critical grain sizes of Watanabe[3] (0.30μm) and Lu[5] (0.5μm). They showed signs of aging after only 24 hours. They exhibited an increase in the grain boundary activation energy as shown in table 1 and figure 2b. They also showed an increase of surface monoclinic zirconia after the grain growth anneal. (The percent monoclinic zirconia in the interior did not increase.) No microcracking was observed before or after aging the larger grain size samples.

Table 1 and figure 2 show that the grain boundary activation energy is more sensitive to aging than x-ray diffraction, as was expected. (This would account for a smaller critical grain size than Lu *et al.*[5]) It is not yet clear whether the change in grain boundary activation energy is a result of monoclinic phase formation, microcracking, or both. The small diameter samples have different grain boundary activation energies and larger changes in activation energy than the larger diameter samples. Since the smaller diameter samples have more surface area, this indicates that the aging process is a surface effect. (The XRD results in table 1 indicate aging upon cooling is a also surface effect.) Figure 2 and table 1 indicate that complex impedance spectroscopy is sensitive to surface effects in bulk samples.

Table 1 also shows a minimum critical grain size effect. The 0.3μm samples did not age while the 0.37 and 0.40μm samples did age. Our minimum grain size for aging is slightly larger than that of Watanabe *et al.*[3]; the difference could result from three factors. One, our samples were prepared differently from those of Watanabe, and may thus differ in impurity content. Assuming a martensitic type of transformation, this could affect surface energy enough to slightly change the critical grain size. Two, Watanabe's samples were aged for 1000 hours while ours were aged 528 hours or less. Again assuming a nucleated martensitic type reaction, this could indicate that the aging process at 300°C in 0.30μm samples is very slow. Three, strength and toughness may be more sensitive to surface aging effects than bulk sample conductivity measurements. All three of the above factors may account for the observed difference in critical grain size. Further experiments with higher relative surface area samples to increase the aging sensitivity of the impedance spectroscopy measurements would be useful. Experiments to explore very long time aging and the effects of impurities would also be useful. Experiments to explore severe aging in Y-TZP using CIS are underway.

Our results are consistent with a nucleated martensitic type reaction on the surface. We observed a critical grain size of $<0.37\mu$m, indicating that small grain size stabilizes Y-TZP through surface energy effects. The aging was also a function of time at the aging temperature indicating a nucleation process may be necessary.

CONCLUSIONS

Complex impedance spectroscopy can be used to monitor the aging process in Y- TZP. The grain boundary activation energy measured by CIS appears to be more sensitive to aging than x-ray diffraction. Aging was observed to be a surface effect. No microcracking was observed at our aging levels. Experiments to examine high level aging using CIS are continuing. The minimum critical grain size for aging was $<0.37\mu m$. This indicates that very small grain size Y-TZP ($\leq 0.3\mu m$) should be stable for structural and electrical applications.

ACKNOWLEDGEMENTS

We wish to thank the ALCOA Foundation for their financial support and G. McManus for his technical assistance.

REFERENCES

[1] T. K. Gupta, J. H. Bechtold, R. C. Kuznicki, L. H. Cadoff, and B. R. Rossing. Stabilization of tetragonal phase in polycrystalline zirconia. *J. Mat. Sci.*, 12:2421, 1977.

[2] N. Bonanos, R. K. Slotwinski, B. C. H. Steel, and E. P. Butler. High ionic conductivity in T-TZP. *J. Mat. Sci. Let.*, 3:245, 1984.

[3] M. Watanabe, S. Iio, and I. Fukuura. Aging behavior of Y-TZP. In *Advances in Ceramics,* Vol 12, page 391, American Ceramic Society, 1983.

[4] T. Sato and M. Shimada. Crystalline phase change in Y-PSZ by low temperature annealing. *Comm. Am. Cer. Soc.*, c–212, 1984.

[5] H. Y. Lu and S. Y. Chen. Low-temperature aging of t-ZrO_2 polycrystals with 3 mol% Y_2O_3. *J. Am. Ceram. Soc.*, 70:537–41, 1987.

[6] J. E. Bauerle. Study of solid electrolyte polarization by a complex admittance method. *J. Phys. Chem. Solids*, 30:2657–2670, 1969.

[7] M. Kleitz, H. Bernard, E. Fernandez, and E. Schouler. Impedance spectroscopy and electrical measurements on stabilized zirconia. In *Advances in Ceramics,* Vol 3, page 310, American Ceramic Society, 1981.

[8] E. J. Kennedy. *Complex Impedance Spectroscopy Characterization of Aging in Y-TZP.* Master's thesis, University of Pittsburgh, 1988.

[9] T. M. Gur, I. D. Raistrick, and R. A. Huggins. AC admittance measurements on stabilized zirconia with porous Pt electrodes. *Solid State Ionics 1*, 251–271, 1980.

[10] M. Ruhle, N. Claussen, and A. H. Heuer. Microstructural studies of Y-TZP. In *Advances in Ceramics,* Vol 12, page 352, American Ceramic Society, 1983.

[11] F. F. Lange, G. L. Dunlop, and B. I. Davies. Degradation during aging of transformation-toughened ZrO_2-Y_2O_3 materials at 250°C. *J. Am. Ceram. Soc.*, 69:237–40, 1986.

[12] F. F. Lange. Transformation toughening. *J. Mat. Sci.*, 17:225, 1982.

Table 1. Aging Effects			
Sample	Aging Time (hours)	% Monoclinic Zirconia on Surface*	Grain Boundary Activation Energy
0.30μm	0-360	<5	1.5 $\pm$ 0.3eV
0.37μm	0	~8	1.14
Large diameter	24	~8	2.71
0.37μm	0	~8	0.86
Small diameter	24	~8	1.67
0.4μm	0	~8	1.29
Large diameter	24	~8	1.37
	48	~8	1.74
0.4μm	0	~8	1.11
Small diameter	24	~8	1.81
	48	~8	2.21

*The % monoclinic ZrO_2 in the interior of all samples was <5%.

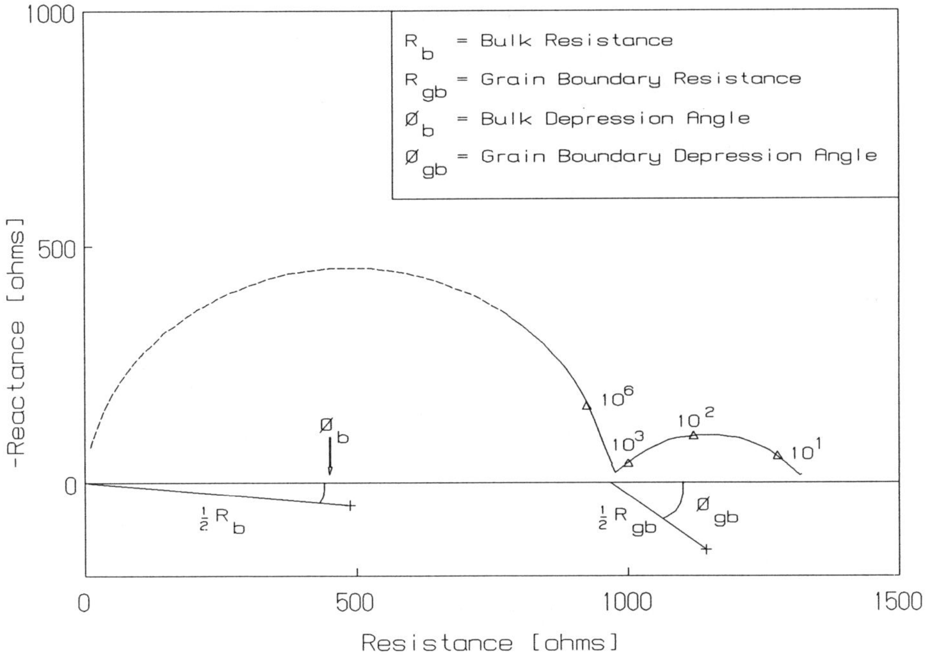

Fig. 1. A typical Cole-Cole plot of a 0.3μm tetragonal zirconia specimen at 800°C. (Most of the individual data points have been removed for clarity; a few representative frequencies are shown.)

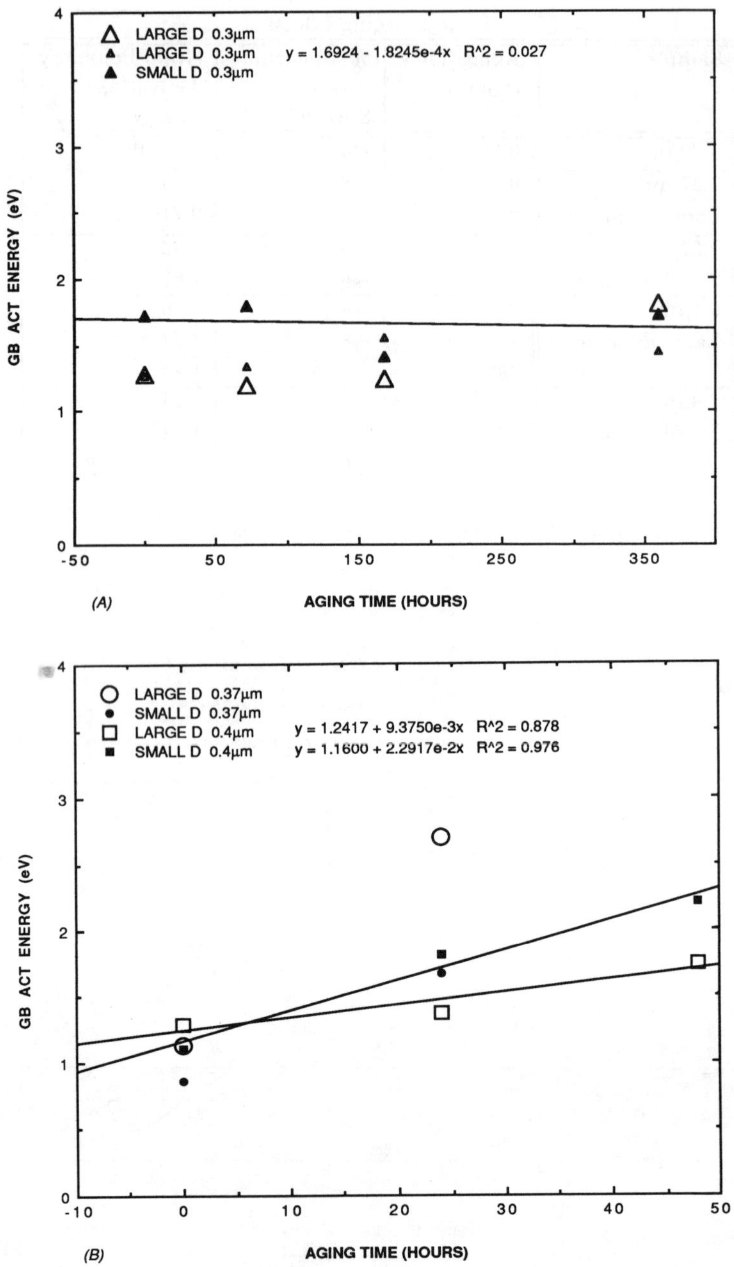

Fig. 2. Grain boundary activation energy as a function of aging time in air. *(A)* 0.3μm grain size samples. *(B)* Larger grain size samples.

Section IV. Electron Microscopy and Microprobe Techniques

ATOMIC RESOLUTION MICROSCOPIES IN MATERIALS

Mehmet Sarikaya

Dept. of Materials Science and Engineering, and
Advanced Materials Technology Program
Washington Technology Center
University of Washington, Seattle, WA 98195

ABSTRACT

High resolution microscopy techniques in the context of their application in materials science research are reviewed here. The principles of three techniques, namely atomic resolution electron microscopy (AREM), field ion microscopy (FIM), and scanning tunnelling microscopy (STM), are discussed briefly with emphasis on the nature of spatial resolution and image formation. The complementary use of these techniques in extracting detailed structural information from materials of current interest are presented and their inherent advantages are emphasized. Finally, a summary of the future directions in applications of these techniques is presented.

1.0 INTRODUCTION

The key to designing and processing of materials with predictable and controlled physical properties is an understanding of the factors that dictate the microstructure. Depending on the property of interest, e.g., electrical, magnetic, optical, transport, and mechanical, the evolution of microstructure may be considered at one or more scales in the range 10^{-2} to 10^{-10}m. The range of scale over which the microstructural details have to be studied by microscopy is but one reason for a variety of complementary analytical techniques being necessary. The dramatic change in the image detail from 1 μm to 0.2 nm brought about by magnification and resolution by transmission electron microscopy is illustrated in Figure (1).[1] In Figure 1(a), the structure of cluster is revealed; in Figures 1(b) and 1(c), particle packing and particle-

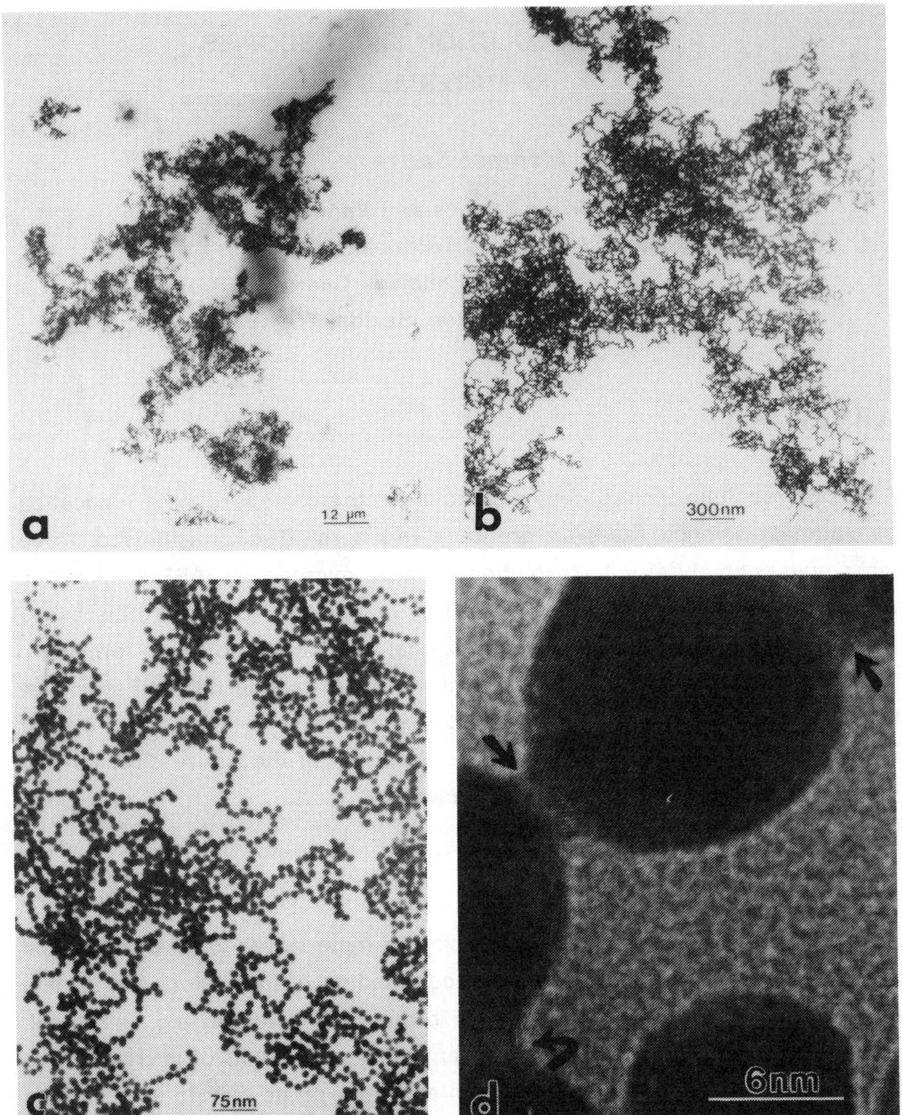

Figure 1. (a) to (c) are BF images of colloidal gold particles (suspended on a carbon film) recorded at various magnifications in a TEM; (d) is an AREM image revealing the contact regions and lattice structures of gold particles (Philips 430T; 300 kV).

particle attachment is revealed; and, finally in Figure 1(d), lattice and defect structure of particles and physical structures of interfaces are revealed at the atomic resolution.

Besides resolution requirements, it is preferable to work with a technique that provides information most directly about the structure↔property relationships being studied, for example, optical microscopy[2a] and spectroscopy[2b] for optical properties, transmission electron microscopy[3-5] for imaging, diffraction, and spectroscopy of matrices, precipitates, defects, and interfaces. In addition, some techniques are specific to surfaces, such as scanning electron microscopy[6] and scanning tunneling microscopy,[7] while others address the microstructures and composition variations of bulk materials, such as transmission electron microscopy,[3] and field ion microscopy[8a,b] --atom probe analyzer.[8a,9] In addition, information on the elemental distributions and chemical and electronic states of atomic species is essential in the understanding of microstructural components and their interrelationships. Finally, investigation with more than one technique may be necessary to eliminate ambiguities and obtain a complete information.

When imaging of microstructures is sought in the most conventional case, the technique of microscopy involving an optical lens is used. The principle of conventional optical microscopy involves a radiation source (e.g., light), a sample (either in a reflection or transmission configuration) and a lens (objective lens). The image is formed and magnified by the objective lens, which receives the transmitted or reflected light from the sample through an aperture with an angle given by a numerical aperture, N.A. The stability of the sample as well as the light source is essential to achieve the instrumental resolution of a particular microscope. The Rayleigh principle states that to distinguish two discrete points in an image, they must be separated by a distance δ, called the resolution, given by:[10]

$$\delta = 0.61\ \lambda / \mathrm{N.A.} \tag{1}$$

Here N.A. is given in radians and λ is the wavelength of the radiation. In the case of optical microscopy, the radiation is incoherent and confined to the visible range (390 nm $< \lambda <$ 770 nm). The smaller the λ and the larger the N.A., the finer the details that can be resolved. There is a limit to the value of N.A. which is dictated by geometry of optics and lens aberration. Image formation in transmission electron microscopy is also based on the same principle, and there is a limit to the size of objective aperture (analogous to N.A.).[12] This is defined by lens defects, specifically by the spherical aberration (see next section), but the simple relationship $\delta \sim \lambda^n$ $(n \leq 1)$ is still applicable.

Recent advances in optical microscopy, however, lead to better than an order of magnitude increase in resolution. In this technique, called near-field scanning optical microscopy,[11] an aperture with a collimated beam emanating from it is brought close to a sample which is at a distance from the aperture less than the aperture diameter. The near-field collimation extends to the sample and the microscopic image of the sample is formed when it is scanned with respect to the aperture. The resolution, better than 500 Å, is related to the aperture size, and not to the wavelength of radiation.

There have been greater advances in revealing structural variations in materials at finer detail, provided mainly by the advances in transmission electron microscopy techniques. These techniques allow microstructural, crystallographic, and compositional analyses of bulk samples to be made directly by imaging, diffraction, and spectroscopy, simultaneously in one combined system.[3-5] It is now possible to observe images of perfect and imperfect crystals at the atomic scale by atomic resolution electron microscopy, AREM.[4,5,12] Improvements have also been made in the direction of achieving small $(d < 5$ nm) electron probe size for diffraction and spectroscopic information from materials including ceramics, metals, polymers, and biological samples.[13]

Concurrent with the advances in AREM, significant improvements have also been made in the design of the instrument and the analysis and interpretation of the results in field ion microscopy (FIM), for atomic resolution imaging, and the atom probe analyzer (AP), for spectroscopy at that resolution.[14] FIM is particularly used for conductive materials; however, recent applications have successfully analyzed polymers,[15] semiconductors,[16] and oxide superconductors.[17]

Finally, the discovery[18,19] of the use of electron tunneling from a needle-shaped probe with an atomically sharp tip, into a flat, plate-shaped sample has led to new pathways to analysis of surfaces of materials at the sub-

Angström level, which was never anticipated.[20] Materials scientists, physicists, chemists, and biologists are now finding improved ways of studying new and traditional samples in atomic detail, where the limit is basically defined by the specimen preparation difficulties.[21-25]

Each of the atomic resolution microscopy techniques has been described separately in detail in some excellent recent reviews.[4,5,7-9,12,14] In this paper these techniques are summarized in a comparative fashion. The overall objective is to give to the materials scientist some collective insight into the advantages and disadvantages which are inherent to each technique so that a broader perspective will be available for future characterization of materials of widely differing nature. In order to accomplish this, the guideline followed is: (i) to present a summary of the principles of each atomic resolution microscopy technique described above; (ii) to illustrate their applications by using some of the latest examples from the literature, specifically on ceramics and semiconductors.

2.0 TECHNIQUES

2.1 Atomic Resolution Electron Microscopy

The fundamental principle of transmission electron microscopy is the same as that of an optical microscope.[10] There are basically three main components in a TEM:[12] the source of electrons which are accelerated by high voltages (typically 100-200 kV); magnetic lenses which vary the path of electrons so that they can be focused; and the sample, which is thin enough (typically 200-1000 Å) to allow electron transparency. Electrons originating from a high brightness filament (such as W, LaB_6, or field emission source, in the order of increasing brightness) are accelerated by a potential created between the cathode and the anode (Figure 2). They travel through a column that is kept in vacuum (typically 10^{-7} torr), and are focused by the condensor lenses on to a thin foil sample. Scattering of electrons takes place through the sample (crystalline and non-crystalline), which then forms a diffraction pattern and an image at the back focal and image planes of the objective lens, respectively. Both the diffraction pattern and the image can be magnified

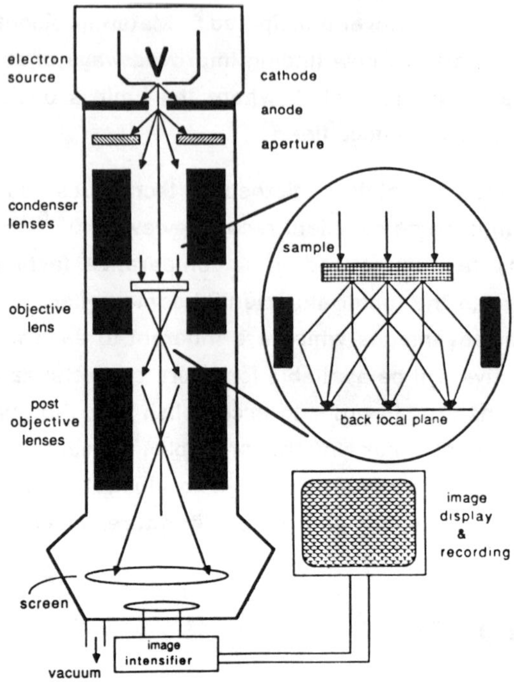

Figure 2. A schematic representation of a TEM column in cross-section. The inset illustrates the formation of diffraction patterns in the parallel illumination mode.

(up to 10^6x) by post objective lenses (usually termed, successively, diffraction, intermediate, and projection lenses) and transferred to a phosphorus screen. There is a sufficient depth of field (>1 μm) to image both the top and the bottom of the sample in focus, and a large enough depth of focus (>50 cm) to enable one to obtain a focused image at the planes of the screen and a photographic plate (or an image intensifier) simultaneously (Figure 2). Image formation is completed, and contrast is increased, by using an objective aperture that is placed in the back focal plane of the objective lens by limiting the scattering angle at which electrons contribute to the image. This type of imaging is called diffraction (or amplitude) contrast imaging, and forms the basis of conventional transmission electron microscopy, CTEM. Bright-field and dark-field images are obtained by placing the objective aperture either on the transmitted beam or one of the diffracted beams, respectively (Figure

3). This produces a reversal of contrast in one image with respect to the other. Under the most general conditions of operation, CTEM allows the morphological and crystallographic analysis of samples at a range of resolution down to 10 Å.[3]

When a plane wave of electrons is incident onto a crystalline TEM foil that is tilted to a zone axis orientation, and many beams (including transmitted and Bragg diffracted beams, as shown in Figure 3c) are used for the image formation, then an interference image is formed. The image, formed by amplitude contrast imaging, consists of an array of periodic contrast related to the crystal structure of the sample. In the simplest treatment, the image is a representation of the projected potential of the crystal in the particular orientation observed. The technique is therefore called atomic resolution electron microscopy, AREM.

A fine example of an AREM image is presented in Figure 4a, which displays an end-on image of a needle-shaped germanium (dc) precipitate formed in an aluminum (fcc) matrix of 1.4 Ge-Al alloy which was solution-treated and aged at 220°C.[26] The lattices of both the precipitate, in [110], and matrix, in [100], are clearly resolved, with the precipitate displaying unusual multiply twinned structure. Such structures have been frequently observed in nanometer-size metallic particles,[27-30] an example of which (from gold) is shown in Figure 4b.[29] Such crystallographic configuration and morphology of germanium precipitates are attributed to the restrictions imposed by symmetry and crystallography of the associated phases.[26] In the case of small particles, multiple twinning is due to the minimization of strain and surface energies.[28-30]

It is essential in AREM to understand a number of factors that may affect the image quality, so that a faithful interpretation can be made. Both the phase and the amplitude of the electron waves vary with increasing thickness of the foil, due to multiple scattering effects. This introduces a shift of atomic columns from their correct positions in the image, and a change in image intensity occurs. A common crystal, such as Si or Al, at a thickness of 10.0 nm can be considered as a weak phase object in which shift and intensity variation in image are minimized. However, in very thin crystals, there is still an image shift introduced by experimental factors, most notably by

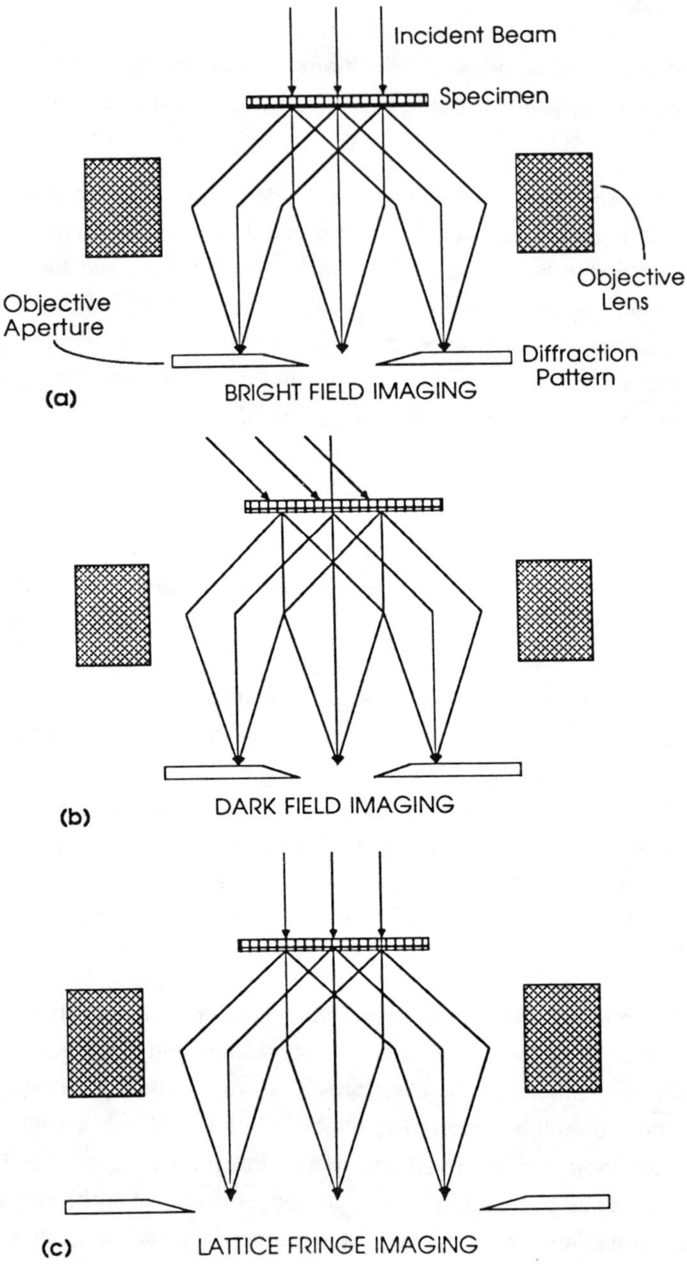

Figure 3. Schematic ray diagrams for (a) bright field, (b) dark field, and (c) lattice fringe imaging.

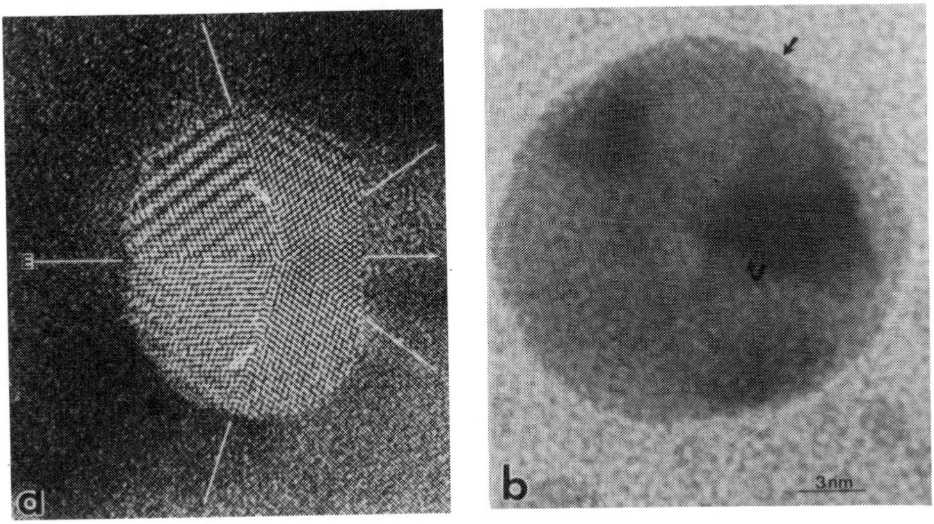

Figure 4. (a) AREM image of pentagonally twinned Ge-precipitate in Al matrix.[26] Beam direction is along <110> Ge and <100> Al; both of the lattices are imaged. Specimen was water-quenched from 540°C and aged 1 hr. at 240°C. (Microscope: Berkeley-ARM/JEOL 1000 operated at 800 kV. (b) AREM image of a single gold particle displaying multiply-twinned structure.[1,30] Particle (colloidally prepared) is suspended on a carbon film (5 nm thickness). (Microscope: University of Washington-Philips 430T, operated at 300 kV.)

spherical aberration, C_s, of the magnetic lenses and defocusing of the objective lens, Δz. This leads to the breakdown of paraxial approximation which gives the point-to-point representation at the Gaussian focal plane. The radius of the disk of confusion in the object plane resulting from the lens aberration is given by

$$\delta_s = C_s \beta^3 \tag{2}$$

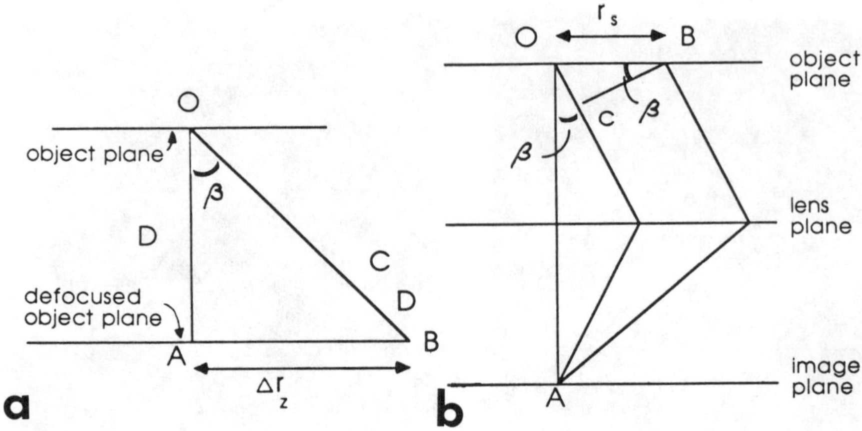

Figure 5. Schematic illustration of phase shift due to (a) defocusing and (b) spherical aberration.

where β is the aperture angle. A phase shift, χ_s, is introduced to those beams which are inclined to the optic axis ($\beta = 0$):

$$\chi_s = 2\pi\frac{\Delta}{\lambda}\ . \tag{3}$$

Here Δ is the path difference, as illustrated in Figure 5a, where,

$$\Delta = r_s \bullet \beta\ . \tag{4}$$

This leads to

$$\chi = \int_0^{\beta} d\chi_s = \int_0^{\beta} \frac{2\pi}{\lambda} C_s \beta^3\, d\beta = \frac{\pi C_s}{2\lambda}\beta^4 \tag{5}$$

In the weak phase object approximation, the contrast is smallest at the Gaussian focus. In AREM, image is focused under the best visual contrast, which involves some defocusing of the objective lens. Defocusing causes a phase shift and a resultant disc of confusion, as illustrated in Figure 5b. When an image is defocused by an amount Δz, which leaves the plane of object unchanged, it has an apparent displacement given by $r_{\Delta z} = \Delta z \sin\beta$.

The path difference is then,

$$\Delta = \frac{\Delta z}{\cos\beta} - \Delta z = \frac{\Delta z}{2}\beta^2 . \tag{6}$$

The phase shift due to defocus is, therefore,

$$\chi_{\Delta z} = \frac{2\pi}{\lambda}\Delta = \frac{\pi\Delta z}{\lambda} \bullet \beta^2 \tag{7}$$

(where $\Delta z < 0$ is underfocus).

The total phase shift, including those due to spherical aberration, χ_s, and defocus, $\chi_{\Delta z}$, is:

$$\chi = \chi_s + \chi_{\Delta z} = \frac{\pi C_s \beta^4}{2\lambda} + \frac{\lambda\Delta z}{\lambda} \bullet \beta^2 \tag{8}$$

The function χ is called contrast transfer function, CTF. The factor, exp(-iχ), has to be multiplied by the diffracted wave function at the back-focal plane of the objective lens, and hence, the image contrast is strongly affected by CTF. CTF is plotted for Philips CM30ST and JEOL JEM 4000EX microscopes in Figure 6a and 6b, respectively. The electron waves corresponding to the flat section of CTF are minimally affected by the phase change due to C_s and Δ_z. This band corresponds, effectively, to a lens without aberration. That is, from the CTF expression, the detrimental effect of spherical aberration is compensated by underfocussing the objective lens.

It is noted from the CTF plots in Figure 6 that sinχ (imaginary part of the wave function) oscillates between +2 and -2. There is a great distortion on those beams corresponding to large β (or corresponding reciprocal vector) and these beams are cut off by objective aperture. The optimum defocus is, therefore, given by (due to Shertzer[31]):

$$\Delta z_{sch} = -(C_s\lambda)^{1/2} \tag{9}$$

It is a usual practice to represent the instrumental resolution of an AREM by the first cut-off point on CTF (shown by arrow in Figure 6a and 6b). This

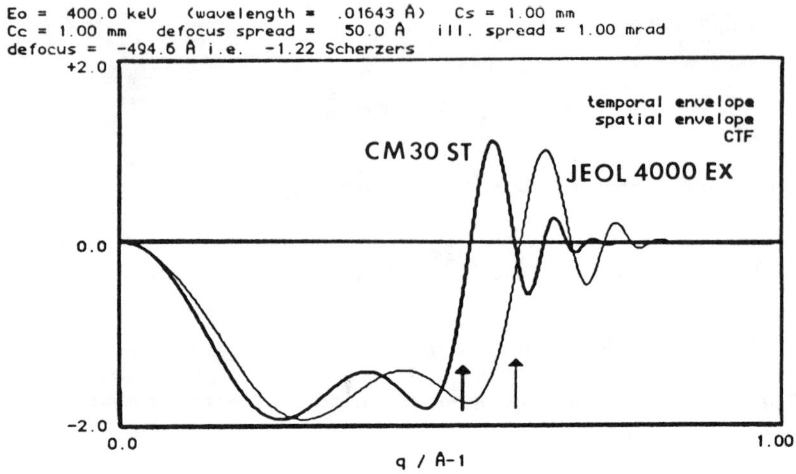

Figure 6. Contrast transfer functions for Philips CM30 ST and JEOL JEM 4000EX.

corresponds to an angle, $\alpha = 1.41\,(C_s\lambda)^{1/2}$, and if $\beta \sim \alpha$, then the corresponding instrumental resolution is given by:[31]

$$\delta = 0.66\, C_s^{1/4}\, \lambda^{3/4} \tag{10}$$

Fine details of an atomic resolution image may be directly related to the structural features of the sample under the stringent conditions of sample, instrument, and electron-optical alignments. These conditions specifically include the use of highly coherent electron-beam, thin and flat specimens imaged in a zone-axis orientation, correct defocus value of the objective lens, finely aligned incident electron beam, and mechanical stability of the microscope column, as each of these parameters significantly affect the image resolution, and subsequent interpretation.[32-35] It is not straightforward to make a direct interpretation of AREM images since image details can vary strongly, for example, by the above parameters. In addition, it is difficult to perform the experiments when imaging a particular crystal at the optimum conditions. In most cases, it is necessary to produce simulated images which take into account the dynamical effects, and compare them with experimental ones to achieve a more quantitative interpretation.[4,5,12,32-35]

2.2 Field Ion Microscopy (FIM)

Formation of an image in the FIM is different than in the TEM and does not require the use of a lens. The instrument is schematically illustrated in Figure 7 and the image formation takes place as follows. As an imaging gas (at 10^{-3} torr) is ionized near a positively charged, cryogenically cooled tip of a specimen at a certain voltage, ions are accelerated radially onto a channel plate/fluorescent screen assembly to form an image of the sample.[36-38] The mechanism by which the imaging gas is ionized is known as field ionization and is based on the electron tunneling process[37] (electron tunneling is qualitatively described in the next section). At the atomic protrusions on the tip (Figure 7) local field strength is raised, which preferentially ionizes atoms at these positions, and corresponding bright spots form on the image. The visible image formed on the fluorescent screen is a projection of the tip of the sample and roughly resembles a stereographic projection of the crystalline sample in the direction of the specimen axis. When the tip radius is less than 500 Å and the tip is at the liquid nitrogen temperature (77 K), atomic structure of the surface becomes resolvable. Magnification of the image at the fluorescent screen is determined by the ratio of the distance between the tip to the screen over the tip radius (typically 5 cm/5 nm $\sim 10^6$ times).[14]

A typical FIM image of a sample tip (as shown in Figure 8) consists of a regular array of bright spots, each of which corresponds to an individual atom at the tip surface. There is no simple correlation between these image spots and the chemical nature of the corresponding atoms. This correlation is possible with a time-of-flight mass spectrometer attached beyond the screen, an arrangement called an atom probe analyzer.[39,40] In this case, the image of the tip is viewed in the FIM mode on a phosphorous screen containing a hole at the center (called the probe hole). In order to identify atomic species on the surface, the sample is manipulated so that the desired area coincides with the probe hole through which the species, which are field-evaporated at the tip with a voltage pulse, pass to enter the drift tube of the spectrometer. An ion detector at the end of the drift tube supplies the timing pulse for the measurement of the flight time of the ions. The ratio of the mass charge, M/n, is calculated from the time of flight, *t*, tube length *d*, and applied voltage, *V*, using the relationship[39]

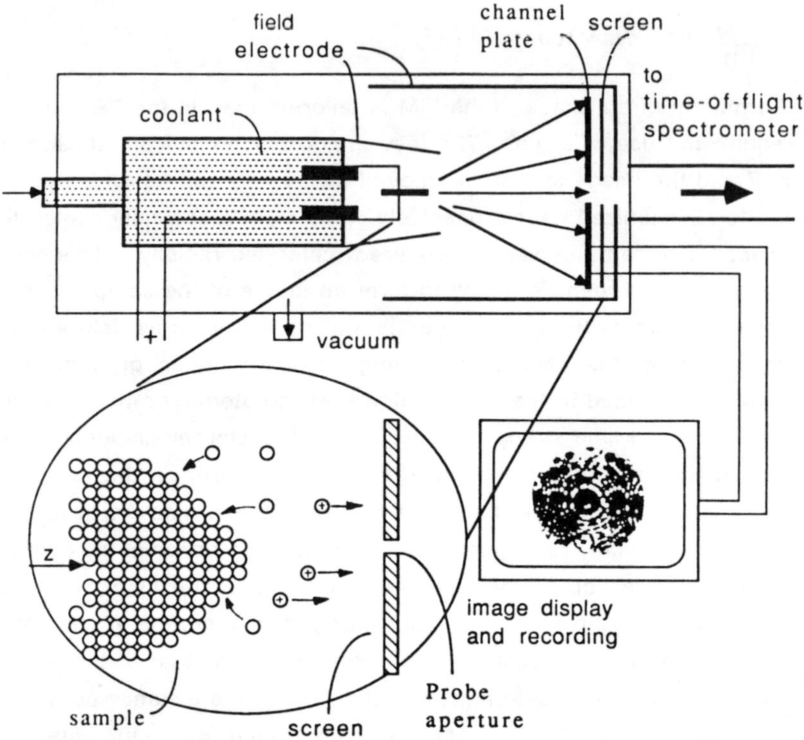

Figure 7. Schematic cross-section of an FIM. AP analyzer is fitted to the plate beyond the phosphorous screen. Inset illustrates the ion evaporation at the tip of a sample.

$$M/n = CVt^2/d^2 \quad (C{:}\ (1930/d^2))\ . \tag{11}$$

FIM and AP analysis techniques have some major limitations. These include the problems with specimen preparations (for non-metallic samples, especially ceramics), specimen fracture due to the high fields applied, and sampling of small amount of analyzed material.[14] Despite these difficulties both in the imaging and spectroscopy modes, FIM-AP analysis provides atomic resolution. Individual atoms, point defects, and point defect clusters can be imaged in the FIM from materials including metals,[8-9,14,39-41] organic compounds,[15] semiconductors,[16] and superconducting oxides.[17]

The spatial resolution obtainable in the FIM image depends on the size and sharpness of the tip, temperature of the specimen, and the effective size

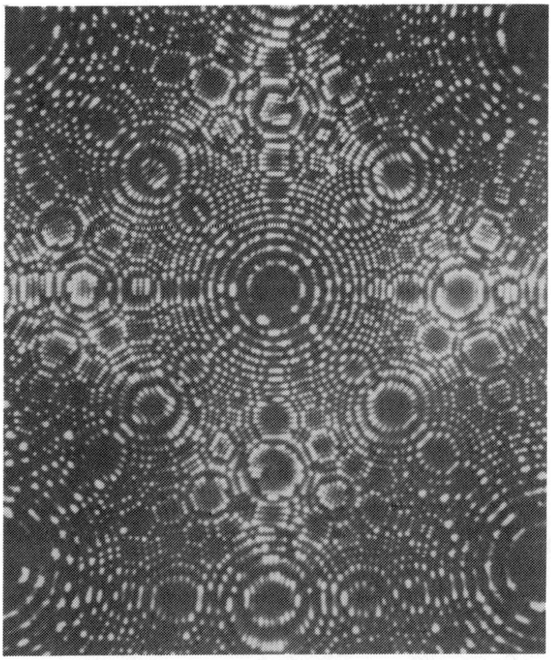

Figure 8. FIM image of Pt showing crystallographic poles. The image of the semi-spherical tip resembles the stereographic projection in this orientation. (Courtesy of E. R. Müller).

of the ionizing gas. The principal limit on the resolution originates from the transverse velocity at the moment of ionization of an imaging gas (H, He, or Ne) with a small atomic diameter (<2.5Å). This may give rise to a lateral spread in the ion beams originating from the individual image points.[40] The field evaporated surface of a well conductive tip of sample is essentially a smooth section taken through the lattice at the tip, and the lattice points which are seen on the screen (Figure 8) are ions from those atoms whose centers intersect or fall closest to this section. As seen in this figure there is roughly a one-to-one correlation between the poles on a stereographic projection centered around the tip axis and the centers of the circular contours formed by protruding atoms in the crystal around these poles.

In atom probe analysis (which is a mass spectroscopy technique), the spectroscopic resolution is superb; individual atoms, isotopes, ions, and ion

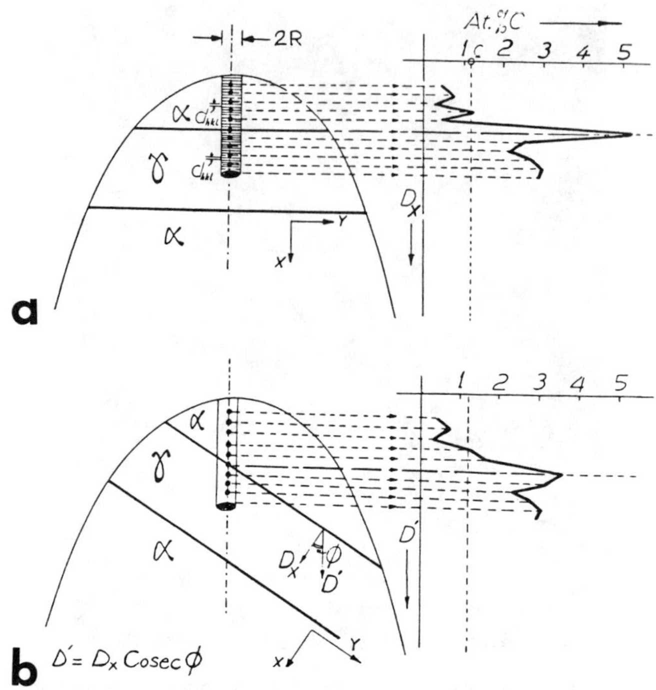

Figure 9. Possible configurations of a thin film of second phase at the tip of an FIM sample: (a) ideal configuration for AP analysis; (b) more frequent configuration plots indicate the expected concentration profile of the segregated atom, C, at the interface between the phases α and γ.[42]

clusters can be detected.[38-41] The spatial resolution has two components: lateral and transverse. The former one is defined roughly by the size of the probe hole (i.e., 2R as shown in Figure 9, which is about 2 to 8 nm). The transverse resolution can be as small as the separation of atomic planes, i.e., d_{hkl} (Figure 9). Concentration profiles through the sample can be obtained layer by layer at the atomic scale as in the case of the ideal geometry as shown in Figure 9a where a narrow (few nm) film of a second phase is located perpendicularly to the tip axis. In this case, even a monolayer of segregating species at the interface between the two phases (α and γ in Figure 9a) is detected. In a narrow region at the interface, the concentration profile, therefore, displays a sharp peak. The figure shows the carbon profile

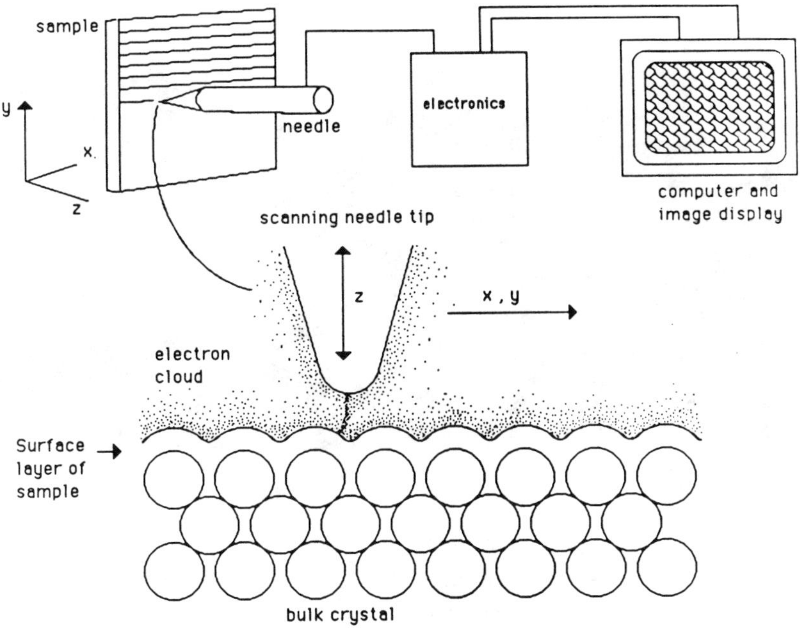

Figure 10. Schematic illustration of the principle of image formation in STM.

across an austenite (γ, fcc)/martensite (α, bct) interface in a low carbon steel.[42] In a more realistic geometry of the microstructure at the tip, when the interfacial film of the second phase is in an oblique orientation (defined by φ) with the tip axis (Figure 9b), the spatial resolution at the interface is effectively deteriorated by cosec φ. The concentration profile displays a wider and lower peak at the interface. The "transverse" spatial resolution for interface concentration at the limit ($\varphi \sim 90°$) becomes the lateral resolution (i.e., the size of the probe hole).

2.3 Scanning Tunneling Microscopy

The basic difference between the STM and other microscopes is that there are no free particles used for imaging, and so there is no need for light, ion, or electron sources, optical or magnetic lenses, and a fluorescent

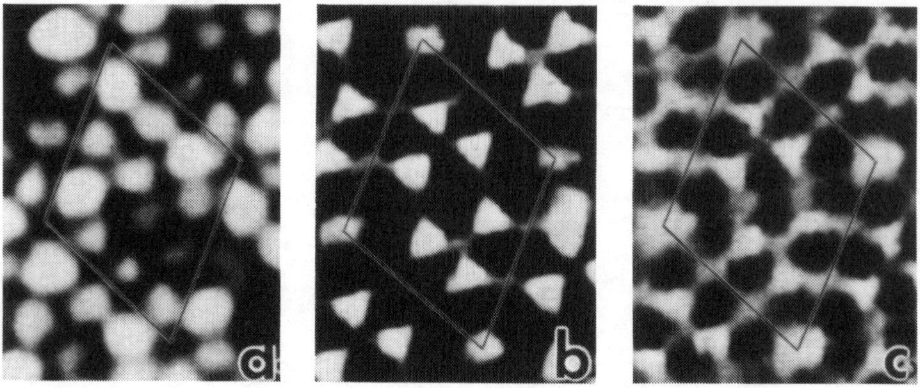

Figure 11. Tunneling image of Si(111) -(7x7) surface with a unit cell of 27 Å edge length. Images reveal filled surface states. (a) Adatom state at -0.35 V; (b) Dangling bond state at -0.8 V; (c) Back bone state at -1.7 V.[22]

screen. In this sense, STM is like FIM; a microscope without a lens. However, there are basic differences between STM and FIM. The bound electrons already existing in the sample serve as the source of radiation. In this case, the sample acts as an electrode, an atomically flat plate, and it is separated by an insulator (vacuum, air, or liquid)[7,18,21-23,43-46] from a probe, which is an atomically sharp needle-like electrode (Figure 10). The tip is pushed towards the sample until the electron clouds of each electrode come into contact.[7,18] A tunneling current is established when a small voltage is applied between the scanning tip and the stationary sample.

The tunneling current is extremely sensitive to the distance between the tip and the surface of the sample as the density of the electron cloud fades exponentially with distance. The tunneling current is established only between the extreme atom at the tip and uppermost atomic layer of the sample plate. This gives the unsurpassed resolution to the microscope both as modulation in the lateral direction (<1.0 Å) and as corrugation in the vertical direction ($\sim$0.1Å). This sensitivity is used to measure the horizontal and vertical positions of the atoms on the surface of the sample. The tip follows the contour above the surface of atoms as the tunneling current is maintained with the electron cloud.

The movement of the probe tip above the surface of the sample is ensured by three mutually perpendicular piezoelectric arms (x, y, and z). Scans in x and y directions are kept at constant velocity electronically. The movement in the z-direction is dictated by the atomic structure of the surface of the sample. By scanning the probe over the surface of sample (10 cm in the image corresponds to 10 Å on the sample, i.e., 10^8x magnification), a three-dimensional image of the surface is obtained. In order to maintain the tunneling current, the microscope is operated either in the constant current or constant voltage mode. The effect of bias voltage on the image contrast of Si (7x7) surface is illustrated in Figure 11.[22] The line forming a parallelopipe outlines the (7x7) unit cell on the surface, which is asymmetric due to differences of electronic structure in each half of the unit cell. The value of different negative bias voltage used reveals different surface states of binding.[22]

3.0 APPLICATIONS

3.1 In Situ Studies of Decomposition of $Mg(OH)_2 \rightarrow MgO$ by AREM

This study demonstrates the use of AREM to study phase transformations in ceramic systems. The example presented here is on the decomposition of $Mg(OH)_2$ (brucite), which has been extensively studied because of the technical interest.[47-49] Several models for the transformation mechanism have been proposed, all relying on indirect measurements.[47-49] The following paragraphs contain a description of a study performed directly with the TEM on the decomposition of single crystal $Mg(OH)_2$, with the aim of improving the understanding of the nature of the reaction.[50]

It is known that single crystal $Mg(OH)_2$ (CdI_2 type structure, space group $P\bar{3}ml$, and $a_o = 3.147$ Å) forms MgO (NaCl type structure, space group, $Fm\bar{3}m$, $a_o = 4.213$ Å) pseudomorphically upon thermal decomposition. Evaporation of chemical water during the decomposition leaves pores in the final product which constitute 55% of the structure. The decomposition reaction in the TEM was accomplished *in situ* by directly heating the $Mg(OH)_2$ single crystal.

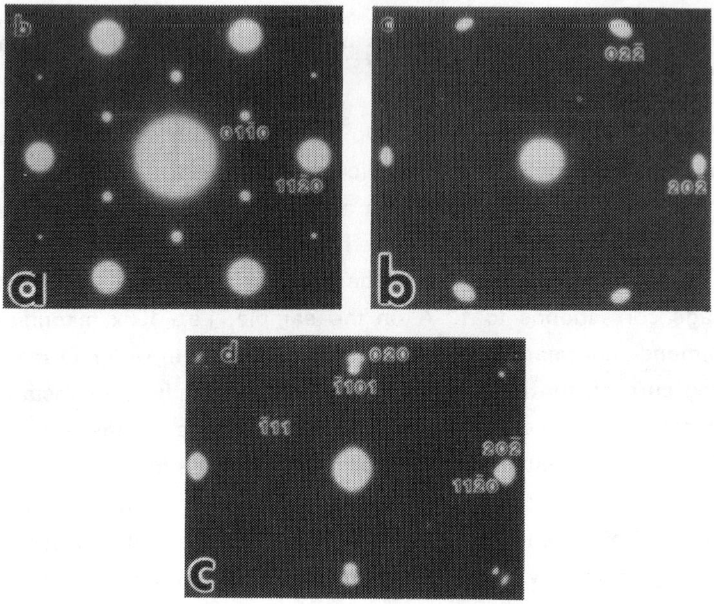

Figure 12. Selected area electron diffraction patterns from (a) $Mg(OH)_2$ undecomposed foil in [0001], (b) from the MgO formed after decomposition, in [111]. (c) Superposition of $[1\bar{1}02]_{Mg(OH)_2}$ and [101] MgO patterns in partially decomposed crystal.[50]

This yielded a topotactic reaction with the following unique crystallographic orientation relationships, as determined by selected area electron diffraction (Figure 12):

$$(111)_{MgO} \;//\; (0001)_{Mg(OH)_2}$$

and

$$[10\bar{1}]_{MgO} \;//\; [11\bar{2}0]_{Mg(OH)_2} \,.$$

It should be noted that, although the $Mg(OH)_2$ pattern (Fig. 12a) exhibits well-defined single crystalline spots, the MgO pattern (Fig. 12b) shows broader spots with a small-angle arc. The latter is attributed to the small size of the MgO crystallites, which have small angular misorientations. In order to better understand the nature of decomposition, the shape, size, and distribution of MgO crystallites were determined directly by studying the product material in the AREM.

Figure 13. AREM image showing misorientation and overlap of individual crystallites of MgO. Notice apparent missing columns of ions (arrow).[50]

It was necessary to tilt the crystallites to <100> direction of MgO, i.e., by about 55° inclined to the [0001] direction of the precursor phase. Figure 13 presents such an image, displaying lattices of MgO crystallites in the <100> projection with 2.1 Å fringe spacing. It is clearly resolved that cube-shaped MgO crystallites form a staggered array with pores dispersed between them. Because of fairly good alignment, the lattices of several crystallites are imaged simultaneously showing clear edges. A small misorientation up to 2.5° between the crystallites, possibly because of the presence of pores, has also been observed frequently.

It is shown in this study that, from the microstructural point of view, the decomposition product is similar to side-band structure with an average wavelength of about 2.5 nm. However, the earlier suggestion[51] that there is

Figure 14. A BF TEM image of $YBa_2Cu_3O_{7-x}$ needle. Transformation twins, in an oblique orientation to needle axis, are clearly seen.[17a]

a connection between the <100> square wave modulation of MgO and a concentration modulation in the (0001) planes of $Mg(OH)_2$ is disputed.[50]

3.2 *Field Ion Microscopy and Imaging Atom Probe Analysis of Superconducting $YBa_2Cu_3O_{7-x}$*

The following example presents a summary of a first attempt to perform FIM-atom probe analysis of the $YBa_2Cu_3O_{7-x}$ superconducting phase.[17] The objective in this work was to study the distribution of atomic species in the orthorhombic $YBa_2Cu_3O_{7-x}$ phase at the atomic scale.[17] The study was performed to understand stability and homogeneity of the samples as these were discussed in the literature[52-56] to be possible sources to influence superconducting properties.

Bar-shaped samples with 0.4 mm square cross-sections were cut from the preheated samples and then electropolished in a 10% perchloric acid-ethylene glycol monobutyl ether solution to form sharp tips.[17] An example

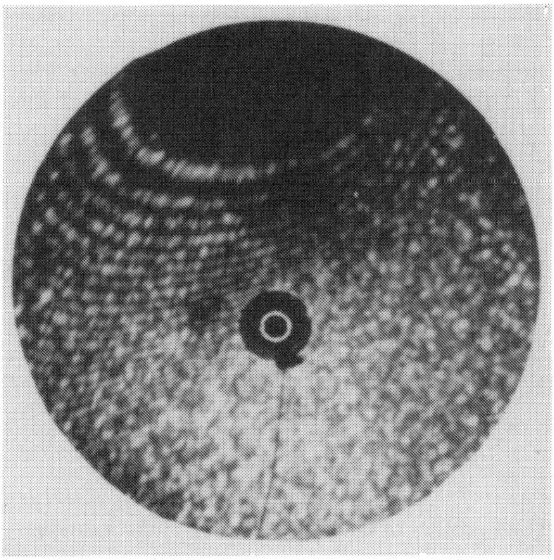

Figure 15. FIM image of the tip of a $YBa_2Cu_3O_{7-x}$ sample. The white circle at the center shows aperture to mass spectrometer covering an area of 2 to 2.5 nm in diameter.[17a]

of such a tip is shown in a TEM image in Figure 14 which displays transformation twins, spaced by about 1000 mm, with the twin normal at approximately 30° to the tip axis. The twins form to accommodate stresses created during the transformation of high temperature tetragonal phase to low temperature (superconducting) orthorhombic phase. The FIM image in Figure 15 displays a ring contrast centered around a pole which is parallel to [001] axis of the orthorhombic lattice. The bright spots in the image are attributed to the metallic atoms and, therefore, the height of the steps would be $\frac{c}{6}[001]$ or about 0.2 mm.

Atomic mass spectra were obtained through a 2 m drift tube using a 2.0-2.5 nm probe hole in an AP analyzer.[17] It is found that field evaporation took place in a layer-by-layer fashion so that composition fluctuation along the field evaporation direction could be measured. The results of a concentration scan of about 6.0 nm long is plotted in Figure 16. All the profiles

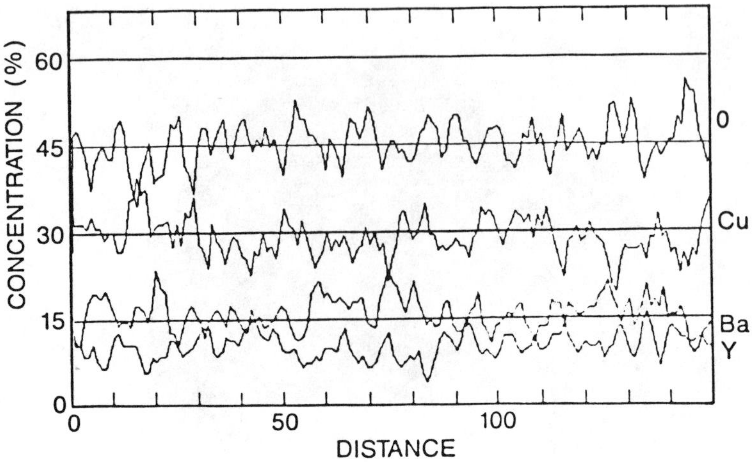

Figure 16. Concentration profiles of O, Cu, Ba, and Y along a random direction. Each unit of analysis is 50 ions which corresponds to a field evaporation depth of about 0.4 nm. Profiles were smoothed by three-point averaging. Scan length is 60 nm.[17a]

display fluctuations. The fluctuations in oxygen appear to be more periodic[17] (about 2.5 nm wavelength) and statistically significant.

Similar fluctuations were also observed by AREM[52] and STM,[53] and on a larger scale by XRD on single crystals[54] and by electron microdiffraction[55] in polycrystalline samples. These fluctuations, measured in terms of change in lattice parameter difference,[55] i.e., $\Delta a / a$ (where $\Delta a = b - a$; b and a are lattice repeat distances in [010] and [100] directions, respectively) were attributed to local variations in the ordering of oxygen. It is concluded, therefore, that the superconducting $YBa_2Cu_3O_{7-x}$ samples are at a metastable state, causing the glassy superconducting behavior.[56] In future studies, it would be useful to have further results of AP analysis taken along [100] and [010] and [110] axes, the last one being perpendicular to the twin plane so that a better transverse resolution is obtained. This would help to determine the oxygen fluctuation near twin and grain boundaries and the understanding of the resulting effects of these defects on superconducting behavior of $YBa_2Cu_3O_{7-x}$.

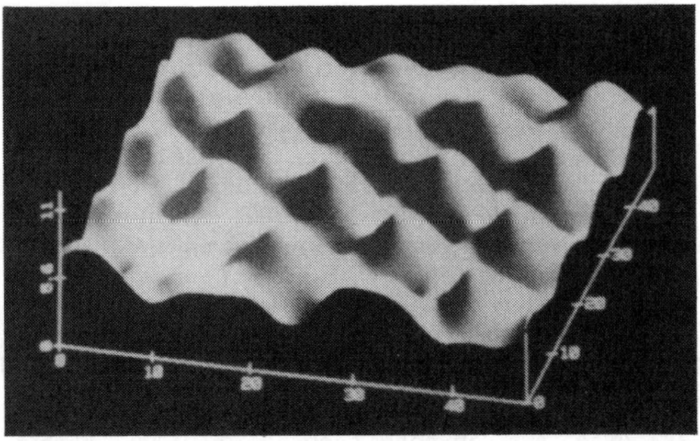

Figure 17. STM image of TaS_2 recorded at room temperature in the constant current mode (tunneling current 2.1 nA) with a platinum-iridium alloy (90%-10%) tip biased at +26 mV relative to the sample. The sample surface was covered with oil, although similar images were also obtained in air. The raw digital data was low-pass-filtered before display. The area markings are in Å.[57]

3.3 Surface Electronic Properties and Effect of Doping Studied by Scanning Tunnelling Microscopy

It is essential to understand the fundamental nature of surfaces in order to utilize their properties. Various properties of surfaces can be simultaneously studied by STM technique with high sensitivies. The example on TaS_2 given below is from a recent paper[57] which illustrates a STM study of structure and electronic properties of surfaces.

Tantalum disulphide, TaS_2, is a member of a group of layered materials, containing transition metals, which exhibit interesting electrical, magnetic, and catalytic properties.[58] The structure of TaS_2 is hcp with a_o and c_o, 3.39 Å and 5.90 Å, respectively, and consists of weakly bonded, S-Ta-S closed packed layers. In the octahedrally coordinated TaS_2, a phase transformation involving commensurate-incommensurate structures occurs upon heating and on doping with an impurity atom. In its charge density wave (CDW) state, both a periodic distortion of the TaS_2 lattice and a change in electron density within S-Ta-S layers take place.[59] This study involves distinguishing these two effects.

In the single crystal form, cleaving TaS_2 produces atomically flat (0001) faces which contain only sulphur atoms. A typical STM image from TaS_2 is shown in Figure 17 which exhibits a hexagonal coordination of the positions of the peaks.[59] However, the lattice formed in the STM image, with peak spanning and corrugation of 11.8 ± 0.2 Å and 5.7 ± 0.5 Å, respectively, do not correspond to the lattice formed by the sulphur atoms on (0001) planes where the atom positions and corrugations are 3.5 Å and 1.0 Å, respectively. These images, which have commonly been observed, have been related to the periodic variation of charge associated with the CDW state in TaS_2.[43,57]

In order to change the electronic properties of TaS_2, it is doped with Ti, which is known to decrease the electron density by one electron per Ti atom.[57] Correspondingly, STM images also change. For sample composition of $Ti_xTa_{1-x}S_2$, low additions of Ti, i.e., $x=0.06$, the CDW is still dominant in the STM image (Figure 18a) with peaks and corrugations, 11.6 ± 0.2 Å and 3.6 ± 0.5 Å, respectively. Some secondary peaks of 3.3 ± 0.1 Å apart can also be resolved in this image which corresponds to the atomic positions in the sulphur lattice. For $x=0.1$ doping, the atomic lattice is clearly defined where peak-to-peak separation is 3.4 ± 0.1 Å (Figure 18b). Even in this case the CDW is still recognizable with a weak periodic modulation (11.6 Å peak separation and 1.4 ± 0.4 Å corrugation). It is concluded that the trend in the decrease of the corrugation distance is significant and it is related to the Ti doping (about 1 Å for $x \cong 0.03$).

It is concluded that the origin of the effects related to doping which modify the STM images are still unknown. Although Ti in TaS_2 causes a lattice distortion and decreases conduction electron density, these two effects perturb the CDW in different ways. The increases in the observed atomic structure (at 295 K) with increasing Ti doping is related to a decrease in the CDW amplitude associated with the incommensurate state (of CDW with the lattice) which otherwise occurs at higher temperatures (~350 K).[59] However, an increase in the CDW wavelength, the expected response to a decrease in the conduction electron density,[60] is not observed in the STM. Nonetheless,

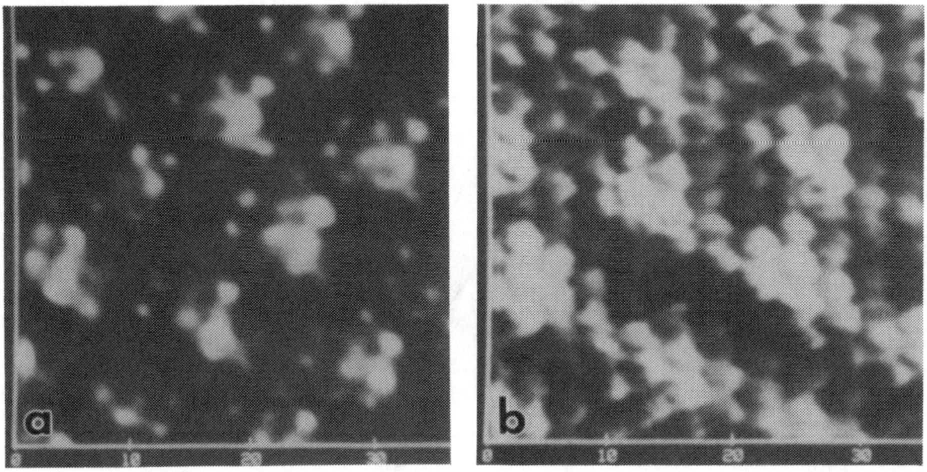

Figure 18. Grey-scale images of $Ti_xTa_{1-x}S_2$ recorded in the constant current mode. The sample surfaces were covered with oil to improve the imaging stability.[17] a) x=0.06; tip bias= +8 mV, tunneling current=2.1 nA. The grey-scale in this image (black minimum to white maximum) is 4.3 Å. b) x=0.1; tip bias voltage +9 mV, tunneling current =2.1 nA. The grey-scale in this image, 2.9 Å, is significantly smaller than in *a*. The *x,y*-axes markings are in Å.[57]

this study demonstrates how chemical modifications can be combined with tunneling microscopy to separate the electronic and structural components of STM surface images.

4.0 PRESENT STATUS AND FUTURE DIRECTIONS

Among the techniques discussed, AREM is the only one which provides bulk information at modes of analysis in imaging, diffraction, and spectroscopy directly from a wide spectrum of thin-foil and particulate materials.[61] The

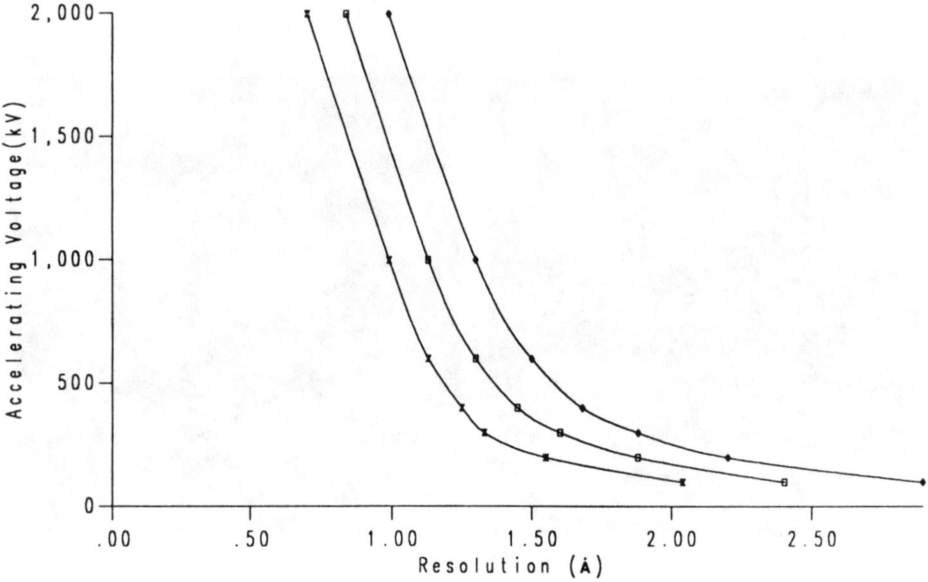

Figure 19. A plot of accelerating voltage versus instrumental resolution (using equation 10) in atomic electron resolution microscopy. C_s values in plot A are taken from ref. 5. B and C correspond to half and quarter values of those C_s corresponding to curve A.

challenge in AREM is the improvement of resolution to an even higher value so that images of interference fringes (corresponding to lattice atomic columns) with greater detail, specific to atomic species, can be obtained. From the resolution formula in equation (10), this is possible by decreasing both λ and C_s values. The effect of the decrease of λ (by an increase of accelerating voltage) on the resolution is given in Figure 19, with the C_s values corresponding to the existing instruments which operate at the given voltages (C_s values were taken from reference 5). Although there is an improvement at the medium voltages, resolution achieves a certain plateau at about 1.4 Å, corresponding to 1.0 MeV accelerating voltage. It is known that increasing voltage increases the radiation damage to the samples by increasing the probability of knock-on phenomenon in which atoms in the lattice of the sample are displaced by the highly energetic incoming electrons.[62] This is especially detrimental to ceramic and polymeric material.[62] The optimum value of accelerating voltage, therefore, is in the 300 to 400 kV level, and

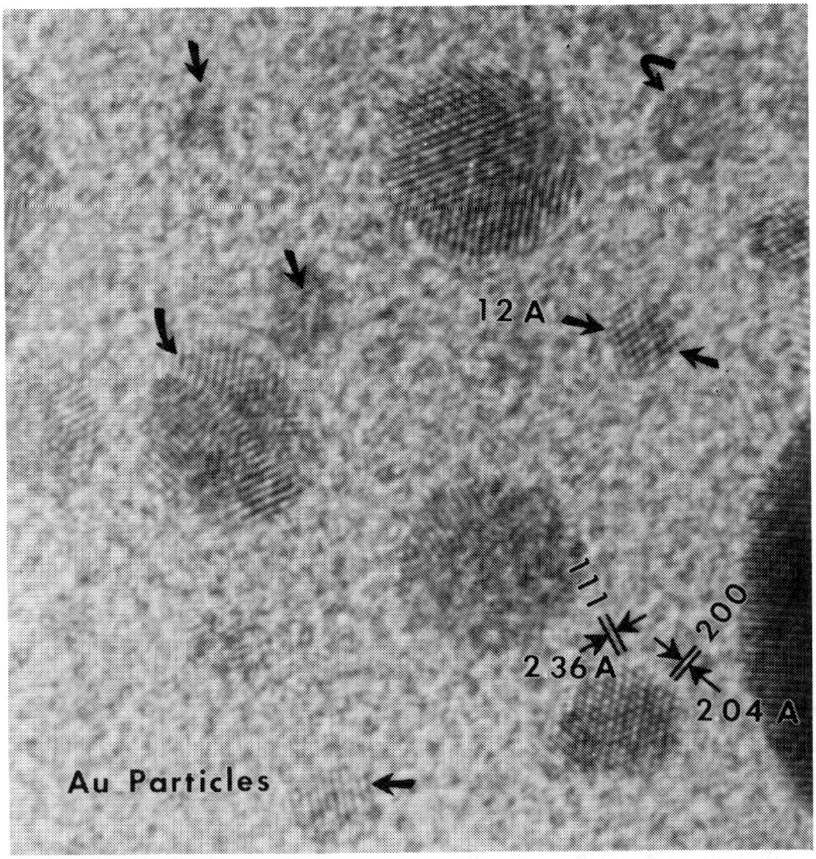

Figure 20. AREM image of ultrafine gold particles[1] (Philips 430T; 300 kV).

this is in fact the voltage range at which the present instruments operate. The example presented in Figure 20 illustrates that the lattices of metallic particles with diameter as small as 12 Å can be resolved with such a medium voltage TEM.[1] Higher voltages also require larger lenses to provide the required field to bend the electrons; correspondingly C_S values increase. If the lens stabilities improve through the use of "new" lens material (such as superconducting lenses) then smaller C_S values may be obtained in the future. The hypothetical curves in Figure 19 give a marginal resolution improvement, which ranges 0.8 to 1.2 Å within 300-400 kV. Consequently, 1.0 Å seems to be the ultimate image resolution achievable in an AREM.

In conjunction with improved lens design, electron probe size can also be made smaller (down to fraction of a nanometer) with adequate brightness. High brightness is achieved through the use of a field emission gun system which also provides the high spatial resolution in analytical electron microscopy techniques performed in an AREM. Among AEM techniques, microdiffraction and convergent beam electron diffraction, CBED, techniques,[63] provide crystallographic information from small structural features. Both energy dispersive x-ray spectroscopy, EDS, and electron energy loss spectroscopy, EELS, provide compositional information. When operated in the parallel detection mode,[64] in the EELS technique, which also provides chemical information, both the counting statistics increase and the acquisition times decrease. These are both crucial factors to be ensured in studying the electron beam-sensitive materials.

The uniqueness of the capabilities, in performing atomic resolution imaging, diffraction, and spectroscopy in a multipurpose AREM[32] can be combined with the possibilities of performing in-situ experiments. These include cooling/heating and deformation. The in-situ experiments are performed, respectively, to study phase transformations and associated phenomena and mechanical response of materials to deformation and crack propagation at the atomic resolution. An example of a heating experiment is illustrated in Figure 21, which presents a sequential imaging of nanometer-sized gold particles during sintering upon electron beam heating.[65] In these studies, the neck formation between the particles as the first step in sintering to achieve densification, as hitherto been assumed,[66] has not been observed. These images, reveal that atoms "simply" form a bridge between the particles which then form coherent or semicoherent interfaces. In addition, close observation of the internal structure during the observation reveals that the lattice of each particle is in continuous motion, like a "semi-liquid," as the particles approach each other and their surfaces achieve a low energy configuration during densification (this work is presently in progress).[65]

In field ion microscopy and atom probe analysis, it is now possible to map the surfaces of samples atom by atom (even in three-dimension by sequential field evaporation) for each individual species.[14] This is done by using each ion, as it is stripped from the tip, with proper correlation procedures.[41] Transverse spatial resolution in quantitative spectroscopic analysis

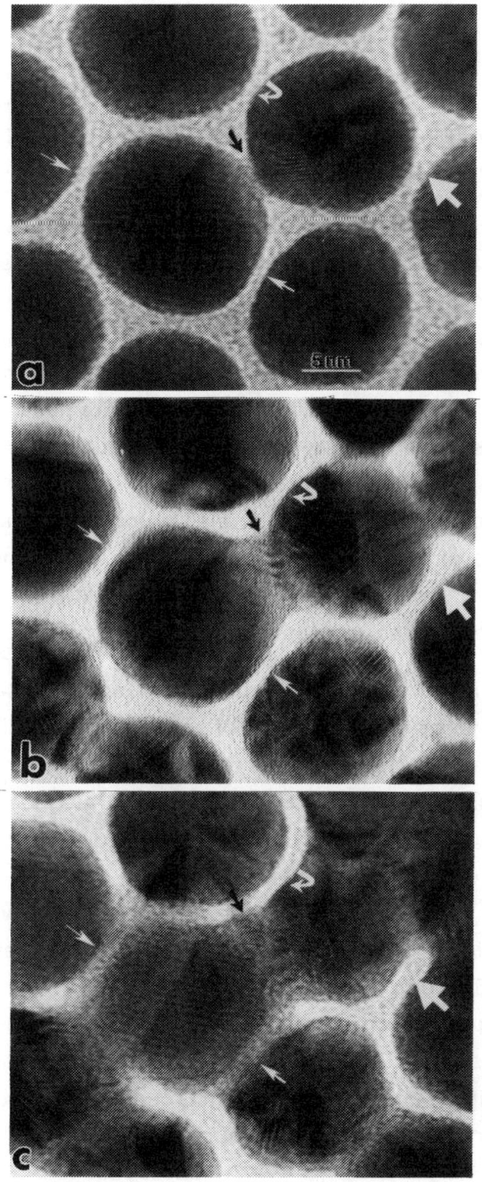

Figure 21. In-situ sintering of gold particles (150-180 Å dia.)[65] in an AREM. (a), (b), and (c) are sequential images recorded during electron beam heating (Philips 430T; 300 kV).

equivalent to atomic steps is achieved, which has not so far been matched by any atomic resolution instrument. The major improvement in the application is expected to be (i) in the sample preparation of ceramics, composites, and semiconducting materials (e.g., by sputtering or ion-milling), and, (ii) in imaging by close control of pulse application during evaporation.

Scanning tunneling microscopy has achieved[19] unprecedented resolution and it is now a unique tool in surface analysis. Consequently, it is almost impossible to predict and list possibl improvements and new applications of STM techniques in the surface characterization of materials as new applications appear almost on a weekly basis. Since there is no requirement of high vacuum for operation, apart from the requirement of clean and flat surfaces, STM has already been used for application in various media, including water, ice, gas, and oil. Surfaces of materials of widely different nature, such as superconducting materials, biological membranes, catalysis, and ceramics have already been imaged[7,18-25,45] with somewhat lower resolution, which, however, is expected to increase with the improvements in the operational procedures of scanning and image display.[7]

Applications in purely ionic crystals, such as most ceramic materials, have been limited because of the lack of electrical conduction. With a small resolution sacrifice, however, STM has recently been used for fracture surface analysis of ceramics, in which case the surfaces are coated with a highly conductive metal. An example is illustrated in Figure 22 which presents an image of a fractured NaCl surface, cleaved along the (100) plane in a three-point-bend test.[67] After the fracture, the sample surface was coated with a 10 Å thick gold layer at 77 K. The resolution in the lateral direction is estimated to be 15 Å, and in the corrugation to be 10 Å. The image in Figure 22 displays a large area of scan, 3000 Å x 3000 Å, with a height difference of 2500 Å. For reasons of unknown nature, extensive regions of clear cleavage fracture (on the right) and regions of high roughness, up to 500 Å high (on the left) are observed on the fracture surface.[67]

The challenges in the STEM field would be in the use of flexible tips to follow surface contours of small particles and fractured surfaces, and recognize various (adsorbed surface and bulk) atoms and ions on the surfaces. There have already been some recent reports in this direction.[68]

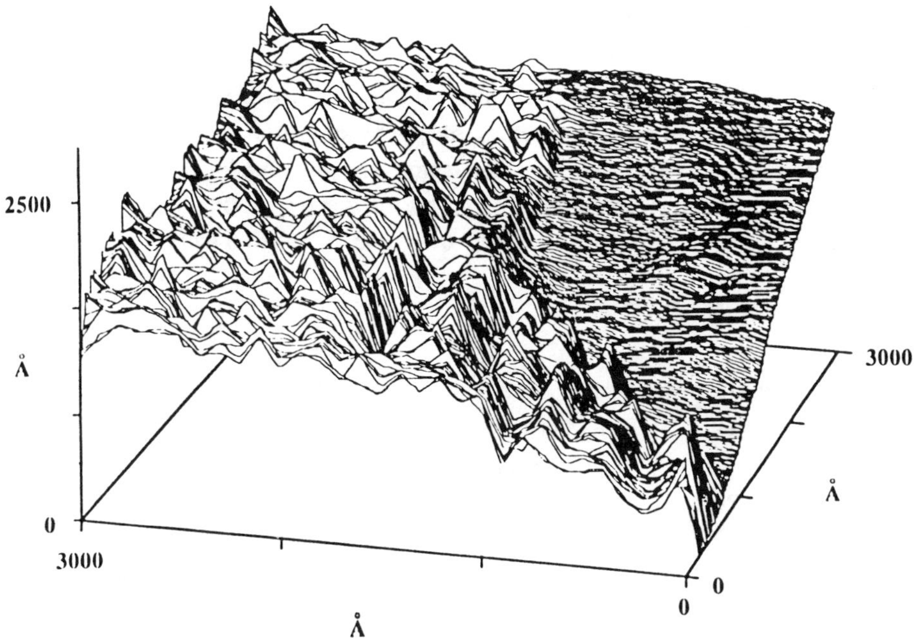

Figure 22. STM scan of (100) fracture surface of NaCl tested under three-point-bend.[67] Imaging was possible after coating the surface with a 10 Å thin layer of gold.

In summary, with the existing and foreseeable technology, the AREM imaging will still be confined to information with spatial resolution not better than 1.0 Å. However, AREM will continue to be the most widely used technique for materials scientists and biologists because it is a unique instrument providing information from inorganic and organic bulk materials by revealing their structural features through all modes of analysis. STM, with its simplicity in design and use in all media, provides information confined to the surface, but at sub-angström resolution in samples of almost any nature. It should be realistic to expect a "marriage" of these two techniques used in a complementary fashion to achieve maximum information in one combined system.[69]

5.0 ACKNOWLEDGMENTS

The author is grateful to Drs. Ulrich Dahmen (Lawrence Berkeley Laboratory), Sidney S. Brenner (University of Pittsburgh), and Thomas Dickinson (Washington State University), for discussion and the use of their research results. Author's own work is sponsored by the Air Force Office of Scientific Research under Grants Nos. AFOSR-87-0114 and AFOSR-88-0135.

6.0 REFERENCES

1. J. Liu, Ph.D. thesis in progress, University of Washington (1988).
2. a. C. Viney, "Optical Microscopy for Studying Molecular Ordering in Polymers," in ***Encyclopedia of Eng. Materials***, N. P. Cheremisinoff (ed.), Part A, Vol. 1, Marcel Dekker, N.Y., 505-536 (1988).
   b. R. Y. Tsien and M. Poenie, "Fluorescence Ratio Imaging of Intracelluiar Ionic Signals," *Trends Biochem. Sci.*, **11**, 450-455 (1986).
3. a. Hirsch, P. B., A. Havie, R. B. Nicholson, D. W. Pashley, and M. J. Whelan, ***Electron Microscopy of Thin Crystals***; Krieger: New York, 1977.
   b. G. Thomas and M. J. Goringe, ***Transmission Electron Microscopy of Materials***; John Wiley and Sons: New York, 1979.
   c. L. E. Murr, ***Electron Optical Applications in Materials Science***; McGraw-Hill: New York, 1970.
   d. L. Reimer, ***Transmission Electron Microscopy***; Springer-Verlag: Berlin, 1984.
4. a. O. L. Krivanek, "High Resolution Imaging of Grain Boundaries and Interfaces," *Chemica Scripta*, **14**, 213-218 (1979).
   b. R. Gronsky, "High Resolution Transmission Electron Microscopy," In ***Treatises of Material Science and Technology, 19B***; Academic: New York, 1983; pp. 225-257.
   c. M. Sarikaya, "Characterization of Advanced Materials by High Resolution Transmission Electron Microscopy," in ***Proc. of MRS in "Atomic and Molecular Processing of Electronic and Ceramic Materials;"*** I. A. Aksay, G. L. McVay, T. G. Stoebe, J. F. Wagner, Eds. (Materials Research Society: Pittsburgh, 1987), pp. 211-226.
   d. G. Thomas and R. Ramesh, "Atomic Imaging and Microanalysis of Ceramics", these proceedings.

5. J. M. Cowley and D. J. Smith, "High Resolution Transmission Electron Microscopy," *Acta. Cryst. A.*, **43**, 739- (1987).

6. a. J. I. Goldstein and H. Yakowitz (eds.), ***Practical Scanning Electron Microscopy; Electron and Ion Microanalysis***; Plenum Press: New York, 1975.

   b. D. P. Woodruff and T. A. Delchar, ***Modern Techniques in Surface Science***, Cambridge University Press: New York, 1988.

7. a. C. F. Quate, "Vacuum Tunneling: A New Technique for Microscopy," *Phys. Today*, **39**, 26 (1986).

   b. J. A. Golovchenko, "The Tunneling Microscope: A New Look at the Atomic World," *Science*, **232**, 48-53 (1986).

8. a. E. W. Müller and T. T. Tsung, ***Field Ion Microscopy: Principles and Applications***; (Elsevier: Amsterdam, 1969).

   b. S. S. Brenner, "Applications of Field-Ion Microscopy Techniques to Metallurgical Problems," *Surface Sci.*, **70**, 427-451 (1978).

9. P. J. Turner, B. J. Regan, and M. J. Soothon, "Field Ion-Atom Probe Analysis," *Surface Sci.*, **35**, 336-344 (1973).

10. H. W. Zieler, "What Resolving Power Formula Do You Use?" *The Microscope*, **17**, 249-270 (1969). "

11. M. Isaacson, E. Belzig, A. Harootiunan, and J. A. Lewis, "Near-Field Optical Microscopy," *Annals of the New York Acad. Sci.*, **483**, 448 (1986).

12. J. C. H. Spence, ***Experimental High Resolution Electron Microscopy***; Clarendon Press: Oxford, 1981.

13. J. J. Hren, J. J. Goldstein, and D. C. Joy, eds., ***Introduction to Analytical Electron Microscopy***, Plenum Press, New York (1979).

14. a. S. S. Brenner and M. K. Miller, "Atomic Scale Analysis with the Atom Probe," *JOM*, **35** (3), 54-63 (1983).

    b. M. A. Burke and M. K. Miller, "Comparison of TEM and APFM in Microstructural Characterization and Interpretation," *TEM Tech.*, **8**, 201-210 (1988).

15. O. Nishikawa, "Atom-Probe Study of Semiconductors and Conducting Polymers," ***Proc. of Microbeam Analysis 1986***, A. D. Romig, Jr. and W. F. Chambers (eds.), San Franciso Press, San Francisco, pp. 351-356 (1986).

16. O. Nishikawa, O. Kaneda, M. Shibata, and E. Nomura, "Atom Probe Study of Al-Ga Exchange Reaction at Al-GaAs Interfaces," *Phys. Rev. Lett.*, **53**, 1252-1255 (1984).

17. a. G. L. Kelly and S. S. Brenner, "Field Ion Microscopy and Imaging Atom-Probe Mass Spectroscopy of Superconducting $YBa_2Cu_3O_{7-x}$," *App. Phys. Lett.*, **51** (22), 1851-1853 (1987).

b. S. S. Brenner and G. L. Kelly, "Atom Probe Microanalysis of Superconducting $YBa_2Cu_3O_{7-x}$," *Mat. Res. Symp. Proc. Vol. 99*, Materials Research Society, Pittsburgh (1988).

18. G. Binnig, and H. Rohrer, "Scanning Tunneling Microscopy," *Helv. Phys. Acta*, **55**, 726-735 (1982).

19. G. Binnig and H. Rohrer, "Scanning Tunneling Microscopy--From Birth to Adolescence, Nobel Lecture," *Angew. Chem. Int. Ed. Engl.*, **26**, 606-614 (1987).

20. G. Binnig, H. Rohrer, Ch. Gerber, and E. Weibel, E., "Surface Studies by Scanning Tunneling Microscopy," *Phys. Rev. Lett.*, **49**, 57-61 (1982).

21. R. S. Becker, "Tunneling Microscopy of Silicon and Germanium," *Proc. Nat. Acad. Sci. USA*, **84**, 4667-4670 (1987).

22. R. M. Tromp, R. J. Hamers, and J. E. Demuth, "Quantum States and Atomic Structure of Silicon Surfaces," *Science*, **234**, 304-309 (1986).

23. R. Sonnenfeld and P. K. Hansma, "Atomic Resolution Microscopy in Water," *Science*, **232**, 211-213 (1986).

24. S.-I. Park and C. F. Quate, "Tunneling Microscopy of Graphite in Air," *Appl. Phys. Lett.*, **48**, 112-114 (1986).

25. J. A. N. Zasadzinski, J. Schein, J. Gorley, V. Elwigs, and P. K. Hansma, "Scanning Tunneling Microscopy of Freeze-Fracture Replicas of Biomembranes," *Science*, **239**, 1013-1015 (1988).

26. U. Dahmen and K. H. Westmaccott, "Observation of Pentagonally Twinned Precipitate Needles of Germanium in Aluminum," *Science*, **234**, 875-877 (1986).

27. I. Hanson and A. R. Thölen, "Adhesing and Twin Formation in Fine Particle Systems," *Scand. J. Metall.*, **6**, 27-29 (1979).

28. a. D. J. Smith and L. D. Marks, "High Resolution Studies of Small Particles of Gold and Silver," *J. Cryst. Growth*, **54**, 433-438 (1981).

b. L. B. Marks, "Surface Structure and Energies of Multiply Twinned Particles," *Phil. Mag. A*, **49** (1), 81-93 (1984).

29. S. Iijima, "Fine Particles of Silicon. I. Crystal Growth of Spherical Particles of Si," *Japan. J. Appl. Phys.*, **26** (3), 357-364 (1987).

30. J. Liu, M. Sarikaya, and I. A. Aksay, "Aggregation and Packing of Nanosize Particles," unpublished research (1988).

31. a. O. Schertzer, "Theoretical Resolution Limit of the Electron Microscope," *J. Appl. Phys.*, **20**, 20-22 (1949).

    b. O. Schertzer, "Limitations for the Resolving Power of Electron Microscope," in ***9th Intl. Conf. on E. M., Toronto, Vol. III***; J. M. Sturgess, Ed.; (The Imperial Press, Ltd.: Ontario, 1978), pp. 124-129.

32. H. Hashimoto, "Achievement of Ultrahigh Resolution by 400 kV Analytical Atomic-Resolution Electron Microscopy," *Ultramicroscopy*, **18**, 19-32 (1985).

33. M. A. O'Keffe, "Image Interpretation in High Resolution Microscopy," *EMSA*, 556-557 (1979).

34. L. A. Burshill, "Interpretation of HREM images of crystals," *Ultramicroscopy*, **18**, 1-10 (1985).

35. W. O. Saxton and D. J. Smith, "The Determination of Atomic Positions in High-Resolution Electron Microscopy," *Ultramicroscopy*, **18**, 39-48 (1988). 36

36. E. W. Müller, and K. Bahadur, "Field Ionization of Gases at a Metal Surface and the Resolution of the Field Ion Microscope," *Phys. Rev.*, **102**, 624-631 (1956).

37. R. H. Good, and E. W. Müller, "Field Emission," in ***Handbuch der Physik***; (Springer: Berlin, 1956; Vol. 21, 2nd ed.), pp. 177-231.

38. E. W. Müller, "Observation of Radiation Damage with Field Ion Microscope," in ***Proc. 4th Intl. Symp. Reactivity of Solids***; (Amsterdam, 1960), p. 682-691.

39. E. W. Müller, J. Panitz, S. B. McLane, "The Atom-Probe Field Ion Microscope," *Rev. Sci. Instr.*, **39**, 83-86 (1968).

40. S. S. Brenner and J. T. McKinney, "Construction and Performance of FIM-Atom Probe," *Surface Sci.*, **23**, 88-111 (1978).

41. T. J. Wilkes, G. D. W. Smith, and D. A. Smith, "On the Quantitative Analysis of Field-Ion Micrographs," *Metallography*, **7**, 403-430 (1974).

42. M. Sarikaya, Ph.D. Thesis, University of California, Berkeley; Lawrence Berkeley Laboratory Report #15111.

43. a. G. C. Slough, W. W. McNairy, R. V. Coleman, B. Drake, and P. K. Hansma, "Charge-Density Waves Studied with the Use of a Scanning Tunneling Microscope," *Phys. Rev. B*, **34**, 994-1005 (1986).

b. R. V. Coleman, B. Drake, P. K. Hansma, and G. Slough, "Charge Density Waves Observed with a Tunneling Microscope," *Phys. Rev. B*, **55**, 394-397 (1985).

44. B. Giambattista, W. W. McNairy, C. G. Slough, A. Johnson, L. D. Bell, R. V. Coleman, J. Schneir, R. Sonnenfeld, B. Drake, and P. K. Hansma, "Atomic Resolution of Solid-Liquid Interfaces," *Proc. Nat. Acad. Sci. USA*, **84**, 4671-4674 (1987).

45. a. P. K. Hansma and J. Tersoff, "Scanning Tunneling Microscopy," *J. Appl. Phys.*, **61**, No. 2, R1-R23 (1987).

    b. P. K. Hansma, V. B. Elings, O. Marti, and C. E. Bracker, "Scanning Tunneling Microscopy and Atomic Force Microscopy: Applications to Biology and Technology," *Science*, **242**, 209-216 (1988).

46. X.-L. Wu and C. M. Lieber, "Determination of the Structural and Electronic Properties of Surfaces Using Scanning Tunneling Microscopy Coupled with Chemical Modifications," *J. Am. Chem. Soc.*, **110**, 5200 (1988).

47. J. Green, "Calcination of Precipitated $Mg(OH)_2$ to Active MgO in the Production of Refractory and Chemical Grade MgO," *Materials Sci.*, **18**, 637-651 (1983).

48. J. F. Goodman, "The Decomposition of Magnesium Hydroxide in an Electron Microscope," *Proc. Roy. Soc. Lond.*, *247*, 346-352 (1958).

49. D. J. Anderson, and R. F. Horlock, "Thermal Decomposition of Magnesium Hydroxide," *Trans. Faraday Soc.*, **58**, 1993-2004 (1962).

50. U. Dahmen, M. G. Kim, A. W. Searcy, "Microstructural Evolution During the Decomposition of $Mg(OH)_2$," *Ultramicroscopy*, **23**, 365-370 (1987).

51. A. F. Moodie and C. E. Warble, "MgO Morphology and Thermal Decomposition of $Mg(OH)_2$," *J. Crystal Growth*, **74**, 89-100 (1986).

52. Z. Hiroi, M. Takano, Y. Ikeda, Y. Takeda, and Y. Bando, "Coherent Intergrowth Structure of Tetragonal to Orthorhombic Ba_2YCuO_{7-x} Observed by Transmission Electron Microscopy," *Japan. J. Appl. Phys.*, **27**(2), L141-L144 (1988).

53. R. Laiho, L. Heikkilä, and H. Snellman, "Microstructural Investigation of the High-T_c Superconductor $YBa_2Cu_3O_{7-x}$ With a Scanning Tunneling Microscope," *J. Appl. Phys.*, **63**(1), 225-227 (1988).

54. H. You, J. D. Axe, X. B. Kan, S. C. Moss, J. L. Liu, D. J. Lam, "High Resolution X-Ray Microstructural Study of Single Crystal of $YBa_2Cu_3O_{7-x}$," *Phys. Rev. B*, **37**, 2301 (1988).

55. M. Sarikaya, and E. A. Stern, "Local Structural Variations in $YBa_2Cu_3O_{7-x}$," *Phys. Rev. B*, **37**, 9373-9381 (1988).
56. G. Deutscher and K. A. Müller, "Origin of Superconductive Glassy State and Extrinsic Critical Currents in High-T_c Oxides," *Phys. Rev. Lett.*, **59**(15), 1745-1747 (1987).
57. X.-L. Wu, P. Zhou, and C. M. Lieber, "Surface Electronic Properties Probed with Tunneling Microscopy and Chemical Doping," *Nature*, **335**, 55-57 (1988).
58. J. A. Wilson, F. J. DiSalvo, and S. Mahajan, "Charge Density Waves and Superlattices in the Metallic Layered Transition Metal Dichalcogenides," *Adv. Phys.*, **24**, 117-201 (1975).
59. F. J. Di Salvo, in ***Electron-Phonon Interactions and Phase Transitions***; T. Riste, Ed.; (Plenum: New York, 1977).
60. F. J. Di Salvo, J. A. Wilson, B. G. Bagley, and J. V. Waszczak, "Effect of Doping on Charge-Density Waves in Layered Compounds," *Phys. Rev. B*, **12**, 2220-2235 (1975).
61. P. R. Buseck, ***High Resolution Electron Microscopy***, North-Holland, Amsterdam (1988); also *Ultramicroscopy*, **18** (1-4), 1988.
62. Reference 13. Chapter 16 by R. M. Glaeser, "Radiation Damage with Biological Specimens and Organic Materials," pp. 423-436; and Chapter 17, by L. W. Hobbs, "Radiation Effects in Analysis of Organic Specimens by TEM," pp. 437-480.
63. J. M. Cowley and J. C. H. Spence, "Nanobeam Diffraction in STEM," *Ultramicroscopy*, **6**, 359-366 (1981).
64. O. L. Krivanek, C. C. Ahn, and R. B. Keeney, "Parallel Detection Electron Spectrometer Using Quadrupole Lenses," in ***Frontiers of Electron Microscopy in Materials Science***, S. Bradley and W. King (Eds.), North-Holland, Amsterdam, 103-116 (1982).
65. J. Liu, M. Sarikaya, and I. A. Aksay, "In-Situ Densification of Clusters of Nanosize Particles," unpublished research.
66. W. D. Kingery, H. K. Bowen, and D. R. Uhlman (Eds.), ***Introduction to Ceramics***, 2nd Ed., John Wiley and Sons, New York, Chapter 10, 448-515 (1976).
67. T. Dickinson, private communication.
68. a) E. Ganz, K. Sattler, and J. Clarke, "One-Dimensional Random Walk of a Five-Atom Silver Cluster on Graphite Observed by Scanning Tun-

neling Microscopy;" b) R. M. Feenstra and P. Martensson, "Scanning Tunneling Microscopy and Spectroscopy of Antimony Overlayers on the GaAs(100) Surface," MRS Fall Meeting, Nov. 28-Dec. 1988, Boston.

69. J. C. H. Spence, M. Kuwabara, and W. Lo, "A Scanning Tunneling Microscope for Use in the Philips EM 4000 Electron Microscope in the Reflection Mode," in ***Proc. of 46th Ann. Meet. Electron Microsc. Soc. of Amer.***, G. W. Bailey (Ed.), San Francisco Press, San Francisco, 446-447 (1988).

ATOMIC IMAGING AND MICROANALYSIS OF CERAMICS AT THE NATIONAL CENTER FOR ELECTRON MICROSCOPY, BERKELEY, CALIFORNIA

Gareth Thomas and R. Ramesh*
Department of Materials Science & Mineral Engineering
University of California at Berkeley and National Center for Electron Microscopy, Materials and Chemical Division
Lawrence Berkeley Laboratory, 1 Cyclotron Road, Berkeley, CA 94720

*now with Bell Communications Research, Red Bank, NJ 07701

ABSTRACT

A brief summary is presented of recent research using atomic resolution imaging and microanalytical procedures that are being applied to a number of ceramic systems at the National Center for Electron Microscopy. This paper is not intended as a comprehensive review but rather to illustrate representative problems for which these techniques can provide information toward understanding often complex materials.

INTRODUCTION

The inherent complexity of the crystal structure and microstructure of ceramics, as well as the interdependence of the final properties on microstructure and processing, mean that detailed characterization of the effect of processing variables on the structure and microstructure is imperative. Thus, in order to understand and improve ceramics, it is necessary to establish these relationships iteratively. One common theme in this research methodology is the characterization of the grain boundaries, and the structural and microcompositional changes that take place near such interfaces. Electron microscopy has become firmly established as a powerful analytical tool for the characterization of the structure and microstructure in ceramics. Due to the various types of interactions of electrons with materials, the electron microscope is unique in that it can provide morphological, structural, and compositional information with high spatial resolution. **Fig. 1** shows the various types of information that can be obtained from a modern transmission electron microscope. However, one has to be aware of some of the inherent problems associated with electron microscopy of ceramic materials. These are (1) specimen preparation for electron microscopy and the possible changes in the nature of the sample due to thin foil effects; (2) radiation damage, especially knock-on damage; and (3) image resolution of anions.

IMAGING AND X-RAY SPECTROSCOPY

It has long been the goal of electron microscopists to achieve interpretable atomic resolution. For crystals sufficiently thin, the weak phase object approximation can be made, which implies that the image is basically a projection of the lattice potential. Hence, contrast of atomic columns is proportional to the projected atomic number along that column. The resolution is fundamentally a function of the electron wavelength and lens aberrations, especially spherical aberration, quantified by the relationship: $R = 0.66\ C_s^{1/4}\lambda^{3/4}$, which is plotted in **Fig. 2** [1]. As shown in this figure, to achieve 2Å point resolution using an objective lens with C_s = 3 mm, one would have to operate in the range of 600-1000 kV. Thus, with increasing energy, the reduction in the wavelength enables interpretable atomic resolution to be achieved when approaching 1 MeV. In addition, the effect of chromatic aberration is reduced by increasing the accelerating voltage, since this parameter varies as $(kV)^{-1}$. Another resolution limit is set by the ultimate cut-off spectra due to instrumental instabilities.

The effects of aberrations and defocus of the objective lens must be taken into account when considering the ability of the lens to transfer sufficient "in-phase" information scattered within the lens aperture that can provide such phase contrast to yield interpretable structure images. The phase shift between a scattered and undeviated electron beam is given by:

$$\chi(\alpha) = \pi/\lambda[(C_s\bullet\alpha^4)/2 + D\bullet\alpha^2] = \pi\bullet\lambda\bullet g^2(C_s\bullet\lambda^2 g^2/2) + D \quad [1]$$

for Bragg scattering where $\lambda\bullet g = 2\theta$, $\alpha = 2\theta$, and D = objective lens defocus. The lens thus transfers information as a complex quantity $i\bullet\chi(g)$, which can be represented as a diagram of (χ) versus g for varying C_s and D. Damping terms can be included to account for chromatic errors, beam divergence, mechanical noise, etc. This plot is known as the contrast transfer function (CTF) and an example is shown in **Fig. 3a**. The phase contrast at high resolution is the imaginary part of the CTF. It can be seen that the objective lens underfocus offsets the loss in resolution due to C_s.

The value of the optimum defocus is known as the Scherzer defocus, obtained by differentiating the CTF and setting $\chi(g) = -\pi/2$ and has the result: $D_{Sch} = -(C_s \bullet \lambda)^{1/2}$. From **Fig. 3a**, it is seen that the resolution of the Berkeley ARM (at 1000 kV) of ≈1.4 Å is the best currently available in the world, along with the ±40° specimen tilt. The resolution of the 400 kV microscopes, **Fig. 3a**, which are based on the design of the ARM is about 1.7 Å. Of course, when a specimen is introduced into the problem, then one must also consider parameters such as thickness, as illustrated by the example in **Fig. 3b**. In this figure, computed images for orthorhombic ZrO_2 are shown for different values of foil thickness and defocus [2]. It is to be noted that the detail available is very different from one image to another. These considerations show that atomic resolution images cannot be interpreted by mere inspection. This must be done via the use of known operating parameters of the microscope and specimen, which involves sophisticated computing in addition to expert microscopy.

At the National Center for Electron Microscopy (NCEM), the atomic resolution microscope (ARM) has been designed to meet the above requirements over its entire voltage range of 400 kV to 1000 kV at constant $C_s \bullet \lambda$. The key to this variable voltage atomic resolution performance is the top entry specimen stage that, in addition to ±40° tilting, can be moved up and down in order to maintain a constant value of $C_s \bullet \lambda$ for a given value of objective lens current [3]. Additionally, this enables tuning of the voltage to below the threshold for knock-on damage, especially for beam sensitive materials such as ceramic oxides and glasses. The key to successful interpretation of atomic resolution images is the computing and image simulation facilities that are available at the Center. Such facilities provide very rapid feedback to the researcher, enabling quick and effective optimization of the imaging conditions.

Image processing is carried out using the hardware and software (SEMPER) facilities described by Saxton et al. [4]. For image simulation, a new software program, NCEMSS (National Center for Electron Microscopy Simulation Software), provides an interactive milieu via a mouse input. Details of the computing facilities are given in earlier references [5,6]. In association with the image processing computer, a second computer system has been designed to acquire and enhance electron microscope images in real time at video rates from three microscopes. This real-time system allows the user to make necessary optical adjustments, e.g., astigmatism, in less than a minute, thus limiting the delay resulting from improper imaging conditions.

Analytical electron microscopy forms the other important facet to modern electron microscopy since it provides spectroscopic and diffraction information from the regions being examined. With the use of energy dispersive x-ray spectroscopy (EDXS) and electron energy loss spectroscopy (EELS), it is possible to obtain quantitative compositional information for all elements above Z=5, with a spatial resolution of about 100 Å. Likewise, microdiffraction and convergent beam electron diffraction provide structural information with a spatial resolution approaching 40 Å (determined by the probe size) for a LaB_6 filament. However, this can be improved by the use of a higher coherence source such as the field emission gun (FEG), which, unfortunately, is not yet widely available. Atomic level resolution can be indirectly obtained by using techniques such as ALCHEMI [7]. Since the optimum microscope requirements are not the same for these different techniques, it is essential that each microscope is dedicated to a specific collection of tasks, e.g., atomic resolution imaging, microanalysis, magnetic imaging. THE NCEM is such a dedicated instrument facility.

In the following, examples from on-going research at Berkeley are presented which illustrate the application of atomic resolution imaging, image simulation and processing, and microanalysis toward understanding modern ceramics.

A. High T_c Oxide Superconductors

The recent discovery of superconductivity above 100 K in Bi-Ca-Sr-Cu-O has further fueled the excitement prevailing in this area of research since the discovery of "warm" ceramic superconductors such as Y-Ba-Cu-O [8]. The superconducting phase in this system, described by the general composition of $Bi_2Sr_2Cu_nCa_{n-1}O_y$, exhibits polytypoid-like behaviour [9], similar to other well-known oxide ceramics. The T_c is known to increase with "n", the number of Cu-O layers. The stacking sequence in these complex oxides has been revealed by atomic resolution imaging [10]. In **Fig. 4a**, a {110} ARM image of the n=3 (T_c = 110K) polytypoid is shown. Since the projection of the atomic potential is imaged, the images of the Bi atoms (Z=83) are the largest, while that of Ca are the smallest. Note that this can also be used to obtain atomic discrimination, i.e., replacement of Ca by Bi (their ionic radii are almost the same) can be directly identified. The SEMPER processed image in **Fig 4b** shows another interesting detail of the structure. The contrast corresponding to the central Cu-O layer (indicated by arrow) is quite different from that of the other two rows, suggesting a deficiency of oxygen along that "chain." Image simulations confirm this inference.

In **Figs. 5a and 5b** the {110} projected potential and a series of simulated images are shown respectively for the perfect crystal. In **Figs. 5c and 5d** the same are shown for a crystal in which the oxygen atoms from the central Cu-O row are removed. In the simulated image in **Fig. 5d**, the contrast of the central Cu-O row is different from that of the other two rows, as seen in the processed image in **Fig. 4b**.

High-resolution imaging, in conjunction with EDXS microanalysis, has also shown that in these solid state processed samples, the grain boundaries are Cu and Ca deficient, thus causing the lower T_c n = 1 or 2 polytypoid to form. **Fig. 6a** is a lattice fringe image showing the polytypoid structures adjacent to the grain boundary. The T_c values decrease toward the grain boundary due to the change in the composition and T_c for each polytypoid. **Fig. 6b** is a typical lattice fringe image of a sample with PbO added, showing a uniform n=3 structure up to the grain boundary. The resistivity plot of the leaded samples are quite different from those of the unleaded samples [9].

Similar effects have been observed in Mg-SiAlONs, **Fig. 7**, in which the cation:anion ratio changes close to the grain boundary, causing the formation of different polytypes [11]. In another example, Be-Si-N [12], the formation of different polytypic structures occurs by the propagation of a stacking fault through the lattice. **Fig. 8** shows a typical example where HREM, in conjunction with image simulation, has been used to identify different polytypoid structures. While the effect of such changes near grain boundaries in MgSiAlONs on their properties is not clear, it is very likely that the low T_c polytypoids adjacent to the grain boundaries can affect intergranular "connectivity" [13].

Another aspect of concern is the nature of the grain boundary plane itself. In the case of Y-Ba-Cu-O, high-resolution imaging of the grain boundaries has shown that approximately 70% of the grain boundaries' planes are of the (001) type. **Fig. 9a** shows one such typical grain boundary, in which the upper grain is in the {100} zone and the lower is in the {441} zone [14]. **Fig. 9b** shows a SEMPER processed image in which only the upper grain has been processed. The interesting feature in this image is that the grain boundary plane is a BaO plane. It is well known that BaO is an insulator and, hence, is likely to be a barrier to the movement of the super-current from one grain to another, especially if this plane is an inherent feature of the structure of the grain boundary. In the case of the BCSCO ceramics, the grain boundary plane is generally a Bi-O plane.

These alkali earth-based ceramics are typical examples in which the effect of the electron beam on the integrity of the sample is pronounced. **Fig. 10** is an example of the BCSCO sample irradiated with an 800 kV electron beam. The edge of the foil is completely damaged due to knock-on damage. This is one problem that needs to be kept in mind when imaging such materials. The operator needs to be extremely efficient and fast in order to acquire the images. On the other hand, dynamic acquisition of the images on a magnetic tape could turn out to be useful.

B. Mullite-Zirconia Ceramics

Incorporation of zirconia particles in a mullite matrix has been found to beneficially improve the fracture toughness. There has been considerable debate over the structure of mullite, especially regarding the ordering of vacancies [15,16]. The may be due in part to the large range of solid solubility in the Al_2O_3-SiO_2 system.

Similarly, several different structural variants of zirconia, including ordered structures, have been observed, e.g., the orthorhombic form show in **Fig. 3b** [2]. Atomic resolution imaging, in conjunction with image simulations and matching, has been utilized to understand the structure of these two phases. **Fig. 11** shows the structural model of mullite that was used in the simulations. EDXS microanalysis showed small mounts of Ti in solid solution (due to the reaction sintering). The Al_2O_3-SiO_2-TiO_2 phase diagram shows a shift in the composition of mullite from the ideal 3:2 (Al_2O_3-SiO_2) to the 2:1 composition.

For the samples used in this study, the ratio was found to be 1.7:1. **Figs. 12a and 12b** show simulated images of a 29 Å mullite crystal for defocus values of 0 Å and 200 Å, respectively. The corresponding ARM image is shown in **Fig. 12c.** The match between the experimental and simulated images is very good, thus verifying the structure model. It may be noted that there is no evidence for any ordering, although other orientations must be explored before this matter can be unequivocally settled. **Figs. 13a and 13b** show simulated images for an 87 Å thick foil at defocus values of 0 Å and 100 Å; and the match with the experimental image, **Fig. 13c,** is very good. The drastic change in the nature of the images, with thickness and defocus, makes it imperative that image interpretation be carried out only with the assistance of thorough image simulation and matching. This is illustrated by the strong influence on the image details of a small tilt of the sample from the exact zone axis.

This influence can be recognized by obtaining the Fourier transform of the image using the SEMPER program. **Fig. 14a** shows the actual image of mullite, with the corresponding optical diffraction pattern shown in **Fig. 14b.** In this pattern, it can be seen that the intensity of the diffraction spots is not the same, indicating a small tilt of the foil. However, this can be corrected in the computer to produce the zone axis image. The results of this are shown in **Figs. 14c and 14d. Fig 14c** is the image of the same region as in **Fig. 14a,** but with the sample in the exact zone. The corresponding optical diffraction pattern is shown in **Fig. 14d.** This example clearly illustrates the critical need for a large tilt range (e.g., ±40°), to enable the operator to tilt the sample into the exact zone axis with the assistance of on-line computer analysis. It also illustrates the capability of currently existent software and hardware to assist the operator in obtaining the optimum information from the microscope. With the advent of dynamic microscope-computer interactions in the near future, such experiments will be possible as the operator is working at the microscope, rather than after the experiment. This will enable the correction of imaging conditions to be done instantaneously.

C. Crystallization of Glasses in Y-SiAlONs

Much research has been carried out on SiAlONs with small amounts of sintering additives such as Y_2O_3/Al_2O_3, MgO, etc., to enhance densification [17-21]. This leads to liquid phase sintering, consequent to which the liquid phase does not completely crystallize. **Figs. 15a-e** show different intergranular configurations after liquid phase sintering. The most desirable microstructure is that in "D", where all the liquid phase has crystallized; or that in "E" where all additives are taken into solution in the matrix, provided no polytypes (or polytypoids) form, such as in **Fig. 6a, Fig. 7,** and **Fig. 8.** The amorphous phase, which forms due to the solidification of the liquid phase, is detrimental to the high-temperature properties of the ceramic. Electron microscopy and microanalysis have been instrumental in understanding these intergranular phases.

Since this problem was observed in the case of many additives, a research program addressing the crystallization characteristics of a ceramic of composition equal to that of the amorphous phase was started at Berkeley in 1980 [see Ref. 22]. These compositions are shown in the isothermal section of the $AlN-SiO_2-Y_2O_3$ phase diagram in **Fig. 16.** At 1000°C dendritic nucleation occurs on heterogeneous sites with the formation of an orthorhombic "y" phase polymorphic with the $Y_2Si_2O_7$ phase. HREM and atomic resolution imaging have been used to examine the details of the different polytypes or polytypoids that form during the crystallization process. It has been observed that the crystallization proceeds via the formation of an intermediate δ_1 phase, which is heavily faulted. The side-arms in this growth structure are close to equilibrium $Y_2Si_2O_7$ and grow epitaxially as one of the twin variants. The details of this structure are shown in the HREM image in **Fig. 17a** [23]. This image shows the nature of the interface between the δ_1-δ_2 dendrites. The flat interfaces are twin boundaries with strain contrast only at the ledges.

The atomic configurations of the different phases appearing upon crystallization are not very different. In such cases, direct imaging along with image matching is extremely useful in identifying the phase. In **Fig. 17b**, an ARM image of the δ_1 $Y_2Si_2O_7$ phase is shown along with the simulated image [24].

The match between the experimental and simulated image is very good except for the slight skew of the image, probably due to a slight specimen tilt. These crystallization processes proceed with the rejection of Al and N ahead of the transformation interface, as evidenced by EDXS microanalysis. Upon prolonged annealing, it is possible to achieve complete crystallization. As in the case of mullite, the image details in this system are sensitive to small misorientations, and interpretation is possible only for extremely thin sections. However, in the case of these crystallized glasses, the thinnest areas are generally found near the amorphous regions. Thus, the image details are convoluted with the contribution from the amorphous phase.

The crystalline details in the thinnest part of the image can be deconvoluted using the SEMPER program, the results of which are shown in **Figs. 18a-d** [25]. In the region of the sample near the amorphous phase, the structural details cannot be readily identified, as illustrated in **Fig. 18a.** In the thicker parts of the sample, **Fig. 18b**, more structural information is available, although not directly interpretable as projections of the atomic potential. In addition, details of the symmetry of the crystal are not obvious from this image. The structural information in the image in **Fig. 18a** can be obtained by deconvolution of the amorphous background by image processing.

Fig. 18c shows the SEMPER processed image. The processing was carried out by obtaining the Fourier transform, followed by filtering using apertures around each intensity maximum in the Fourier transform. This filtered Fourier transform was subsequently reversed to obtain the filtered images in **Figs. 18c-d.** In **Fig. 18e** the processed image from the thin region is overlaid on the simulated image and shows an excellent match. The capability of such image processing techniques for deconvoluting the effect of the influence of an amorphous phase, whether it is an inherent part of the sample or produced due to the sample preparation method, is clearly illustrated in this example.

D. Al-O-Ns

Another related area of research is that of Al-O-Ns. It has been suggested that the study of the much simpler system Al_2O_3-AlN would provide more information regarding the processing of related ceramics, the SiAlONs. The Al_2O_3 and AIN-rich ends of this system are themselves potential candidates for high technology applications [26]. For example, the AIN-rich compositions are being considered as candidates for electronic packaging applications due to the high thermal conductivity and good mechanical properties of AIN. However, densification of AIN has been a difficult problem. Addition of small amounts of Al_2O_3 to AIN leads to the formation of a series of polytypoids due to the insertion of a chemically distinct layer into another chemically distinct structure.

These phenomena also occur in the alkali earth ceramic superconductors, as noted earlier, **Figs. 6a-b.** AIN has the 2H wurtzite structure, shown in **Fig. 19a,** the cation:anion ratio being 1. As this value decreases, due to the addition of Al_2O_3, more of the tetrahedral sites in the wurtzite structure are occupied. However, this leads to configurations where adjacent tetrahedra share a common base, causing the occurrence of rather short anion-anion distances. This is avoided by changing the cation arrangement locally to cubic, thus creating a stacking fault, shown schematically in **Fig. 19b.** The local cubic layer has the composition of MX_2, where M is the cation and X is the anion. The spacing of these faults is determined by the overall composition, i.e., the amount of Al_2O_3. Thus, the repeat sequence of these polytypoids can be used as an accurate local composition scale, i.e., the image periodicity directly reflects the M/X ratio.

Figs. 20a-c [27,28] show results of a typical set of experiments in which HREM, CBED, and x-ray microanalysis with an ultra-thin window detector (UTW) have been used to identify the structural details of the polytypoids in this system. **Fig. 20a** is the lattice fringe image of the 32H polytypoid, along with the SAD pattern in the inset. X-ray microanalysis data, **Fig. 20b,**

obtained with experimental k-factors and after correction for absorption, confirms the cation:anion ratio to be 16:17. The atomic resolution image in **Fig. 20c**, however, shows the same 32H region to consist of intergrowths of the 21R and the 27R structures. Atomic resolution imaging, in this case, is useful in obtaining supporting evidence for the microanalysis data since the composition can be directly inferred from the stacking sequence in the HREM image, as summarized in **Table 1**. Microanalytical quantification of oxygen and nitrogen has to be done with care. In **Fig. 20b**, one typical example is presented that illustrates the limitations of using system-generated theoretical k-factors [29].

Table 1. Polytypoid structures generated with different cation:anion (M/X) ratios. These are observed in Al-O-N as well as Si-Al-O-N systems.

n	M/X	Polytype	c(Å)
4	4/5	8H	23.04
5	5/6	15R	41.85
6	6/7	12H	32.88
7	7/8	21R	57.12
8	8/9	16H	40.70
9	9/10	27R	72.09
11	11/12	33R	86.46
12	12/13	24H	Not measured
13	13/14	39R	101.40
16	16/17	32H	82.82
∞	1	2H	4.98

The quantified data, using theoretical k-factors, is very different from the stoichiometric composition. The correct stoichiometry is not obtained even using experimental k-factors. It is necessary to apply the absorption corrections to the x-ray intensity of the nitrogen peak in order to obtain the correct composition. Thus, it is essential to know the foil thickness with a fair degree of precision. CBED is very useful in this aspect since the foil thickness within the probe diameter can be obtained. Alternatively, the thickness can be estimated by energy loss spectroscopy. Such absorption corrections become more and more critical as the concentration of heavy elements in the matrix increases; for example, the BCSCO superconductors.

One simple method to carry out microanalysis in such cases is to obtain spectra from points along a line that is normal to the edge of the foil. In this case, depending upon the wedge angle, the thickness will progressively increase. The exact location of the probe can be determined by the carbon contamination spot on the foil. The quantified data for the different points can be plotted against the distance of that data point from the edge. The composition, for an infinitely thin sample, can then be obtained by extrapolating to zero (i.e., the edge of the foil).

SUMMARY

It is realized that in this short review one cannot cover the whole range of electron microscopy techniques as applied to modern ceramics. However, it is hoped that the examples shown here are representative of the significance of modern electron microscopy methods of atomic resolution imaging, diffraction, and spectroscopy in the task of characterizing and understanding typical ceramic materials. The ARM at Berkeley is currently unique in providing the highest resolution [(1.6 Å) over a ±40° tilt range] of any instrument in the world. It is located at the National Center for Electron Microscopy at the Lawrence Berkeley Laboratory, which is a user-operated facility open to all qualified electron microscopists funded by the U. S. Department of Energy.

ACKNOWLEDGMENTS

The authors wish to acknowledge the technical contributions of and discussions with Drs. M. A. O'Keefe, K. M. Krishnan, C. J. D. Hetherington, R. Kilaas, and R. S. Rai. The continued attention to the high performance of the ARM by C. Nelson is also acknowledged. The research work on electronic ceramics and superconductors is supported by the Director, Office of Energy Research, Office of Basic Energy Sciences, Materials Science Division of the U. S. Department of Energy under Contract No. DE-AC03-76SF00098. The work on structural ceramics is supported by the National Science Foundation through grant No. DMR 83-1317239.

REFERENCES

1. G. Thomas and M. J. Goringe, "Transmission Electron Microscopy of Materials." John Wiley and Sons, New York (1979).

2. T. A. Bielicki, U. Dahmen, K. H. Westmacott and G. Thomas, in Zirconia '86, Proc. of Int. Conf., ed. S. Somiya, Tokyo, Japan (in press).
3. R. Gronsky and G. Thomas, in Proc. of 41th Ann. Meeting of EMSA, ed. G. W. Bailey. Claitor's, Baton Rouge, p. 310 (1983).
4. W. O. Saxton and T. L. Koch, J. of Microscopy, 127:69-75 (1982); and W. O. Saxton, T. J. Pitt and M. Horner, Ultramic., 4:343 (1979).
5. R. Kilaas, in Proc. of 45th Ann. Meeting of EMSA, ed. G. W. Bailey, San Francisco Press, San Francisco, CA, p. 66-69 (1987).
6. M. A. O'Keefe and R. Kilaas, in Proc. of 6th Pfefferkorn Conf., Niagara Falls, NY, Scan Mic. Suppl. 2, p. 225 (1988).
7. K. M. Krishnan, P. Rez, and G. Thomas, Acta Cryst. B41, 396 (1985).
8. J. G. Bednorz and K. A. Muller, Z. Phys., B64, 189 (1986).
9. R. Ramesh, G. Thomas, S. M. Green, C. Jiang, Yu Mei, M. L. Rudee and H. L. Luo, Phys. Rev., 38:7070 (1988).
10. C. J. D. Hetherington, R. Ramesh, M. A. O'Keefe, R. Kilaas, G. Thomas, S. M. Green, and H. L. Luo, Appl. Phys. Lett., 53:1016 (1988).
11. D. R. Clarke, J. of Amer. Cer. Soc., 63:208 (1980).
12. T. M. Shaw and G. Thomas, J. of Solid State Chem., 33:63 (1980).
13. R. Ramesh, G. Thomas, S. M. Green, C. Jiang, Y. Mei and H. L. Luo, in Proc. of 1988 Applied Superconductivity Conf. San Francisco, CA (1988).
14. H. W. Zandbergen, R. Gronsky and G. Thomas, J. de Microscopie et du Spectroscopie Electronique (1988) in press.
15. R. Sadanaga, M. Tokonami and Y. Takeuchi, Acta Cryst., 15:65 (1962).
16. D. Schryvers, K. Srikrishna, M. A. O'Keefe and G. Thomas, J. of Materials Research (in press).
17. K. H. Jack, in Proc. of NATO Advanced Study Institute, ed. F. L. Riley, Noordhoff Publishers (1977).
18. D. R. Clarke and G. Thomas, J. of Amer. Cer. Soc., 61:114 (1978).
19. R. E. Loehman, J. of Amer. Cer. Soc., 62:49 (1979).
20. G. Thomas, C. Ahn, and J. Weiss, Comm. of Amer. Cer. Soc., 65:185 (1982).
21. T. M. Shaw, G. Thomas and R. E. Loehman, J. of Amer. Cer. Soc., 67:643 (1984).
22. G. Thomas, in Ceramic Microstructures, ed. J. A. Pask and A. G. Evans, Plenum Publishers, pp. 55-72 (1988).
23. T. R. Dinger, R. S. Rai and G. Thomas, J. of Amer. Cer. Soc. (1988) in press.

24. R. S. Rai, M. A. O'Keefe and G. Thomas, in Proc. of MRS Symp., Vol. 82, p. 429 (1987).
25. M. A. O'Keefe, R. S. Rai and G. Thomas, presented at the MRS Symp. on High-Resolution Electron Microscopy, Boston, MA, (1988).
26. N. D. Corbin and J. W. McCauley, in Proc. of the Soc. of Photo-Optical Inst. Engineers, 297:19 (1981).
27. K. M. Krishnan, R. S. Rai, G. Thomas, N. D. Corbin and J. W. McCauley, in Proc. of MRS Symp., Vol. 60, 211 (1986).
28. K. M. Krishnan and G. Thomas, in Proc. of Microbeam Analysis 1987, ed. R. H. Geiss, San Francisco Press, San Francisco, CA pp. 222-224.
29. K. M. Krishnan, unpublished work.

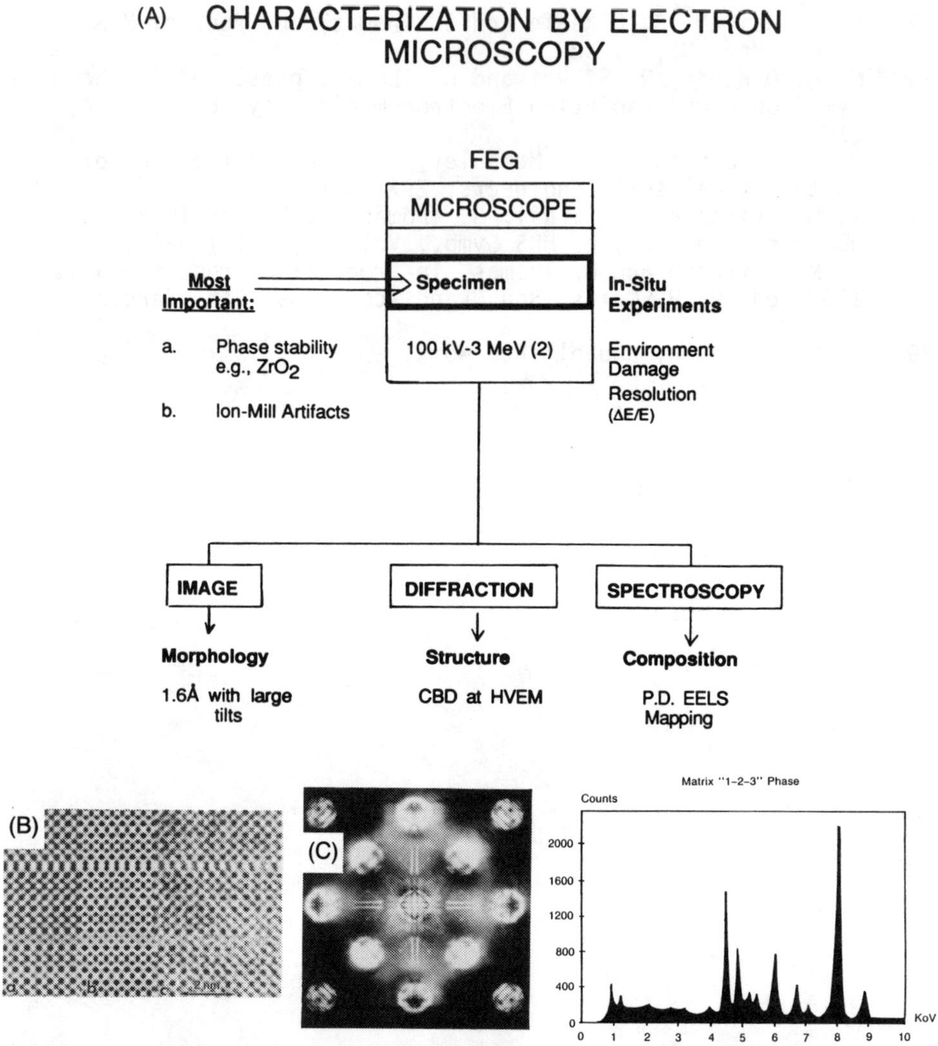

Figure 1: A schematic showing the different types of information that can be obtained from a modern transmission electron microscope, and some of the problems associated with the preparation and imaging of thin foils prepared from the bulk.

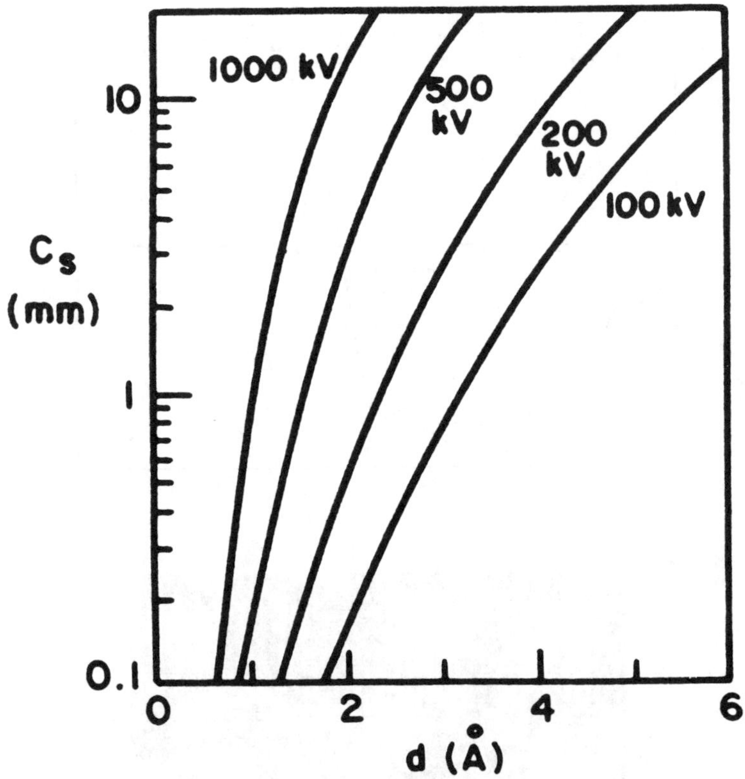

Figure 2: A plot of C_S, the spherical aberration coefficient versus resolution, for different accelerating voltages. Note that for a typical C_S value of 3mm, to obtain a resolution of 2Å, the accelerating voltage has to be in the range of 500-1000kV. (Courtesy of J. Wiley and Sons, New York).

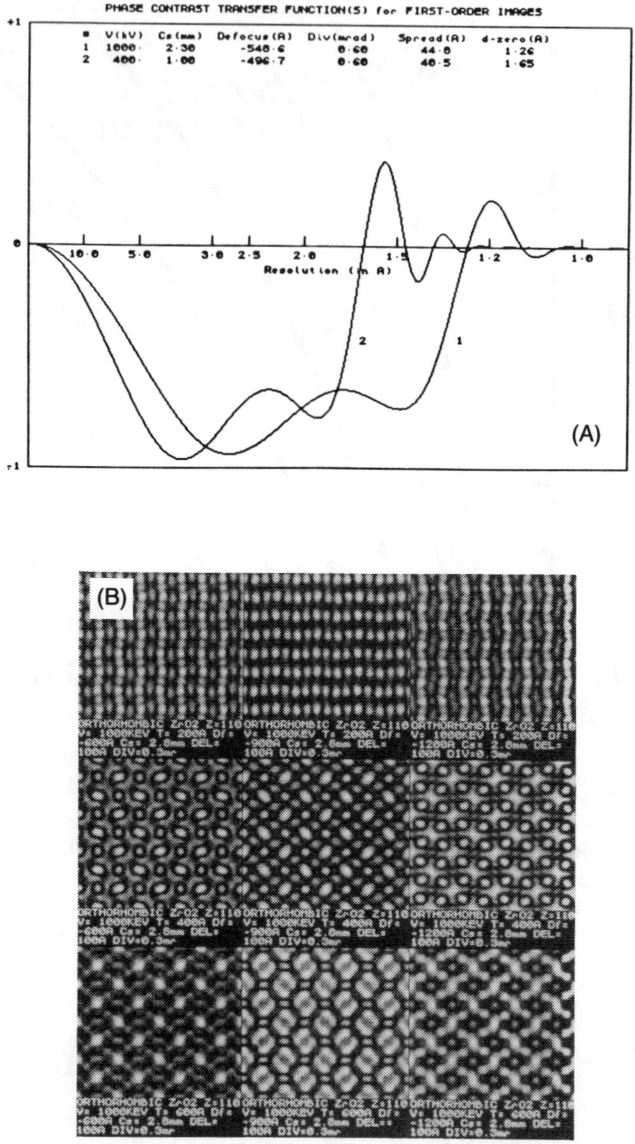

Figure 3: (a) Plots of the contrast transfer function at Scherzer defocus for two accelerating voltages, 400kV and 1MeV ;(b) computer generated images of orthorhombic ZrO_2 for varying specimen thickness and objective defocus. Notice the large variations in the image detail. (Courtesy R.Kilaas).

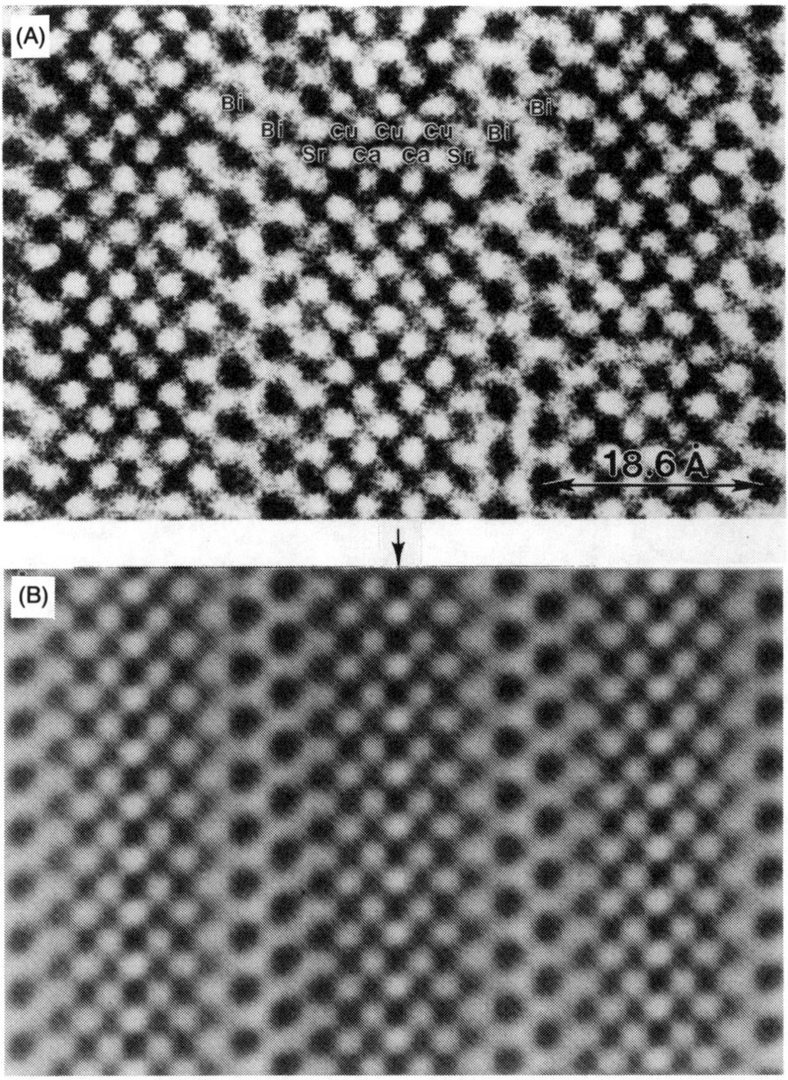

$Bi_2Sr_2Ca_2Cu_3O_y$ **(2-2-2-3)** $T_c = 110K$ **c=37.2Å**

Figure 4: (a) [110] ARM image of the n=3[T_c=110K] polytypoid in the Bi-Ca-Sr-Cu-O system ; (b)image in (a) after SEMPER processing. Note the intensity difference in the central Cu-O row compared to the other two rows.(Courtesy Appl. Phys. Lett.)

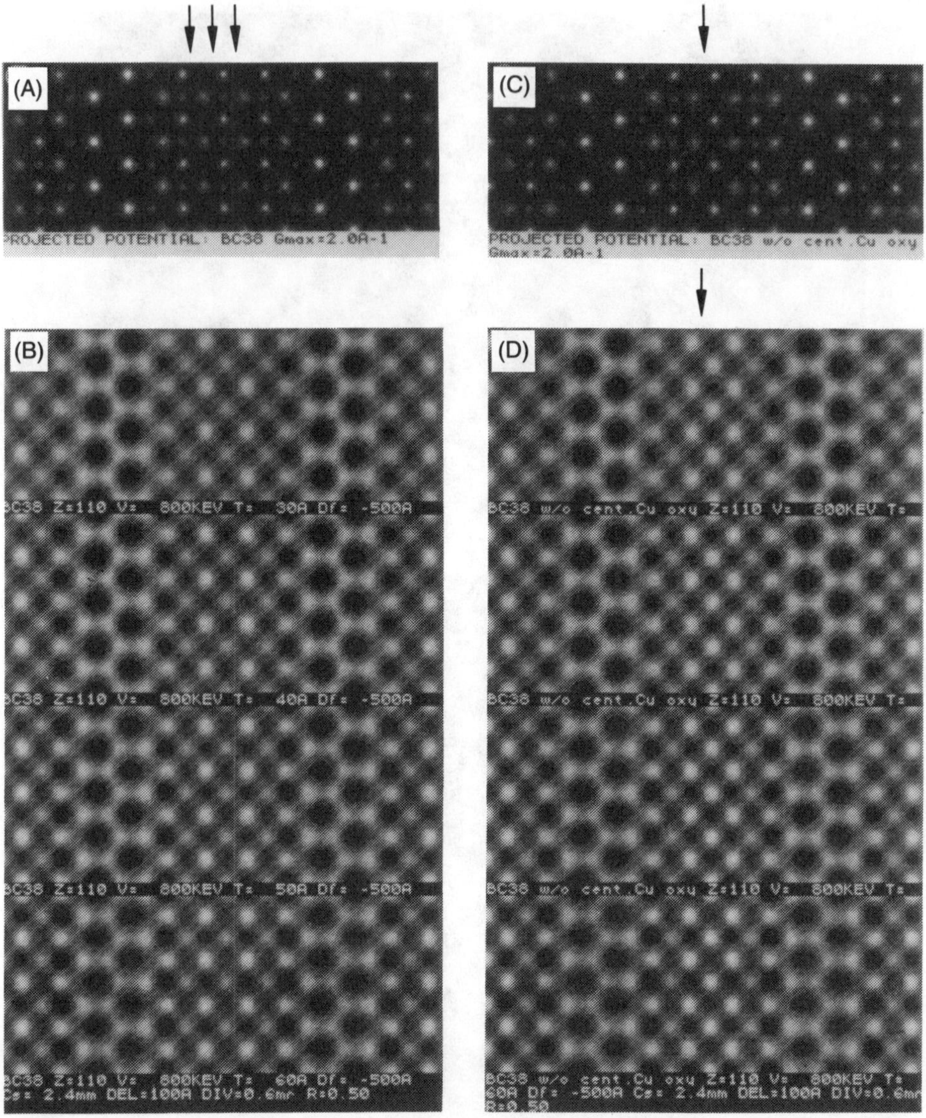

Figure 5: (a) [110] projected potential for the perfect n=3 structure ; (b) simulated images corresponding to (a) ; (c) projected potential for a structure where the oxygen from the central Cu-O row has been removed ; (d) simulated images corresponding to the structure in (c) showing the intensity difference between the central Cu-O row and the other two rows. (Courtesy Appl. Phys. Lett.).

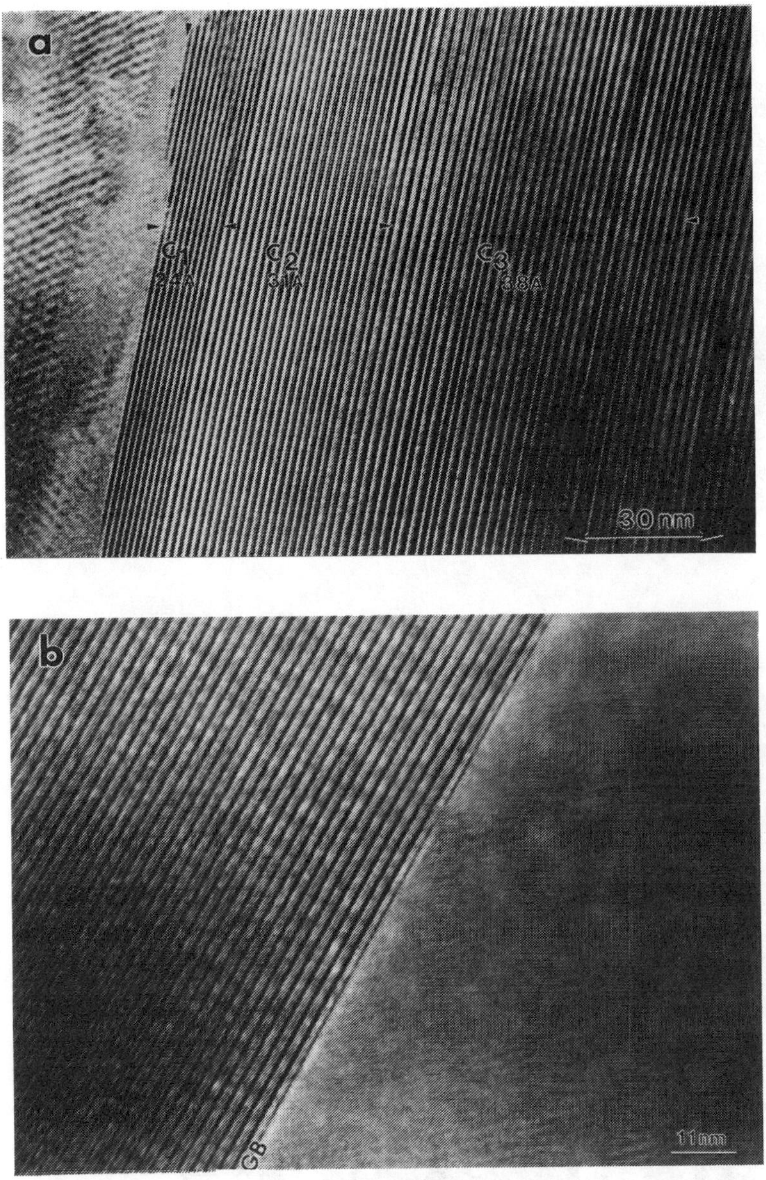

Figure 6: (a) Lattice fringe image near the grain boundary of the un-leaded sample, showing the change in composition (and T_c) from that of n=3 to n=1 ; (b) lattice fringe image in the case of a leaded sample, showing the uniform composition of n=3 up to the grain boundary.(Courtesy Phys. Rev. B).

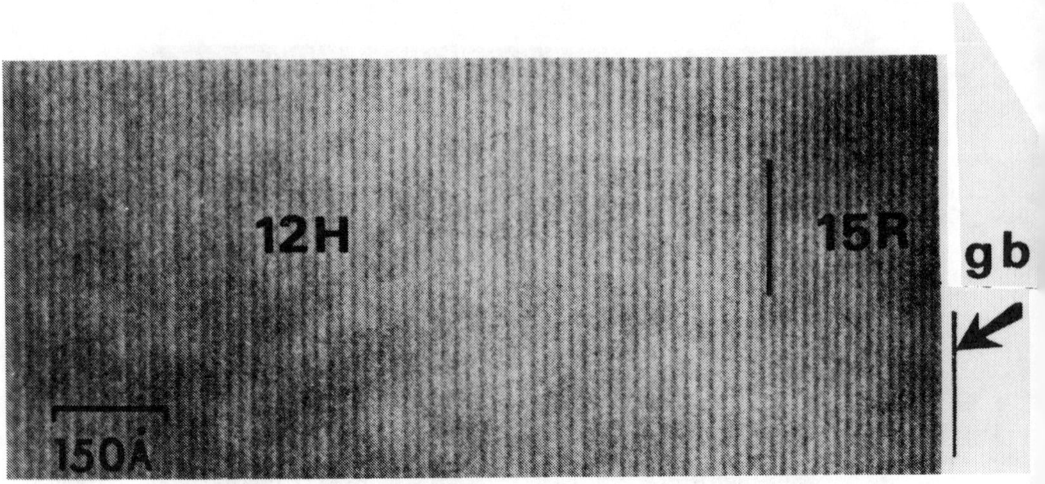

Figure 7: Lattice fringe image showing polytype formation adjacent to a grain boundary in MgSiAlON. (Courtesy Jl. Amer. Cer. Soc.).

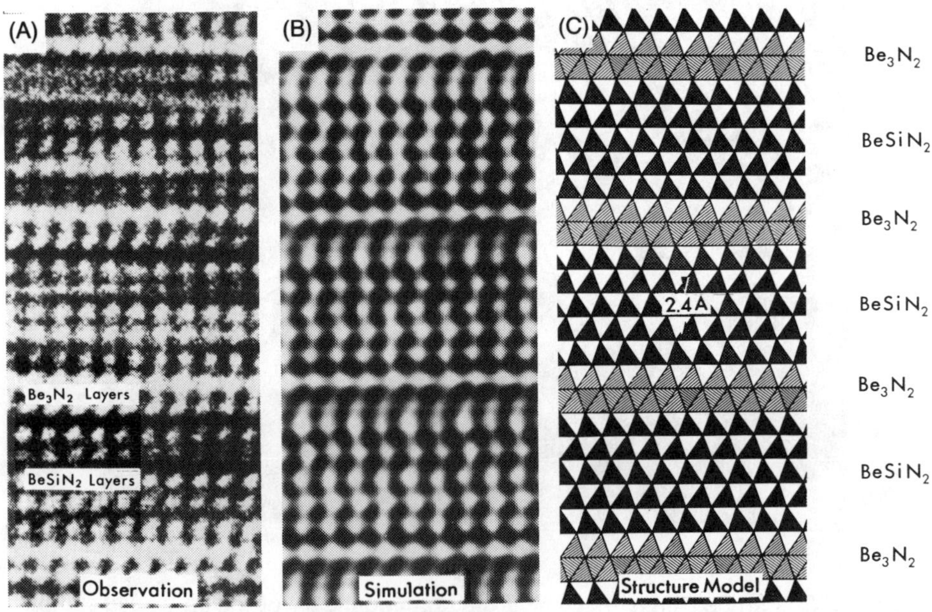

Figure 8: High resolution image, simulated image and structural model illustrating the formation of different polytypes in Be-Si-N.

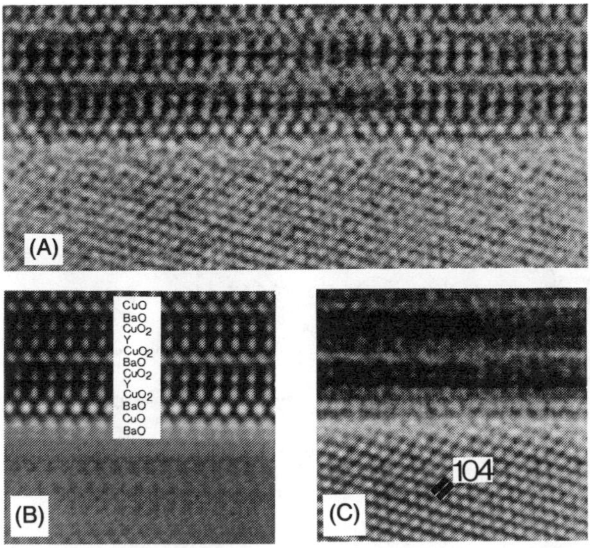

Figure 9: (a) HREM image of a grain boundary in Y-Ba-Cu-O ceramic ; (b) SEMPER processed image in which the upper grain in the [100] zone axis has been processed to reveal the grain boundary plane to be a Ba-O plane.(Courtesy H.Zandbergen).

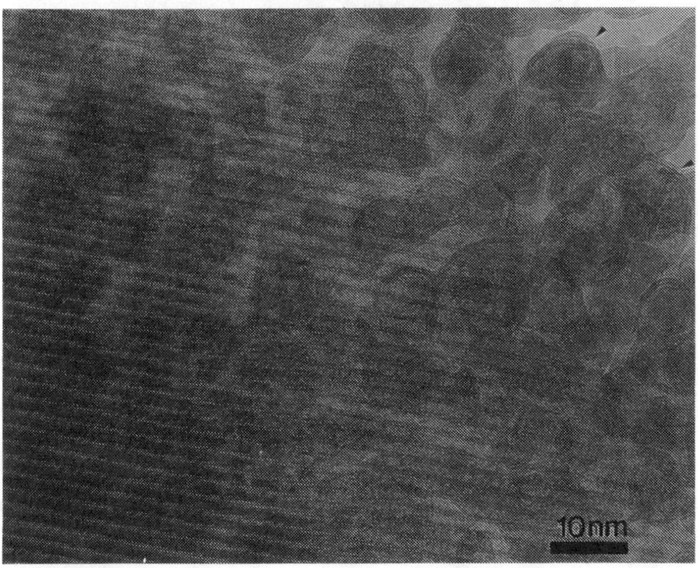

Figure 10: HREM image of a Bi-Ca-Sr-Cu-O thin foil whose edge has been damaged due to electron beam and Argon ion beam irradiation.

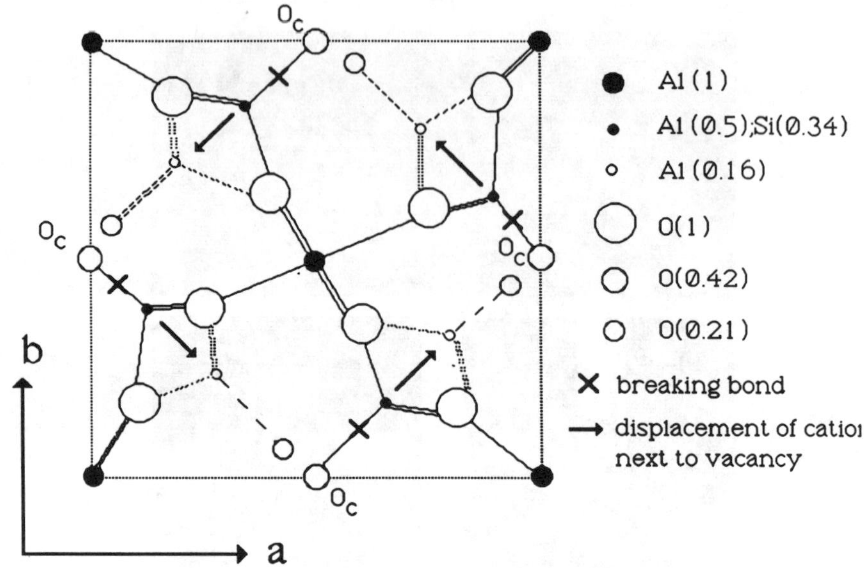

Figure 11: Structural model of mullite in the [001] orientation.

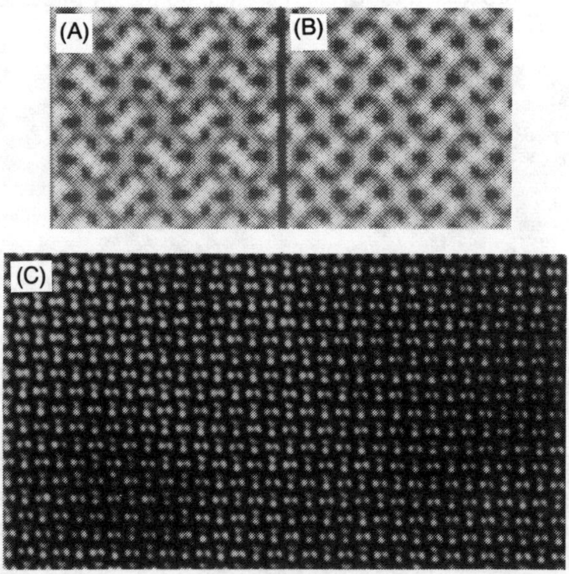

Figure 12: (a&b) Simulated images of mullite for a foil thickness of 29Å and for defocus values of 0Å and -100Å respectively ; (c) ARM image corresponding to the simulated images in (a&b).(Courtesy Jl. of Materials Research).

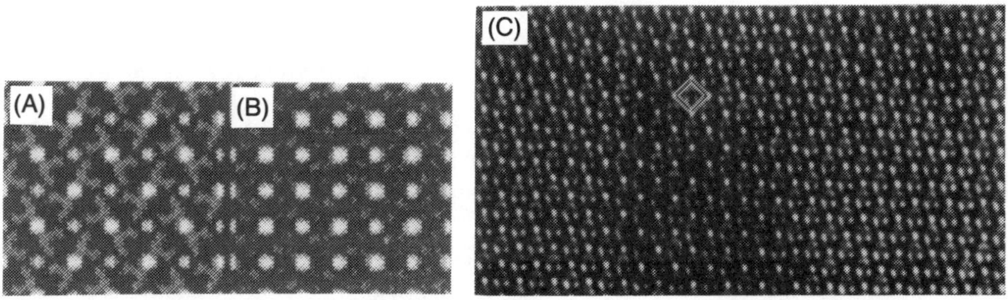

Figure 13: (a&b) Simulated images of mullite for a foil thickness of 87Å and for defocus values of 0Å and 100Å respectively ; (c) ARM image of comparable foil thickness, showing the image details to match well with those in the simulated images.(Courtesy Jl. of Materials Research).

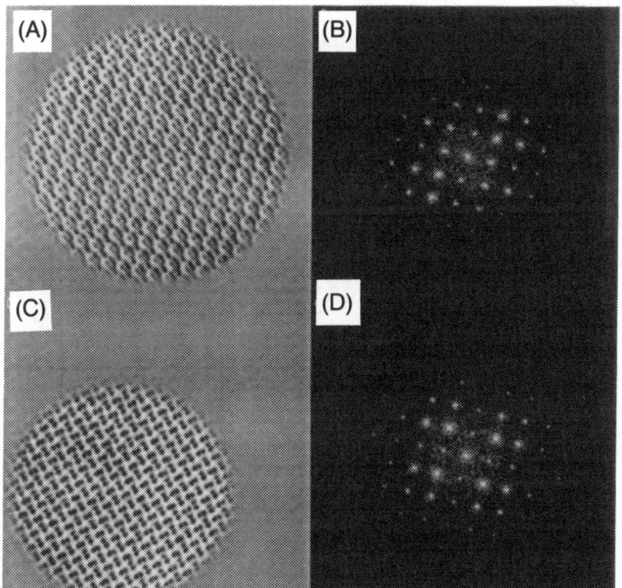

Figure 14: (a) ARM image of mullite, whose Fourier transform is shown in (b). Note the small tilt from the exact zone axis. (c) image of the same region as in (a), but after processing by tilting the foil in the computer, the Fourier transform for which is shown in (d).(Courtesy Jl. of Materials Research).

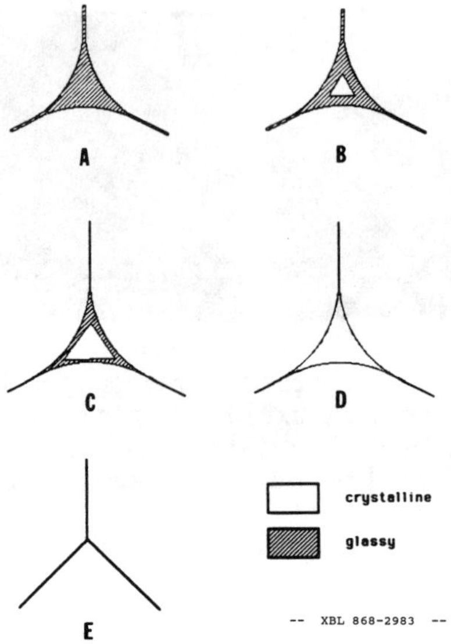

Figure 15: (A-E) Schematic diagrams of the grain boundary microstructures that can evolve in liquid phase sintered ceramics.

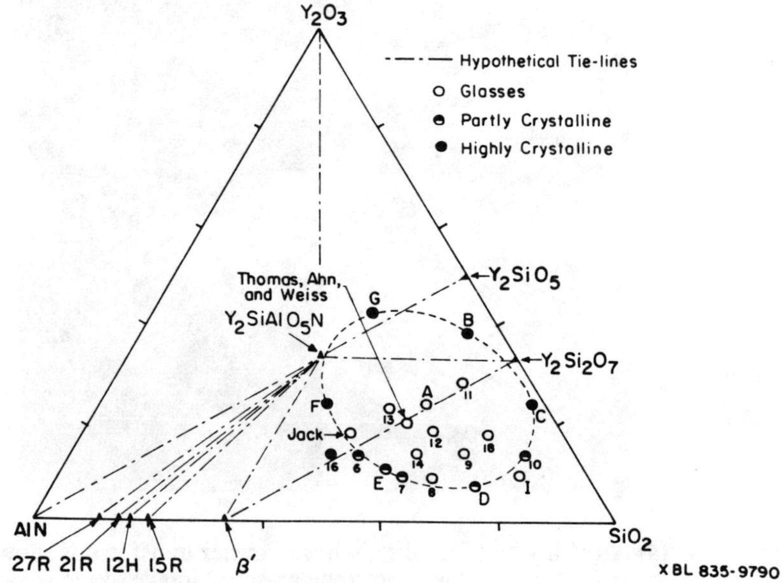

Figure 16: Isothermal section of the AlN-SiO_2-Y_2O_3 phase diagram.(Courtesy Jl. Amer. Cer. Soc.).

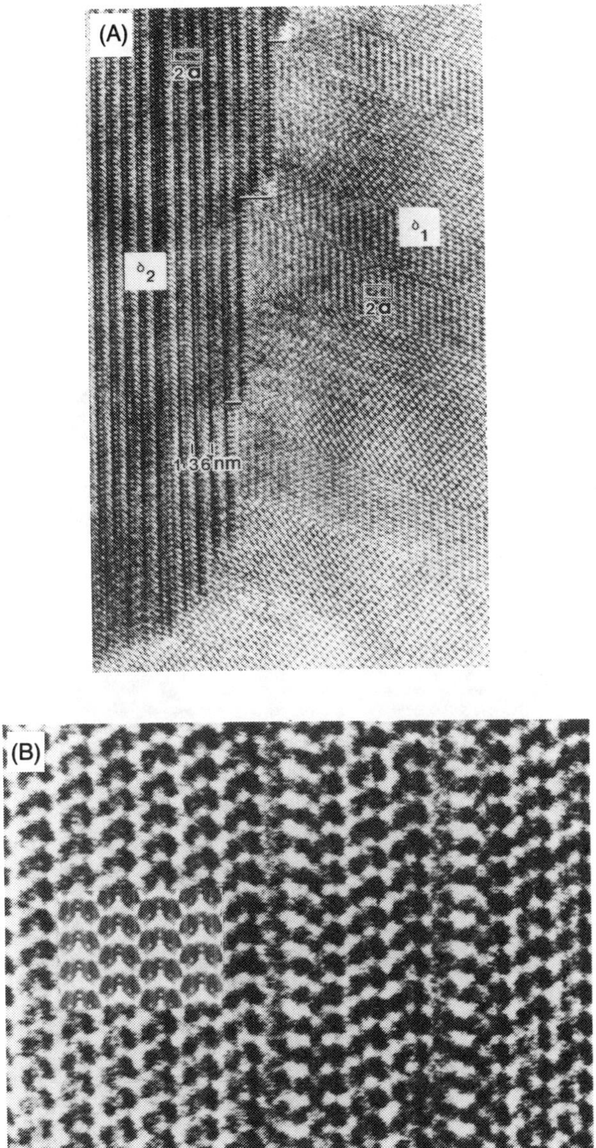

Figure 17: (a) HREM image of the interface between the δ_1-δ_2 dendrites showing flat interfaces, with strain contrast only at the ledges.(Courtesy Jl. of Amer. Cer. Soc.) ; (b) atomic resolution image of the δ_1 $Y_2Si_2O_7$ phase along with the simulated image, showing a good fit (notice the slight skew, probably due to a small tilt of the sample).

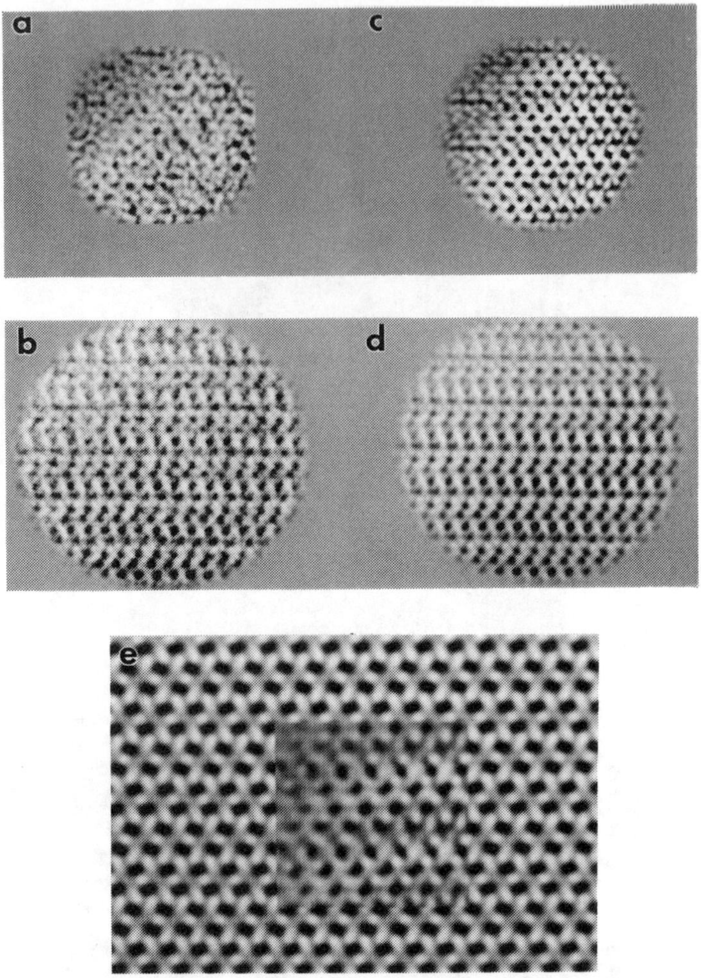

Figure 18: (a) Un-processed image of the $Y_2Si_2O_7$ phase near the amorphous edge, showing no discernible structural detail ; (b) un-processed image of the same crystal as in (a), but in a thicker region, showing some structural detail, but lacking in the symmetry details ; (c) processed image corresponding to (a) showing the intrinsic structure present in the image as well as the correct symmetry of the crystal ; (d) processed image corresponding to (b), showing again the influence of sample misorientation on the thicker regions ; (e) processed image from the thin region, overlaid on the simulated image, showing an excellent match (Courtesy Materials Research Society and M.A.O'Keefe).

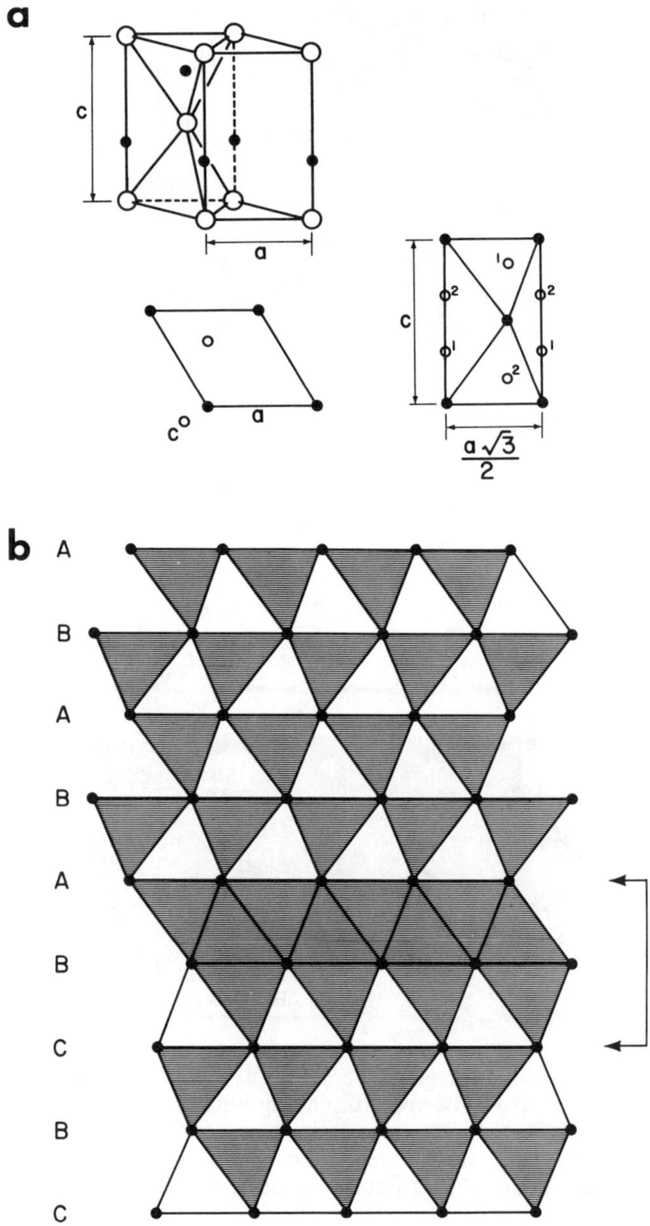

Figure 19: (a) The 2H wurtzite structure of AlN ; (b) structural model showing the formation of a stacking fault in the wurtzite structure.

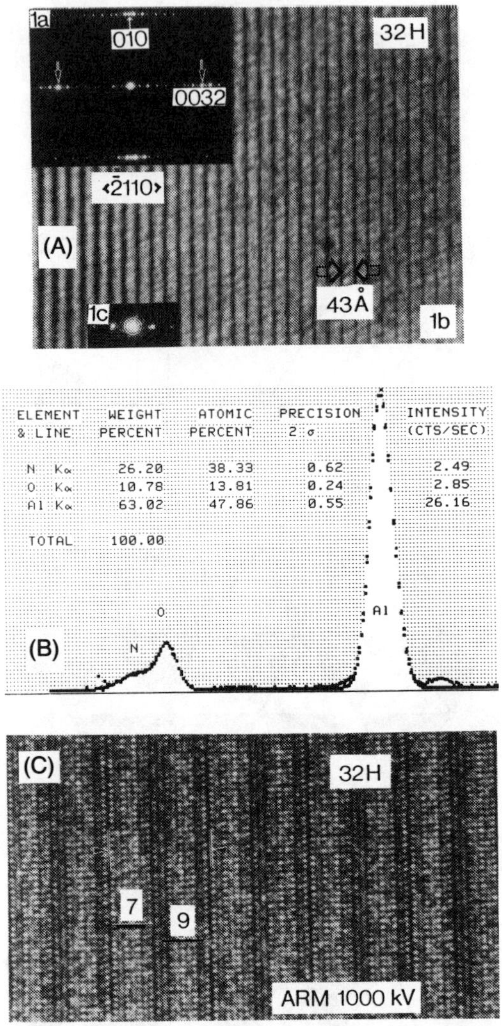

Figure 20: (a) Lattice fringe image of the 32H polytypoid along with the SAD pattern in the inset ; (b) EDX spectrum along with the quantified data that determines the cation : anion ratio to be 16:17 ; (c) atomic resolution image showing the actual structure of the 32H polytypoid to be an intergrowth of the 21R and the 27R structures.(Courtesy San Francisco Press).

SOME THEORETICAL AND EXPERIMENTAL CONSIDERATIONS IN SCANNING TUNNELING MICROSCOPY OF CERAMIC MATERIALS

Dawn A. Bonnell
The University of Pennsylvania
Philadelphia, PA 19104

ABSTRACT

The application of STM and STS to the high spatial resolution electronic structural analysis of ceramic materials is discussed. Examples of the two extremes in imaging conditions are given. SiC is used to illustrate the case where a high density of surface states pins the sample Fermi level and the data can be interpreted in the same manner as those from narrow band gap semiconductors. STS results on ZnO demonstrate that images can also be obtained on samples where the Fermi level in not pinned. In this case, difficulties in image interpretation remain. Techniques for imaging insulators are also reviewed.

INTRODUCTION

In the last few years, scanning tunneling microscopy (STM) has developed from a difficult measurement performed in a handful of labs around the world into a widely accepted method of surface characterization. The excitement over this surface analysis method stems from the fact that information is obtained about bonding in solids at unprecedented levels of spatial resolution. Three-dimensional atomic images have been produced of semiconductors (Si, GaAs) [1,2], semimetals (graphite) [3], and even metals (gold) [4]. Imaging and scanning tunneling spectroscopy (STS) have been used to investigate surface reconstructions, adsorbate reactions, diffusion processes, charge density wave formation, etc [5]. Currently, there are perhaps as many as 150 microscopes in operation around the world. This increase in STM related research, along with the recent availability of commercial microscopes, suggests that the technique will be utilized in investigations of an increasing variety of physical phenomenon involving a wider range of materials. This activity raises the question as to which materials can, and which cannot, be characterized by STM.

The issue addressed in this paper is the extent to which STM and STS can be applied to ceramic materials. The fundamental process of electron tunneling into (or out of) the sample implies that some degree of electronic conduction in the sample is required. However, the limits on sample conductivity have not yet been specified in terms of materials parameters such as energy band gap, carrier mobility, or carrier number density. We will consider here STM analysis of inorganic materials with wider band gaps and lower conductivities than materials previously analyzed. The theoretical description of tunneling, as it occurs in the STM configuration, will be followed by a discussion of the implementation of imaging and spectroscopy techniques. Experimental results on single crystal SiC and single crystal and polycrystalline ZnO will be used to illustrate extremes in imaging conditions likely to arise in STM characterization of ceramics in general. Finally, approaches to the analysis of insulating materials will be reviewed.

SCANNING TUNNELING MICROSCOPY AND SPECTROSCOPY

STM is the most recent of a long list of devices based on electron tunneling which includes: the Field Ion Microscope [6], Esaki diodes [7], Zener diodes [8], Josepheson junctions [9], and Inelastic Electron Tunneling Spectroscopy [10]. It was the earlier development of these tunneling devices which provided a theoretical framework that could be almost directly applied to the STM at its invention. To summarize this theory, we begin by considering electron tunneling in the general case. An infinite potential energy barrier exists between two surfaces separated by a large distance, however, the barrier becomes finite when the two surfaces are in close proximity. When a bias is applied between two surfaces within some tens of angstroms of each other, a statistically significant number of electrons with energies <u>lower</u> than that of the barrier will negotiate the energy barrier producing a 'tunneling current'. Although classically forbidden, electron tunneling can be explained within the wave-particle dualism concept of quantum mechanics. The equation for current resulting from electrons tunneling between two planar electrodes in the limit of low temperature and low applied bias is [11]:

$$I = \frac{2\pi e}{h} \sum_{s,t} f(E_s)\,[1 - f(E_t + eV)]\,\left|M_{s,t}\right|^2 \delta(E_s - E_t) \tag{1}$$

where $M_{s,t}$ is the tunneling matrix, f(E) the Fermi function, e the charge of an electron, h Planck's constant, and the subscripts s, and t refer to the electrodes (later specified as sample and tip, respectively.) Tunneling, as it occurs in the STM configuration, is described in figure 1. One electrode is a metallic point probe (the tip) and the other is a planar semiconductor or metal (the sample). Tersoff and Hamann [12] have taken the geometry of the STM and the electronic structure of the tip into consideration and evaluated the tunneling matrix in equation (1) to produce a current equation specific to the STM. Assuming a spherical wave function for the tip structure and using reasonable values for the materials parameters pertaining to the tip, they obtain for a constant applied bias:

$$I = C D_s e^{-1.025 Z \phi^{1/2}} \quad (2)$$

in the limit of low temperature and applied bias, where D_S is the density of states of the sample, ϕ the tunneling barrier, is on first approximation $(\phi_S + \phi_t)/2$, and C is a constant specific to the tip size and material.

Equation (1) emphasizes that the contrast in the image is a convolution of the electronic structure of the sample, the electronic structure of the tip, and the tunneling barrier function (related to the surface work function). It is only when the latter two contributions are constant that the variation of contrast in the image can be attributed to electronic properties of the sample. Fortunately, these conditions can be easily met on many materials; therefore, equation (2), in which these assumptions are implicit, is a reasonable approximation.

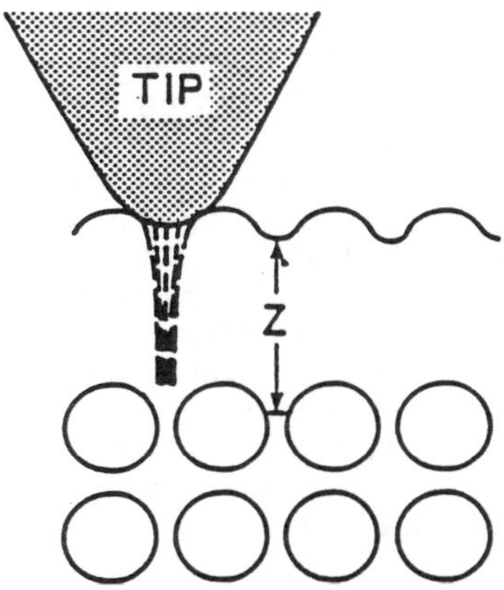

Figure 1. Schematic of tunneling between tip and sample in the STM configuration. In 'constant current' imaging Z is held constant as the tip is scanned over the surface. A series scans produces a real space image of constant charge density.

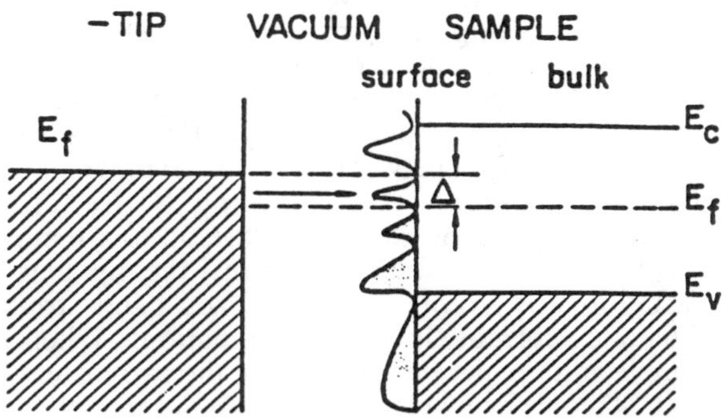

Figure 2. Idealized representation of the electronic structure of the sample and tip. E_f is the Fermi energy, E_v the valence band edge, and E_c the conduction band edge. States on the sample surface may be due to surface reactions or extensions of midgap bulk states. Shaded areas indicate occupied states. A bias between the sample and tip displaces the Fermi levels with respect to each other, allowing of electrons within the energy window to traverse the vacuum gap.

To produce images, the STM can be operated in two modes. In 'constant current' imaging a feedback mechanism is enabled which maintains a constant current while a constant bias is applied between the sample and tip. As seen in equation (2), these conditions require a constant sample-tip separation. As the tip is scanned over the sample, the vertical position of the tip is altered to maintain that constant separation. The signal required to alter the vertical tip position is the image, which represents a constant charge density contour. Altering the level of the current or the applied bias produces contours of different charge densities. An alternative imaging mode is 'constant height' operation in which constant height and constant applied bias are simultaneously maintained. The variation in current is then monitored to produce the image as the tip scans the sample surface. There are advantages inherent in both modes of operation; the former produces contrast directly related to electron charge density profiles, while the latter provides for faster scan rates not being limited by the response time of the vertical driver. Atomic resolution images are possible only under optimized sample and tip conditions. Larger sample-tip separations and blunt tips have the effect of smearing the localized structure and produce topographic images with somewhat lower resolution.

Tunneling spectroscopy (STS) performed in the STM provides information about the electronic structure of the sample by probing the sample density of states as a function of energy. The concept is similar to traditional sandwich junction spectroscopy (IETS), the primary difference being that a vacuum gap exists between the two electrodes rather than an insulator. There are also two methods of performing STS. The first, referred to as point spectroscopy, involves moving the tip to a feature of interest, disengaging the feedback mechanism, modulating the tip bias and recording the resulting variation in current. The idealized band structures in fig. 2 schematically illustrates current flow in this situation. When a bias is applied between the sample and tip, the Fermi levels of the two materials are displaced with respect to each other allowing electrons within the energy window to transverse the vacuum gap. As the bias is ramped, either positively or negatively, the current varies in response to the changing electron density in the energy window. Ramping in both directions probes both the occupied and unoccupied energy states, with the magnitude of the current at a specific voltage directly related to the density of states of the sample at that energy. Since the constant, C, in equation (2) is a linear function of voltage, spectroscopy data can be related to the sample electronic structure (D_S) in the form of conductance:

$$I/V = C^* D_s \, e^{-1.025 \, Z \, \phi^{1/2}} \qquad (3)$$

and voltage derivatives thereof. The second method of obtaining spectroscopic information involves simultaneously collecting images at various biases. This can be accomplished by modulating the bias at a high frequency (with respect to the time constant of the feedback controller) and recording the current at several discreet values of applied bias or using the same sample and hold method of point spectroscopy at every image point. The simultaneous collection of multiple images is sometimes referred to as current imaging tunneling spectroscopy (CITS) [13]. A comparison of the periodic contrast in such images with atomic charge superposition and total energy calculations indicates the spatial position of energy states with respect to the lattice.

STM has been used extensively in the characterization of the Si (111) surface, to the point that the (7x7) reconstruction of this surface is now universally associated with the technique. This familiar structure can be used to illustrate both point spectroscopy and CITS of semiconductors. An example of point spectroscopy is given in fig. 3 which is taken from a study of Al adsorption on Si (111) by Hamers and Demuth [14]. The electronic structure of the Si (111) 7x7 surface determined by STS is compared to that determined by ultraviolet photoemission (UPS) [15] and inverse photoemission (IPS) [16]. Occupied states at -1.6 eV, -0.85 eV, -0.1 eV, and unoccupied states at +0.5 eV, +1.6 eV are characteristic of the stacking fault adatom structure of the 7x7 reconstruction. These states are evident in STS and in photoemission results. The excellent agreement between the energies of the states identified in the STS spectrum and those determined from photoemission results illustrates the accuracy with which tunneling spectra characterize surface electronic structures. The measurements average over an area larger than a unit cell and each state corresponds to the energy of a particular bond. For example, the state at -0.8 eV arises from the dangling bond whereas the state at -1.6 eV corresponds to the silicon back bond. Naturally, the bonds are localized at different positions with respect to the lattice; therefore, images of these states have different geometries but should have the same translation as the unit cell. Figure 4 is a schematic representation

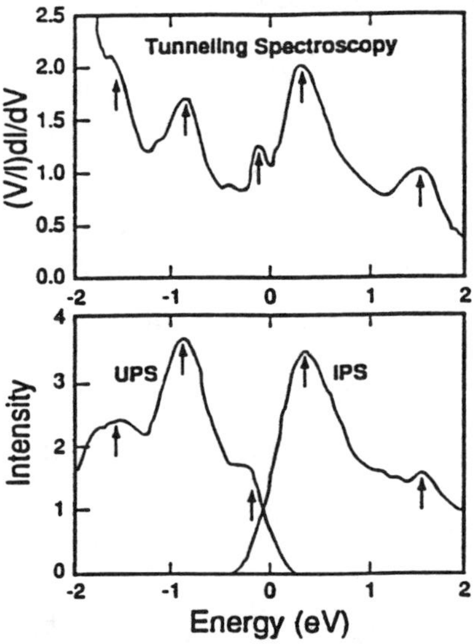

Figure 3. Comparison of density of states by STS point spectroscopy with results of photoemission [15] and inverse photoemission [16] spectra. Agreement in the energies of the states between various methods verifies the accuracy of STS results. (Redrawn after ref. 14)

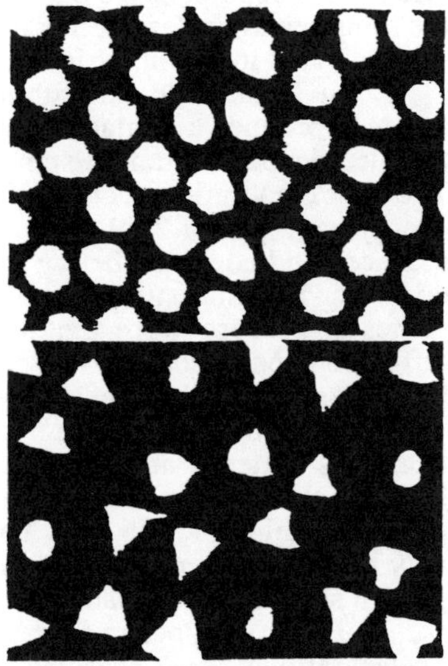

Figure 4. Schematic representation of CITS images obtained at +0.7 and -1.8 V on the Si (111) 7x7 surface. Unoccupied (+) and occupied (-) states are located at different positions in the unit cell, which is outlined.

comparing current images such as those produced by CITS at +0.7 eV and -1.8 eV. Note that the features in the images appear quite different but do have the lateral translation of the unit cell. Image simulations based on theoretical calculations are necessary to interpret atomic resolution images, ie. to relate the features in the image to bond types and atom positions.

STM OF CONDUCTIVE CERAMICS

The term 'ceramic' refers to a broad materials classification which has a variety of definitions depending on the context of its usage, however, large band gaps (> 1.5 eV) and low conductivity are universally associated with ceramic materials. The issue addressed here is the point at which these properties preclude the use of STM as a characterization tool. An important consideration is the electron density of the sample surface which will be high in the case of a reconstructed surface or when midgap bulk states extend to the surface and low when the sample is unreconstructed and does not contain a large density of defect states. A high density of surface states will pin the Fermi level of the sample and allow spectroscopy to be performed as shown in fig. 2. The Fermi level might not be pinned in a sample with a low density of surface states. In this case, electron occupation of the energy bands will vary in response to the field between the sample and tip (applied bias) and band bending will occur. The first case is demonstrated in the analysis of SiC, the second in the analysis of ZnO.

SiC can be doped with numerous elements, the most common being Al and N to produce a partially compensated crystal. Impurities, such as Fe, which produce deep level defect states can also be incorporated in the SiC structure. The bulk electronic structure of the 6H modification is semiconducting with an energy band gap of 2.89 eV with the possibility of defect states both above the valence band edge and below the conduction band edge depending on impurity content. For this study, highly doped SiC single crystals produced by the Acheson process were examined. The crystals were confirmed by x-ray diffraction and Laue back scatter techniques to be of the 6H modification with the basal plane parallel to the surface. Crystals were cleaned in HF and transferred immediately to the vacuum. Constant current image and electronic spectrum from this material are shown in fig. 5 [17]. Images were easily acquired at biases in the range of + 1.5 V. Steps are evident in the image with heights of 0.4 nm and spacing of 1.45 nm. The morphology of these steps indicates that they may be growth steps, which is consistent with the observation of macroscopic growth steps on the surface.

Electronic spectra from the same surface were acquired at several sample-tip separations. Peak energies were independent of sample-tip separation. STS spectra, presented in the form of dynamic/static conductance ratios, indicate the existence of three midgap states, two are occupied (-0.8 eV, -0.45 eV) and one is unoccupied (+0.3 eV). These energies are consistent with those expected from defect states produced by N and Al doping and could therefore be bulk states that extend to the surface. Alternatively, these states could be due to a reaction of fluorine (from the cleaning procedure) with the SiC surface. It was not possible to distinguish between the two possible sources of states in the absence of simultaneous chemical analysis. It is, however, the presence of these states that pins the Fermi level and allows spectroscopy results of SiC to be interpreted in the same manner as those of Si, in which the Fermi level is also pinned. Implicit in this interpretation of spectroscopy data is the assumption that the voltage drop occurs in the vacuum, or that any voltage drop that occurs in the sample is small and constant or a very slowly varying function of applied bias.

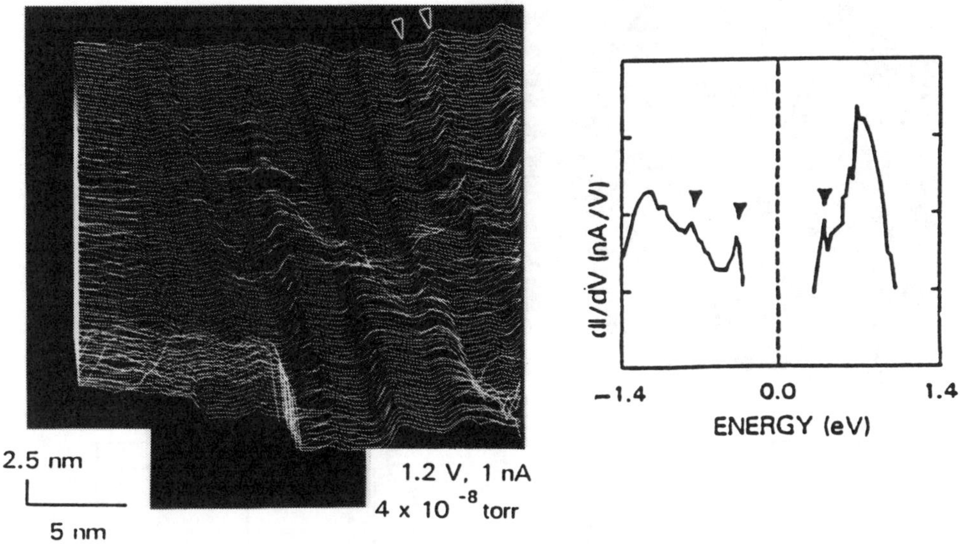

Figure 5. Constant current image and STS measurements of 6H SiC basal plane examined under a vacuum of 10^{-6} torr. Arrows mark the smallest steps which are about 0.4 nm in height and spaced laterally 1.45 nm. Surface states are observed at -0.8, -0.45, and +0.3 eV.

To consider the situation where the Fermi level might not be pinned and, consequently, some voltage drop occurs in the sample, doped polycrystalline ZnO and ZnO single crystals with controlled excess metal content were analyzed. Free sintered surfaces of the polycrystal and cleaved surfaces of the single crystal were analyzed. Both samples are n-type semiconductors with a band gap of 3.34 eV. Defect states are possible for singly and doubly ionized Zn interstitials and oxygen vacancies, in addition to those due to solid solution impurities in the polycrystalline sample. In contrast to imaging the SiC, stable images of polycrystalline ZnO were produced only in a restricted range of applied biases, primarily with the sample biased negatively relative to the tip. Characteristic topographic structure, shown in fig. 6, is large regions which are flat to within a fraction of a nanometer separated by 1-3 nm steps. STS results from the single crystal show that the tunnel junction exhibits diode behavior with increased conductivity under forward biasing conditions. The valence band edge is indicated by the sudden decrease in conductance between -1 V and -2 V, but the conduction band edge is not observed. The shift in position of the valence band edge as a function of sample-tip separation indicates that the band energies vary with the magnitude of the field between the sample and tip (ie. a Schottky barrier is developed). It appears that band bending on the order of 1 eV is possible.

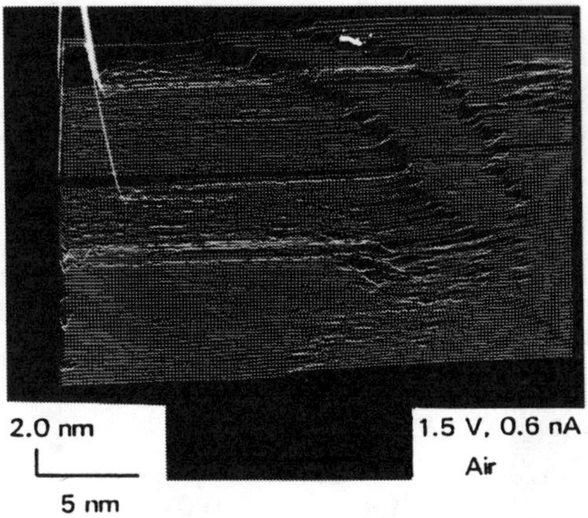

Figure 6. Constant current image of polycrystalline ZnO 0.1 CoO, 0.1 MnO, 0.56 Bi_2O_3, 150 ppm Al and tunneling spectrum of single crystal ZnO with approximately $1.5x10^{16}$ /cm ^{3} excess metal. Voltages correspond to different sample tip separations. The I/V characteristics in the inset illustrate the rectification of the junction.

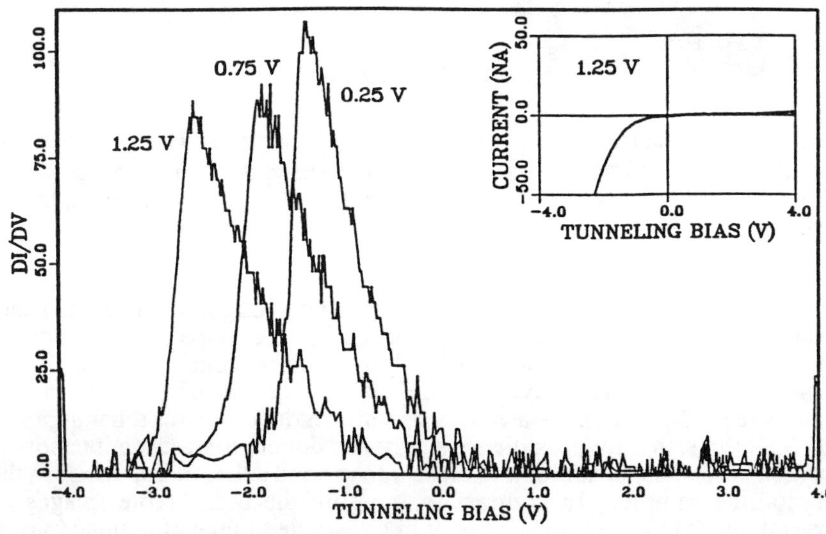

This variation of sample electronic structure during analysis has important implications to the physics of STM imaging. Band bending, as it occurs with a negative sample bias, is schematically illustrated in figure 6. Note that the equipotential lines, which describe the voltage drop in the sample, reflect the geometry of the tip producing spreading resistance. Using the current density equations as developed by Flores and Garcia [18] incorporating the difference in electron and hole mobilities and the data from fig. 6, one can show that the voltage drop in this ZnO sample for sample negative polarity is 10 times that for the sample positive polarity. This calculation provides a quantification of the image instability described above. Perhaps more important than the magnitude of the voltage drop in the sample is the depth associated with depletion region. An electron (or hole) must tunnel through the spreading resistance region to an allowed states at its energy, as well as through the vacuum gap, as illustrated in fig. 7. Under these conditions, charge carrier diffusion processes, which are ignored in the development of equations (1) - (3), must be taken into consideration. It is also possible in this more complicated process for the electrons to lose energy in an inelastic process or undergo band-to-band tunneling. The intriguing possibility exists that, if these various processes can be properly accounted for, the tunnel current could be a probe of subsurface electronic structure by accessing the states within the spreading resistance region.

The STS results on single crystal ZnO have demonstrated that STM images can be obtained even in the event of Schottky barrier formation. These results also illustrate the

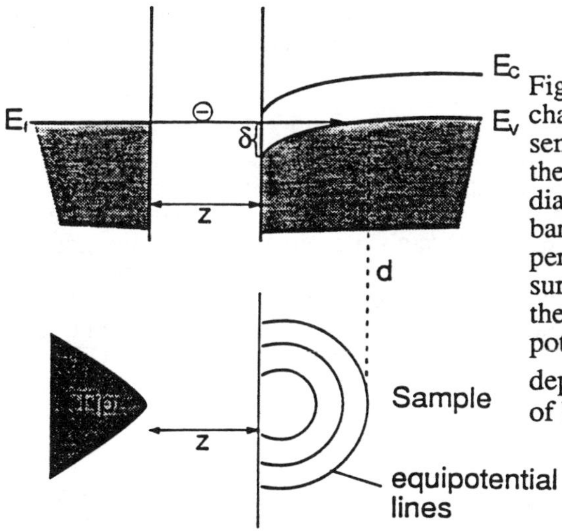

Figure 7. Development of a space charge region in an n-type semiconductor under the field of the STM probe tip. The band diagram illustrates the change in band edge energies along a line perpendicular to the sample surface, while the lateral extent of the band bending is seen in the potential diagram, where d is the depletion depth, δ the degree of band bending.

difficulties associated with unambiguous interpretation of such images. The resolution of these difficulties awaits further work.

STM OF INSULATING CERAMICS

When the charge carrier concentration of the sample is too low to sustain a tunnel current, it will not be possible to perform STM analysis as outlined above. Two approaches have been taken to extend the application of STM to more insulating materials. The first involves increasing the energy of the system so that a higher number of electrons will have sufficient energy to participate in transport processes. This can easily be done through thermally stimulated conductivity where the number of carriers is increased by heating the sample. The approach has proven successful in the analysis of intrinsic or low-doped Si [19] which could not be imaged at room temperature. The inherent disadvantage of the approach is that in situ heating increases thermal drift in the instrument which decreases tne lateral spatial resolution and becomes limiting at about 200°C. It is also possible to increase conductivity through photoexcitation of electrons across the energy band gap. The band gap of the sample determines the wavelength of light focussed on the junction. This approach was demonstrated by using a W halogen lamp and a NeHe laser to illuminate GaAs, thereby increasing its conductivity by about two orders of magnitude [20]. Topographic images could then be acquired and spectroscopy performed.

If the conductivity of the material can not be increased sufficiently by increasing the total energy of the system, then it is still possible to produce topographic images by depositing a thin film of metal on the sample surface. This eliminates the possibility of spectroscopic analysis of the sample because the electronic structure determined at the tunnel junction will be that of the metal film. However, surface geometric structure can be imaged with the limiting resolution dependent on the thickness of the coating. Using this technique it is possible to image features on insulating materials such as fused silica, NaCl, KBr, KI, and oxidized Al [21].

Finally, atomic force microscopy (AFM) may be the tool of choice in geometric structure analysis of some insulators. The technique is similar to STM, but senses atomic forces (for example Van der Waals interactions) between surfaces rather than current [22]. Images produced represent constant force or constant force derivative profiles. Although atomic resolution images of insulators have not yet been produced, there is no fundamental consideration that precludes the possibility. The obvious disadvantage of the technique is that it does not provide for electronic structural analysis. This may be overcome in the future with the development of optical spectroscopy in conjunction with AFM.

SUMMARY

The STM analysis of ceramic materials with band gaps larger than 1.5 eV and low conductivity approaches the limits of applicability of the technique but can, in many cases be accomplished to yield high spatial resolution electronic structural information. In the case where a significant midgap state density can be introduced, the Fermi level will be pinned and spectroscopic data can be interpreted in the same manner as that from narrow band gap, high conductivity semiconductors. Topographic images can also be produced of samples with unpinned Fermi levels, as demonstrated in the analysis of ZnO single crystals. The pertinent questions in this case are what sample conditions result in a depletion depths that prohibit tunneling processes, and can we relate spectroscopy during Schottky barrier formation to subsurface properties.

Quantitative answers to these questions are necessary in order to interpret images of these model structures, which will provide a framework within which to consider more complicated structures such as recently obtained STM images of the WC/Al_2O_3 interface [23]. It is clear, however, that STM can be applied to a wide variety of materials which are not considered conventional semiconductors.

Acknowledgements

It is a pleasure to acknowledge encouragement and support by D. R. Clarke. I am also grateful to R.J. Hamers for helpful discussions and permission to use figure 3, to G. Pike for supplying the ZnO, and to J. McGroarty for technical assistance. This work was supported in part by the National Science Foundation under grant #MSM-8707629 and a grant from IBM Research.

REFERENCES

1. G. Binnig, H. Rohrer, Ch. Gerber, E. Weibel, "7x7 Reconstruction on Si (111) Resolved in Real Space", Phys. Rev. Lett. 50 (1983) 120.
2. R.M. Feenstra, A.P. Fein, "Local state Density and Long Range Screening of Adsorbed Oxygen on the GaAs (110) Surface", Phys. Rev. Lett. 58 (1987) 1668.
3. Sang-Il Park, C.F. Quate, "Tunneling Microscopy of Graphite in Air", App. Phys Lett. 48 (1986) 112.
4. V.M. Hallmark, S. Chiang, "Observation of Atomic Corrugation on Au (111) by STM", Phys. Rev. Lett. 59 (1987) 2879.
5. See for example: Proceedings of the Second International Conference on Scanning Tunneling Microscopy ed. R. M, Feenstra J. Vac. Sci. and Techn. A Vol. 6 March/April (1988).
6. E. Muller, "Electron Microscope Observations from Field Cathodes" Z. Physik 106 (1937) 541.
7. L. Esaki, "New Phenomenon in Narrow Germanium p-n Junctions" Phys. Rev. 109 (1958) 603.
8. E.O. Kane, "Theory of Tunneling" J. Appl. Phys. 32 (1961) 83.
9. P.W. Anderson, A.H. Deyem, Phys. Rev. Lett. 13 (1964) 95.
10. ed. P.K. Hansma Tunneling Spectroscopy-Capabilities, Applications, and New Techniques, Plenum Press NY (1982).
11. J. Bardeen "Tunneling from a Many Particle Point of View", Phys. Rev. Lett. (1961).
12. J. Tersoff, D.R. Hamann "Theory of Scanning Tunneling Microscopy", Phys. Rev. B 31, (1985) 805-813.
13. R.J. Hamers, J. Demuth, "Atomic Structure of Si (111) (3 X 3) Al", J. Vac, Sci. and Technol. A 6 (Mar/Apr 1988) 512-516.
14. R.M. Tromp, R.J. Hamers, J.E. Demuth "Si (001) Dimer Structure Observed with STM" Phys.Rev. Lett. 55 (1985) 1303-1307.

15. F. Himpsel, Th. Fauster, "Probing Valence States with Photoemission and Inverse Photoemission", J. Vac. Sci. and Technol. A 2 (1984) 814-819.
16. Th. Fauster, F. Himpsel, "Momentum Resolved Bremsstrahlung Spectroscopy with a Tunable Photon Detector" J. Vac. Sci. and Technol. A1 (1983) 111.
17. D.A. Bonnell, D.R. Clarke, "Scanning Tunneling Microscopy and Spectroscopy of Ceramics: SiC and ZnO", J. Am. Cer. Soc. 71 (1988) 629-637.
18. F. Flores, N. Garcia, "Voltage Drop in Scanning Tunneling Microscopy for Si", Phys. Rev. B 30 (1984) 2289-91.
19. G. Binnig, H. Rohrer, F. Salvan, Ch. Gerber, A. Bara "Revisiting the 7X7 Reconstruction of Si (111)" Surf. Sci. 157 (1985) L373.
20. G.F.A. Van de Walle, H. Van Kempen, P. Wyder, P. Davidsson, "Scanning Tunneling Microscopy on Photoconductive Semi-insulating GaAs", Appl.Phys.Lett. 50 (1987) 22-24.
21. R.C. Jaklevic, L. Elie, W. Shen, J.T. Chen "Application of STM to Insulating Surfaces" J. Vac.Sci. Technol. A 6 (1099) 448-453.
22. G.Binnig, C.F. Quate, Ch. Gerber "Atomic Force Microscopy" Phys. Rev. Lett. 56 (1986) 1164-66.
23. D.A. Bonnell, "Characterization of Carbides by STM" J.Mat.Sci. and Eng. Dec. (1988).

APPLICATIONS OF THE POTENTIAL MICROPROBE TO ELECTRONIC CERAMIC MATERIALS

W. K. Schubert
Sandia National Laboratories
Albuquerque, NM 87185

ABSTRACT

The potential microprobe consists of a fine tipped tungsten needle mounted on a computer controlled, three axis mechanical/piezo positioner. The probe is mounted to the stage of a standard scanning electron microscope, and allows one to make electrical potential measurements on a very fine spatial scale (≥ 10 nm). The potential microprobe and its operation are discussed. Applications of the technique to polycrystalline silicon, ceramic superconductors, and ZnO varistor materials are presented.

INTRODUCTION

The electronic properties of polycrystalline materials are often dominated by grain boundary effects. Impurity segregation and/or charge trapping at defects in the boundaries may cause the electrical characteristics to be radically different from the single crystal material. These grain boundary related phenomena may be a help or hindrance in tailoring the electronic properties for a specific application.

To understand these phenomena, it is helpful to be able to make electrical measurements on a finer spatial scale than the inhomogeneities (the grain size for instance). Techniques have been developed for probing with several micron spatial resolution. Servo-driven or manually operated micromanipulators coupled with optical microscopes are commonly used in the microelectronics industry, and are capable of such measurements. Other techniques involving rather laborious sample preparation and contacting methods[1,2] have also been developed. Examples are photolithographically defining a fine grid of electrodes to which individual wires must be thermosonically bonded, or directly spot welding fine wires to individual grains. These techniques also are limited by the fact that the electrodes are fixed in position.

At much finer resolution (≤0.1 nm), one can use a scanning tunneling microscope (STM). These instruments however, are intended to resolve atomic step type dimensions and have difficulty dealing with significant surface roughness. Thus for many practical ceramic materials, the range of travel vertically or laterally is too small to allow useful measurements.

The potential microprobe was developed to probe the intermediate spatial range from 10 nm to about 50 microns, and is intended to be used in conjunction with a scanning electron microscope (SEM). It has been used successfully in several applications[3,4], some of which will be reviewed in this paper. Key aspects of the microprobe are discussed below, both with regard to its construction and its capabilities. Finally, three example applications of the instrument to electronic ceramic materials are discussed, each illustrating different capabilities and measurement techniques.

CONSTRUCTION AND USE OF THE POTENTIAL MICROPROBE

The heart of the potential microprobe is a three axis, combination mechanical/piezo stage which is used to precisely position a sharp W needle on the sample. Commercially available, vacuum compatible, miniature stages were modified to allow the piezo stacks to work in series with the normal micrometer drives.[5] The micrometer drives were coupled to the outside of the SEM vacuum chamber via flexible cables and rotary feedthroughs. This set-up provides up to 6 mm range with 5 micron resolution from the micrometer drives and 10 microns of travel with about 10 nm resolution (under ideal conditions) from the piezo elements for all three axes. The finely sharpened W needles (50 to 150 nm tip radius) are normally used as probes for integrated circuit probing stations.[6]

The W probe acts as a moveable contact in two or three terminal current-voltage measurements. A small computer (Hewlett-Packard Model 86) controls the power supplies which drive the piezo elements, the constant current and voltage sources as well as a Keithley 619 electrometer. The entire system is shown schematically in Figure 1. This arrangement allows one to make various electrical measurements on a grid or line of prescibed points, all under computer control while viewing the sample in the SEM. Software controlled surface sensing (i.e. knowing precisely when the needle contacts the surface) can be done by detecting current or potential depending on the particular sample and type of measurement being made. It should be noted that even when measuring very small currents, the current densities at the point of contact can be quite high due to the small contact area, and care must be taken not to damage the sample.

Several things must be considered in evaluating the accuracy of measurements made with the potential microprobe. Spatial accuracy depends primarily on the repeatability and accuracy in placement of the probe needle on the sample surface. A major source of variability in needle placement arises from backlash in the micrometer drives, but this is easily avoided by always approaching the final position from the same direction. The positioning from the piezo drives is as accurate as their programming voltage, and variability here is generally negligible. When measuring profiles on a very fine scale (≤50 nm) one must also account for thermal expansion effects. These can be minimized by controlling the temperature of the SEM chamber and stage a few degrees above room temperature.

Perhaps the most difficult source of positional variability arises from the skating of the W needle on the sample surface. To preserve the fine W tip, it is usually best for the contact angle of the needle with respect to the sample surface to be about 45 degrees. Unfortunately, this situation can lead to skating along the surface as the needle is lowered past the point of initial contact. The amount and direction of skating is very sample dependent. The surface topography determines the direction of skating. Flat surfaces at least make the direction predictable. The distance skated depends on how much pressure is required to make good electrical contact. Insulating surface oxides (such as normally found on Al or Si) must be penetrated to make good contact, and may require some degree of scratching of the surface.

In addition to spatial accuracy, one must evaluate possible errors in the electrical measurements made with the microprobe. There are several possible contributors. When the W needle acts as one terminal in a two terminal configuration, accurate potential measurements depend upon low contact resistance between the needle and the sample. This resistance may be high, particularly in lightly doped semiconducting samples, or on samples where passivating native oxides occur. Contact resistance can also vary considerably with contact pressure. Errors due to contact resistance can be reduced in a three terminal configuration because the full sample current does not pass through the point contact of the probe for the potential measurement. Besides altering the potential distribution, high contact resistance can contribute to poor signal-to-noise ratios.

The electron beam of the SEM itself can contribute to electrical measurement errors or artifacts. The energetic beam can generate electron-hole pairs which may be separated and collected by built in or applied fields. One must insure that such effects are negligible or take steps to avoid them, such as blanking or turning the electron beam off during the measurement. The electron beam can also carbonize surface hydrocarbon layers, creating under certain conditions, a shunting path which could mask the real potential profile.[4] In severe circumstances, this may preclude SEM viewing altogether. Finally, physical damage of the sample surface induced by the electron beam is a possibility.

There is a subtle electrical contribution to errors in high spatial resolution measurements on semiconducting samples. This arises from the depletion layer set up by the needle/semiconductor junction itself. The radius of the approximately hemispherical depletion region depends on the doping density of the sample and the metal/semiconductor barrier height for the particular system of interest. This depletion region could round off sharp features in a potential profile by altering the sample current path just under the probe.

There are also sample requirements for successful potential microprobe measurements. One must be sure the surface measurement yields useful information about the bulk material. Surface damage or contamination which might shunt the underlying material must obviously be avoided. Sample thickness can be critical in systems where grain boundaries dominate the electrical properties. The current path must be near the surface where it can be sensed by the probe. In practice, this means that the sample thickness should be roughly one to two average grain diameters or less. For samples processed by conventional pressing and sintering, this thickness requirement must be met by grinding and polishing. Grain pullout can become quite severe during mechanical thinning operations, making sample preparation difficult. A final chemical etch or thermal anneal may be required to minimize surface damage artifacts in the electrical measurements.

ILLUSTRATIVE APPLICATIONS

High Resistivity Polycrystalline Silicon

Polycrystalline silicon, which has been used to model electronic conduction in polycrystalline ceramics[7], is commonly used in the microelectronics industry. It is usually produced by low pressure chemical vapor deposition. For high value resistor applications, the impurity or doping level is generally controlled by ion implantation and thermal annealing. Charge trapping at the grain boundaries leads to potential barriers whose height and width are determined by the trap density and the doping density.[7] The barrier impedance depends exponentially upon the barrier height. The situation is complicated further by dopant segregation and deactivation at the grain boundaries, and the rapid diffusion of dopant atoms in the grain boundaries as compared with the grain interiors.[8] One method for forming polysilicon resistors is to produce lightly doped regions in the heavily doped n-type polysilicon interconnect lines. To control such a process, one must have a good understanding of the impurity diffusion and activation processes which are occuring. The potential microprobe can easily contribute to such understanding.

The inset to Fig. 2 shows the W needle of the potential microprobe situated over a polysilicon resistor in an actual integrated circuit. A voltage is applied to the resistor through two external leads (not shown), and the W needle is used as a moveable third terminal in a three terminal measurement. The overlying layers in the circuit have been stripped back chemically to expose the polycrystalline silicon surface. In this particular process, the high resistance region was photolithographically defined to be 10 μm long. However, the electrical length of the resistor, determined from the potential profile measured with the microprobe, was found to be only about 6 μm (Fig. 2). Rapid diffusion from the heavily doped regions surrounding the resistor had not been accounted for when designing the circuit and defining the process.

High T_c Ceramic Superconductors

Much of the early development of the new high critical temperature (T_c) superconductors involved the use of conventional pressing and sintering techniques. At Sandia, static magnetization measurements in the 90K YBaCuO system revealed that not all of the resulting pellet was superconducting when sintered at 1100°C in air.[3] Much work was done to empirically determine the optimal thermal anneal parameters for creating the desired phase.

The potential microprobe was used to determine the microstructural location of the high T_c phase. To make the measurement, a small piece of a sintered pellet was mechanically polished to produce a reasonably smooth surface. Gold electrodes were sputter deposited onto the polished surface and gold leads were indium soldered to the electrodes. Optical microscopy and preliminary microprobe measurements indicated the presence of two clearly separated phases in the sample. To characterize the material further, a current of 150 mA was forced between the electrodes. The W needle of the microprobe was connected to the low electrode through the Keithley 619 electrometer in the current measuring mode. In this configuration, we hoped to determine the current path (the measurements were done at room temperature). When the needle touched material in the current path, a portion of the sample current should be diverted through the electrometer. The voltage resolution of the electrometer (1 μV), precluded measuring the potential profile in the highly conductive pellets.

It was quickly determined that all of the current was flowing in thin shells of a soft material surrounding hard grains which were about 15 μm in diameter. Current flow was never detected when the needle contacted the hard grain interiors. However, when placed directly over a grain boundary or when touching the sidewall of a void in the material, currents detected by the electrometer increased by two to three orders of magnitude. Several clearly identifiable grain boundaries were probed with finer spatial resolution (≈40 nm). A typical result of such a measurement is shown in Fig. 3. The noise in the current values arises from the varying contact area each time the needle touches the soft material. In spite of the current noise, the profile clearly indicates that the shells of conducting material are about 0.7 - 0.8 μm thick. Energy dispersive analysis of X rays (EDAX) was performed on this same grain boundary in another SEM, and clearly showed that the grain boundary region was rich in Ba and Cu, and poor in Y as compared to the insulating grain interior. Further EDAX work determined the qualitative composition of the conducting shells to be $YBa_2Cu_3O_y$ (the so-called 123 material), while the grain interior composition was Y_2BaCuO_y (211 material).

Using these complementary techniques, it was deduced that incongruent melting at the high sintering temperature being used at that time was creating a liquid phase (the 123 material) which coated grains of insulating material (211 material). Fractional crystallization thus resulted in a multiphase pellet. The sintering temperature had to be reduced to avoid this occurrence.

ZnO Varistor Materials

As in high resistivity polysilicon, the electrical properties of polycrystalline ZnO are determined by double depletion layer, electrostatic potential barriers which form at the grain boundaries.[9,10] Impurity atoms can be added to the system to both raise the equilibrium barrier height (thus reducing transport across the barrier at low fields), and to accelerate its collapse once a certain field is reached (contributing to highly nonlinear I-V characteristics at breakdown). These effects enable one to dramatically improve the switching behavior of a ZnO varistor.

The medium field ZnO varistor materials used in this study were sintered from chemically prepared powders according to protocols developed by Dosch[11] and Voigt.[12] Measurements were made on samples having the composition 98.94 mol% ZnO, and 0.56 mol% Bi_2O_3, with the remainder being either 0.5 mol% CoO or 0.25 mol% CoO plus 0.25 mol% MnO. The final sintering temperature was 1000°C, resulting in an average grain size of about 25 microns.

To avoid sample thickness related artifacts in the measurements, small pieces were cut out of the initial disk, potted in Crystalbond, and thinned by mechanical polishing to about 40 μm in thickness. Attempts to remove even more material consistently failed due to severe grain pullout. The sample surface was then briefly etched in a 1:1 solution of HF:Cr_2O_3 (5M) to remove polishing damage and to better delineate the grain boundaries. Gold electrodes were evaporated about 250 μm apart on the surface and leads were attached with Ag paint.

Current-voltage curves could not be measured between the Au contacts on these very thin samples, apparently because no continuous path existed. This limited the measurements to two terminal I-V curves taken with the W needle contacting grains just off the periphery of a Au contact. Several grains were carefully chosen which had only one path back to the Au electrode which crossed a single grain boundary (that is all alternative paths crossed two or more grain boundaries). This criteria seemed to insure the observation of small subthreshold currents and sharp turn-on.

Typical results for a single boundary I-V curve measured in this manner are shown in Fig. 4. Several curves are shown which were taken consecutively with varying waiting periods in between measurements. Note that the high α region begins between 3 and 4 volts (the nonlinearity coefficient α is defined as $d[\ln(I)]/d[\ln(V)]$.). This agrees with previous determinations of the turn-on voltage for individual grain boundaries by Einzinger[1,2] measured at fixed current. It also supports models of the high α behavior based on hot electron production of holes, which require that the hot electrons acquire at least the gap energy (3.2 eV in ZnO).[9,10]

The data of Fig. 4 indicate a maximum α value of about 13. Measurements done on the full pellet from which this sample was taken gave a maximum α of 25.[13] The degradation of α in the thin sample may be due to residual damage from the polishing not removed by the final chemical etch. The samples cannot be annealed after thinning because of the potting compound. Contact resistance between the W needle and the ZnO grain could also degrade the measured α value.

The subthreshold (i.e. below turn-on) current in successive runs shows varying degrees of degradation and recovery. Curve 2 was taken within minutes of curve 1, and shows a rather large increase in the current at low applied voltages. The two curves eventually join in the high α region around 4 volts. Curve 3 was taken about 30 minutes after curve 2, and shows substantial recovery of the subthreshold currents toward their initial values (complete recovery was never observed). The same sequence of degradation and recovery of subthreshold currents is repeated in curves 4 (taken within a few minutes of curve 3) and curve 5 (taken 35 minutes after curve 4). This behavior was observed to some degree in every grain boundary studied. The amount of degradation increased when larger currents were forced through the sample, as demonstrated by the lesser degradation between curves 3 and 4 as compared to 1 and 2. If this process was repeated enough times, or the maximum currents were quite high (100 μA), the boundary did not recover and the varistor action disappeared.

Interstitial ion migration has been proposed as a degradation mechanism by several researchers.[14-16] Chiang, et al[14] monitored impurity segregation in severely degraded samples using EDAX in a scanning transmission electron microscope. In fresh samples, Bi was observed to show a high degree of segregation to the grain boundaries, while the Co signal was not enhanced at the grain boundary. After electrical degradation, in boundaries perpendicular to the applied field direction, pile up of both Bi and Co was observed on one side of the boundary. These results imply that drift of these ions may be related to the degradation. Other experiments have implicated interstitial Zn donors (Zn_i) as prime players in the degradation.[15,16] This hypothesis is based on the ability to reduce both the instability and the concentration of Zn interstitials (inferred from deep level transient spectroscopy measurements) by judiciously annealing the sample. An alternative technique is to reduce interstitial migration by introducing impurities which compete for the interstitial sites. Either of these techniques reduces the number of Zn interstitials reaching the grain boundary, where through a series of reactions, they eventually annihilate negatively charged Zn vacancies and lower the potential barrier. This migration into the depletion region would be field assisted during device operation. Gupta and Carlson[17] also propose the reversal of the reactions when the field is removed, and thus recovery of the barrier height.

The subthreshold recovery indicated in Fig. 4 seems consistent with the Zn interstitial migration model. However, the recovery was never observed to be complete. The irreversible degradation and loss of varistor action which occurs with repeated cycling of the grain boundary or the passage of very high currents, indicates more permanent trapping or further reaction which immobilizes the migrating ion. This "permanent" degradation cannot be removed until the sample is annealed at much higher temperatures.[18]

FINAL COMMENTS

The examples discussed above illustrate the capability of the potential microprobe to provide useful information in electrically characterizing inhomogeneous materials. Its range and resolution often make the microprobe the only (or at least simplest) measurement alternative. In its present configuration, the microprobe is relatively inexpensive, taking advantage of equipment which is routinely available in most laboratories. There are refinements and improvements which could be made if space in the SEM chamber is adequate and expense is not a problem. The mechanical stages could be eliminated entirely through the use of commercially available "inchworm" piezo drives.[19] Such drives can provide centimeters of travel with 10 nm resolution all from the piezo elements, thus eliminating the need for rotary vacuum feedthroughs and putting the full range of travel under computer control. Another improvement would be the incorporation of multiple needles, which could eliminate the need to evaporate contacts and attach leads in some cases.

ACKNOWLEDGEMENTS

The author would like to acknowledge C. H. Seager and W. R. Bayless, who played primary roles in designing and building the microprobe. C. H. Seager and J. H. Perry participated in many of the potential microprobe experiments discussed.

REFERENCES

1. R. Einzinger, "Grain Junction Properties of ZnO Varistors", Appl. Surf. Sci. 3, 390-408 (1979).

2. R. Einzinger, "Grain Boundary Properties in ZnO Varistors", Adv. in Cer. 1, 359-374 (1981).

3. D. S. Ginley, E. L. Venturini, C. H. Seager, W. K. Schubert, R. J. Baughman, J. F.Kwak, J. E. Schirber, and B. Morosin, "Grain Boundary Superconductivity in the YBaCuO System", MRS Symposia Extended Abstracts, High-Temperature Superconductors, ed. D. U. Gubser and M. Schluter, EA-11, 201-204 (1988).

4. C. H. Seager and W. K. Schubert, "Direct measurement of majority carrier quasi-fermi levels in Schottky barrier and metal-insulator-semiconductor diodes", J. Appl. Phys. 62, 4313-4316 (1987).

5. The micrometer stages and piezo stacks were obtained from Control Technics Corp., 22600C Lambert St., Suite 909, El Toro, CA 92630.

6. Tungsten probes can be obtained from The Micromanipulator Company, 1120 Industrial Ave., Escondido, CA 92025.

7. G. E. Pike and C. H. Seager, "Electronic Properties of Silicon Grain Boundaries", Adv. in Cer. 1, 53-66 (1981).

8. W. K. Schubert, "Properties of furnace-annealed, high-resistivity, arsenic implanted polycrystalline silicon", J. Mater. Res. 1, 311-321 (1986).

9. G. D. Mahan, L. M. Levinson, and H. R. Phillip, "Theory of conduction in ZnO varistors", J. Appl. Phys. 50, 2799-2812 (1979).

10. G. Pike, "Electronic properties of ZnO varistors: A new model", MRS Symposia Proceedings, Grain Boundaries in Semiconductors, ed. H. J. Leamy, G. E. Pike, and C. H. Seager, 5, 369-379 (1982).

11. R. Dosch, B. Tuttle, and R. Brooks, "Chemical preparation and properties of high-field zinc oxide varistors", J. Mater. Res. 1, 90-99 (1986).

12. J. Voigt, Patent Disclosure SD-4544, S-66621; "A continuous direct oxalate precipitation process for the preparation of high field zinc oxide varistor precursor powders", 1987.

13. B. A. Tuttle, J. A. Voigt, E. J. Binasiewicz, and K. M. Kimball, "Electrical properties of medium field chem-prep varistors", Sandia Report No. SAND87-1824 (1987).

14. Y. Chiang, W. Kingery, and L. Levinson, "Compositional changes adjacent to grain boundaries during electrical degradation of a zinc oxide varistor", J. Appl. Phys. 53, 1765-1768 (1982).

15. A. Rohatgi, S. K. Pang, T. K. Gupta, and W. D. Straub, "The deep level transient spectroscopy studies of a ZnO varistor as a function of annealing", J. Appl. Phys. 63, 5375-5379 (1988).

16. T. K. Gupta and A. C. Miller, "Improved stability of the ZnO varistor via donor and acceptor doping at the grain boundary", J. Mater. Res. 3, 745-754 (1988).

17. T. K. Gupta and W. G. Carlson, "A grain-boundary defect model for instability/stability of a ZnO varistor", J. Mater. Sci. 20, 3487-3500 (1985).

18. G. E. Pike, S. R. Kurtz, P. L. Gourley, H. R. Phillip, and L. M. Levinson, "Electroluminescence in ZnO varistors: Evidence for hole contributions", J. Appl. Phys. 57, 5512-5518 (1985).

19. "Inchworm" is a registered trademark of Burleigh Instruments, Inc., Burleigh Park, Fishers, NY 14453.

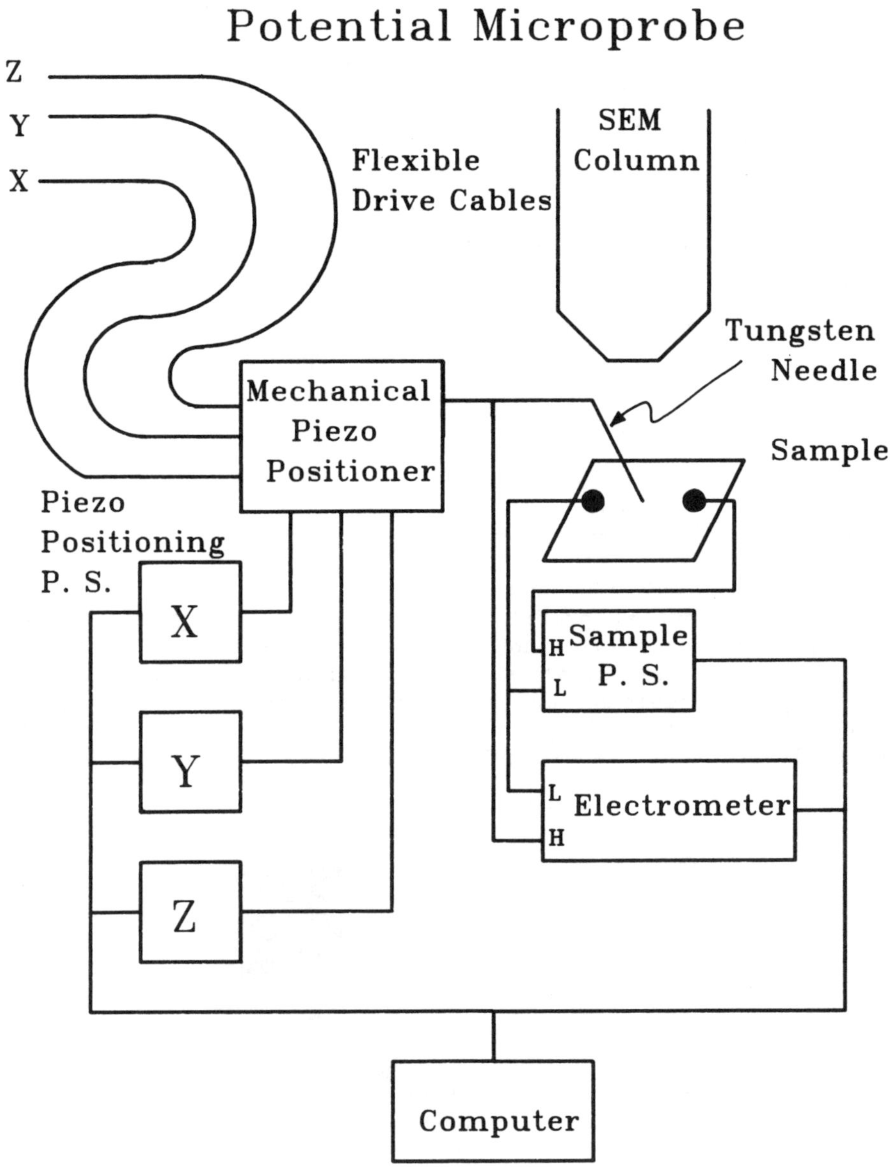

Fig. 1. A schematic representation of the primary components of the potential microprobe, shown in a three-terminal configuration. The sample power supply can be either a constant current or voltage source.

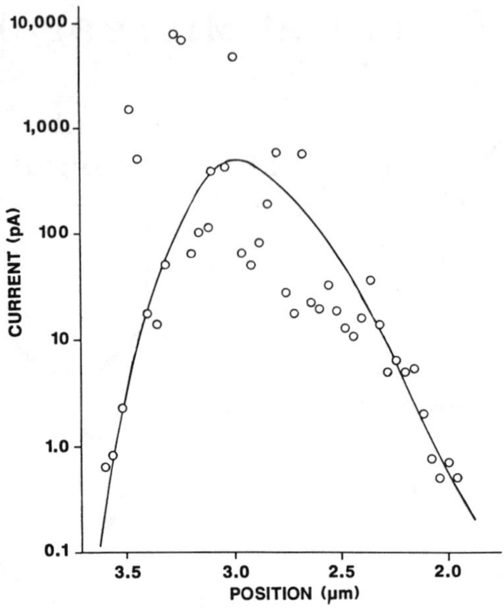

Fig. 2. A potential profile measured on an integrated circuit resistor made from polysilicon, indicating an electrical length of about 6 μm. The inset shows the sharp microprobe needle contacting the resistor at the starting point of the potential profile. The potential rises abruptly to the applied value when the needle contacts the electrically isolated Al crossover.

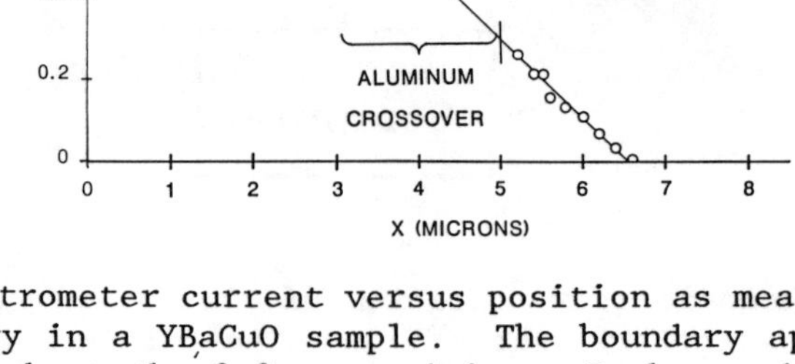

Fig. 3. Electrometer current versus position as measured across a grain boundary in a YBaCuO sample. The boundary appeared in the SEM image at about the 2.9 μm position. Background current noise is < 1 pA.

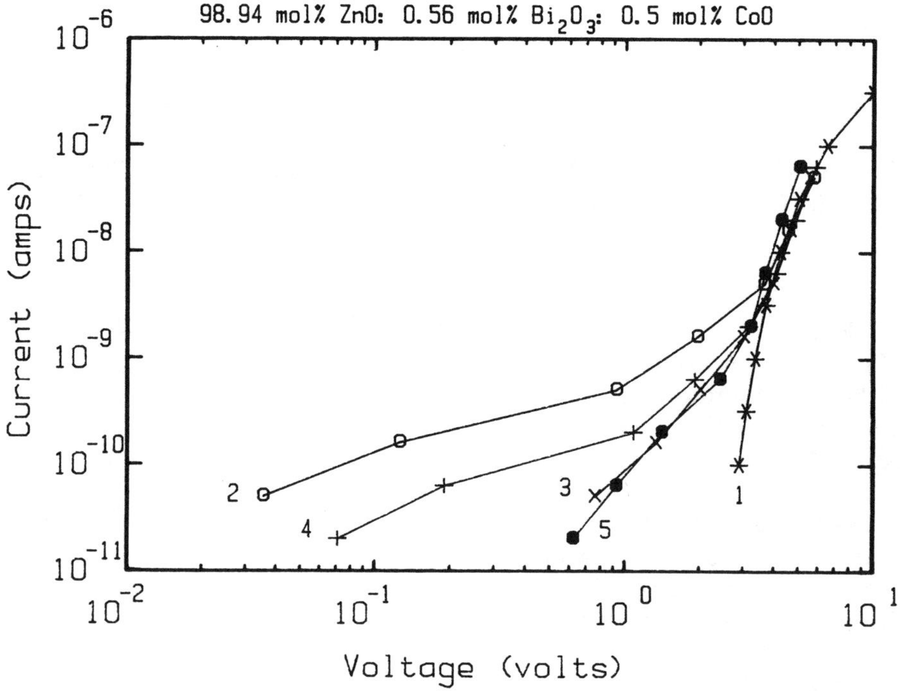

Fig. 4. I-V curves measured on a single ZnO grain boundary in a two-terminal configuration. The measurements were made in the numbered sequence as described in the text.

ADVANCED TEM TECHNIQUES FOR CHARACTERIZATIONS OF CERAMIC MATERIALS

M. Rühle
Materials Department
College of Engineering
University of California
Santa Barbara, California 93106

G. Necker and W. Mader
Max-Planck-Institut für Metallforschung
Institut für Werkstoffwissenschaften
Stuttgart 1, Federal Republic of Germany

ABSTRACT

High Resolution Electron Microscopy (HREM) is now widely used to investigate structures and defects in materials. The point-to-point resolution of the electron microscope limits the applicability of the technique. Furthermore, the resolution limits the detection of the exact number of atoms and their location in the complete structure or within a defect. A quantitative evaluation of HREM images requires a comparison to simulated images. The HREM technique is applied for a characterization of the interface between Ag and CdO.

INTRODUCTION

Recently, the resolution of commercially available electron microscopes has been significantly improved. The point resolution, which is defined more accurately in Sect. 3.1, is now better than 0.16nm for instruments with an acceleration voltage of 400kV. A new generation of high-voltage, ultra-high-resolution instruments pushes the resolution limit to about 0.1nm. Considerable advances have also been made in the analysis of micrographs obtained by high-resolution electron microscopy (= HREM).[1–3] Methods and programs have been developed[3,4] which allow the simulation of HREM images of any given atom arrangements. Those simulated images are used to interpret experimentally obtained micrographs. These recent developments enable us to use HREM as an important method for solving problems in materials science. It is expected that atomic structure of different lattice defects, such as phase boundaries, grain boundaries and dislocations, can be determined by HREM. The difficulties in the interpretation of HREM are similar to those in conventional transmission electron microscopy of lattice defects in crystalline materials. The experimentally obtained micrographs do not usually present a direct image of the object.

The extraction of information on the structure of lattice defects from HREM micrographs is complicated.[1,2] Despite this, very useful information has been obtained on the structure of lattice defects using HREM, particularly for semiconductors. With the new class of instruments, it is expected that HREM will also be an important technique for characterization of defects in ceramics and metals.

This paper starts with some comments on high-resolution electron microscopy and then discusses applications of HREM in materials science. The various steps needed for successful working with HREM are described. Specifically, results for the CdO/Ag interface are described in detail.

DIFFERENT TRANSMISSION ELECTRON MICROSCOPY TECHNIQUES

The different possibilities offered by modern transmission electron microscopy is summarized schematically in Fig. 1. Conventional TEM techniques (CTEM) involves bright field (BF) and dark field (DF) imaging and selected area diffraction (SAD). These are used for morphological analysis, phase and defect identification studies. Spectroscopy can be performed with high spatial resolution. The probe size in scanning transmission electron microscopy ranges from ~ 1.5nm to 50nm. Energy dispersion spectroscopy (EDS) uses X-rays emitted from the specimen for a chemical characterization, while characteristic energy losses can be used for identifying qualitatively and quantitatively the light element by energy loss spectroscopy.[5–7] The surrounding of atoms can be probed by extended energy-loss fine structure (EXELFS) studies whereas energy-loss near-edge structure (ELNES) investigations result in information on the excitation state of atoms or ions. High resolution electron microscopy (HREM) studies result in structure images. This paper focuses on the latter studies.

COMMENTS ON DIRECT LATTICE IMAGING OF DISTORTED MATERIALS USING HREM

The geometric beam path through the objection lens of a TEM is shown in Fig. 2a. Beams from the lower side of the object travel both in the direction of the incoming and diffracted beams. All beams are focused by the objective lens in the back focal plane to form the diffraction pattern. In the image plane, the image of the object is produced by interference of the incoming and diffracted beams. Fig. 2b uses wave optics to describe physical processes which contribute to the image formation. From the lower side of the foil, a wave field emerges that can be described by a transmission function q(x, y). For an undistorted lattice, q(x, y) represents a simple, sine-shaped amplitude and intensity distribution. The transmission function of a complex lattice with a high periodicity (e.g., a periodic grain boundary) is very complicated and q(x, y) is a non-periodic function for the distorted region of a crystal.

The intensity distribution in the diffraction pattern is given by the Fourier transformation Q(u, v) of the transmission function q(x, y) where u, v are the coordinates in the diffraction plane (reciprocal space). Since spherical aberration cannot be avoided with rotationally symmetrical electromagnetic lenses[1,2], the beams emerging from an object at a certain angle (Fig. 2a) undergo a phase shift relative to the direct beam. Imaging under a slightly defocusing mode, Δf, leads also to a phase shift which depends on the sign and magnitude of the defocus value, Δf. The influence of the lens errors and the defocus on the amplitude of the diffraction patterns can be described by the contrast transfer function $\chi(u, v)$:

$$\chi(u,v) = \frac{\pi}{\lambda}\left[\Delta f(u^2 + v^2)\lambda^2 - \frac{1}{2} C_s(u^2 + v^2)\lambda^4\right],$$

λ = wave length of electrons, C_S constant of spherical aberration. The dependency of the real and imaginary components of the contrast transfer function as the function of reciprocal distancy ("space frequency") is shown in Fig. 3.

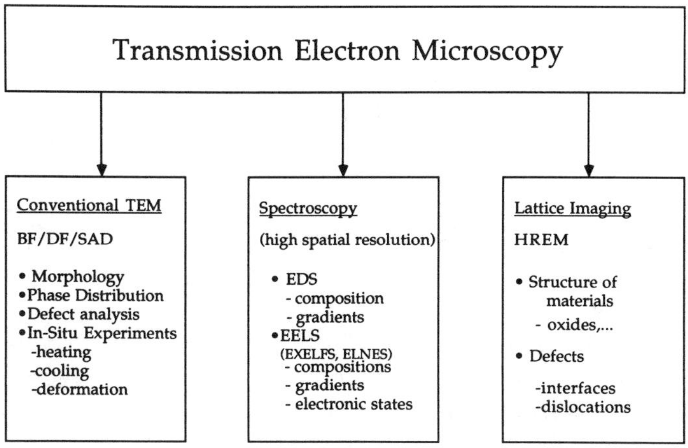

Fig. 1. Schematic illustrations of different TEM techniques (BF = bright field imaging; DF = dark field imaging; EDS = energy dispersive spectroscopy; EELS = electron energy-loss spectroscopy; EXELFS = extended energy-loss fine structure; ELNES = electron energy loss near-edge structure; HREM = high-resolution electron microscopy)

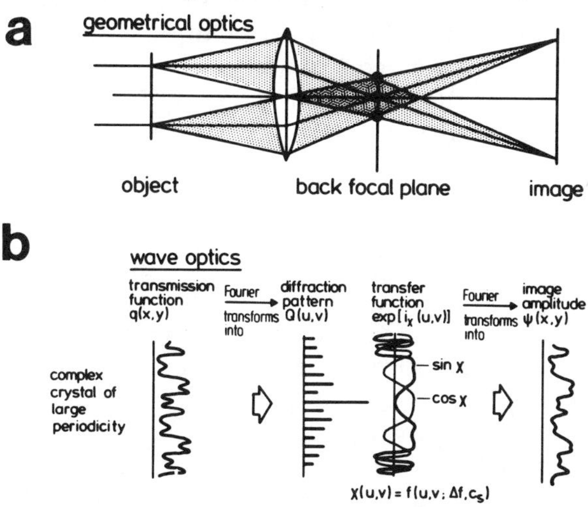

Fig. 2. Image formation by the objective lens of a transmission electron microscope

a) Geometrical optical path diagram
b) Wave optical description (see text for explanation)

The image is formed by a second Fourier transformation of the amplitude distribution in the diffraction pattern multiplied by the contrast transfer function. The amplitude in the image plane, $\psi(x, y)$, is therefore not identical to the wave field in the object plane (transmission function $q(x, y)$). The image is severely modified if scattering to large angle occurs, since the influence of the spherical aberration increases strongly with increasing scattering angle. The modification is most severe if the components in the diffraction pattern coincide with the oscillating part of the contrast transfer function (large values of u, v). If, however, the wave vectors lie within the first wide maximum of the contrast transfer function $\chi(u, v)$, it can be assumed that characteristic features and properties of the object can be directly recognized in the image.[1,2]

For good HREM imaging, the first zero value of the contrast transfer function under the optimum defocusing condition (Scherzer focus) must be at sufficiently large reciprocal lattice spacings so that lattices of many materials can directly be imaged. Good imaging conditions are fulfilled for lattices with large lattice parameters.[8–11] If, however, deviations in the periodicity exist, components of the diffraction pattern appear at large diffraction vectors. It is then most likely that certain Fourier components possess reciprocal lattice distances larger than the Scherzer focus. Lens aberrations and defocusing cannot be neglected.

Resolution

The image contrast obtained in a high-resolution electron microscope (Fig. 2) is produced by interference of the diffracted beams with the transmitted beam. The interference can also be represented as a impulse reaction in the form of a delta function, the location and shape of which is highly dependent on the focusing distance.[1,12] From this description, it can easily be concluded that the resolution is not uniquely defined and that various distances can be adopted to define the resolution: d_1 is defined as the Scherzer resolution, which is determined from the focusing distance Δf at which the influence of negative defocusing compensates as far as possible for the influence of the spherical aberration.[1,2] The information limit, d_2, represents the smallest resolvable distance on an image at an optimum defocusing distance.

For a HREM analysis of crystalline materials, the first step is to establish the number and location of the positions of atom rows. It is assumed that the lattice to be investigated is periodic in the direction of the incoming beam so that the projected potential on the lower surface of the object is formed by the superposition of identical atoms that lie one above the other in the z direction (Fig. 4). To determine the atomic structure of distorted regions (or regions of large lattice periodicity), the resolution is defined below. The evaluation of the number of atomic rows and their exact positions are discussed in the next section. Different zone axes of crystalline materials can be used for the evaluation of crystal structures and of defects if the lattice plane distances are larger than the Scherzer resolution d_1 of the instrument.

If lattice defects must be analyzed by HREM, one must first establish the number of atomic rows that exist in the surroundings of a lattice defect. Naturally, no information can be obtained in any HREM investigation on defects which are not parallel to the direction of the incoming electron beam. HREM images always represent a two-dimensional projection of electron distribution on the exit surface of the transmitted foil. For the image evaluation, it is implied that the atomic rows are exactly parallel to the incoming electron beam.

In order to discuss the possibilities of determining the atomic structure, it is necessary to introduce a parameter d_0 representing the minimum distance of two neighboring atom columns in the region of the lattice defect. Comparing d_0 with d_1 and d_2 enables us to decide whether the distorted structure (or structural unit) in the lattice defects can be resolved in all details or not. Three cases can be distinguished:

a) $d_0 > d_1$. In this case, all atom columns can be observed; artefacts are not introduced into the Scherzer image. For thin foils (typical thicknesses 5 to 10nm), channels in the lattice structure appear as white spots on the image. If certain *a priori* information on the atom configuration is available, the observed HREM micrographs can be directly compared with the atomic structure.[1,2]

b) $d_1 < d_0 < d_2$. The determination of the number of atom columns of a distorted structure requires micrographs taken under a series of defocusing distances which differs typically by small values, ~ 5nm. Artefacts such as additional dark contrasts may appear on the HREM micrographs under certain defocusing conditions. A quantitative evaluation requires the comparison of micrographs taken under specific focusing conditions with computer simulated images in order to differentiate interferences produced by artefacts and real lattice distortions such as additional atom columns, for example. However,

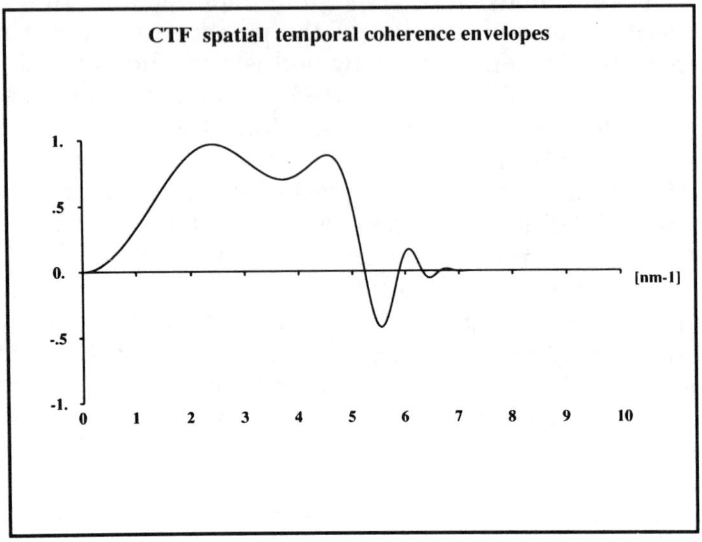

Fig. 3. Contrast transfer function (CTF) of a high-resolution electron microscope

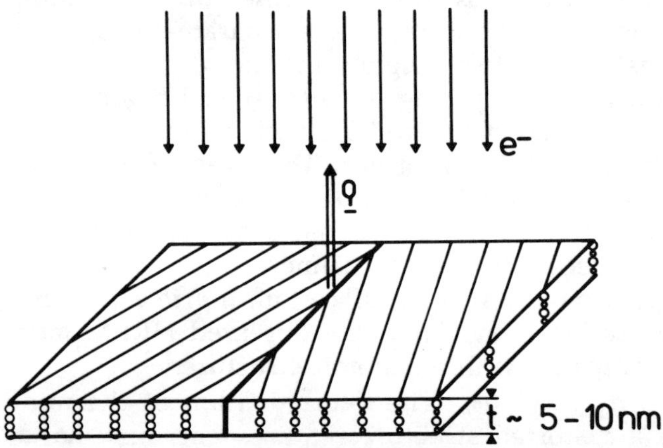

Fig. 4. Direct lattice imaging by HREM. The crystalline specimen must be adjusted so that the direction of the incoming electron beam coincides exactly with the orientation of atomic rows. The schematic drawing includes a grain boundary. HREM can successfully performed for pure tilt boundaries with tilt axis and parallel to the direction of the incoming electrons

this comparison is very difficult and not straightforward, as several different structures can often give the same "image." If sufficient *a priori* information is available, a clear interpretation will be possible.

c) $d_0 > d_2$. The existing atom rows can no longer be resolved as such, and thus only an absolute measurement of the intensity distribution would give the exact number of atom rows which produce a light (or dark) contrast spot on an HREM image. This evaluation would be very difficult, especially since any surface defects introduced during specimen preparation may cause phase shifts which modify drastically the intensity distribution on the HREM images.

From this brief summary of the optimum conditions for observing structures in the vicinity of lattice distortions (grain boundary, dislocations), the following conclusions may be drawn:

a) d_1, the point resolution in the Scherzer focus, must be as small as possible.

b) Analyses of structures surrounding lattice defects are confined to those cases in which strict symmetry exists in the direction of the incoming electron beam. As an example, only pure tilt boundaries without rotation components and without facets perpendicular to the incoming beam can be analyzed since only these defects have a distortion field independent of the z-direction (direction of the incoming electron beam). In interpreting the HREM images, it can then be assumed that the atom rows in the vicinity of the lattice defects are only displaced parallel to each other.

c) The specimen thickness must be as small as possible and less than an effective extinction length of the selected zone axes. In most cases, the foil thickness should be between 5 and 15nm. The exact values of the foil thickness t and the defocusing value Δf are very important parameters which must be precisely determined. Computer simulations for which the experimentally determined parameters t and Δf are required are thus indispensable if details are to be resolved which are of magnitude d_1 or below d_1.

d) Extremely careful adjustment of the electron microscope has to be carried out in order to obtain a good correlation between the image and the projected structure.

Determinationof the Atom Column Locations

After determining the number of atom columns in the vicinity of the lattice defect, the coordinates of the individual atom columns must be determined with the best possible accuracy. The evaluation demands a comparison between experimentally obtained micrographs and computer-simulated images. A "trial and error" method is usually applied. The comparison is made by superimposing the two images and visually comparing the location of the individual contrast features. It is important that the location of all contrast maxima and minima coincide. Very reliable data of atomic coalescence coordinates are obtained for $d_0 > d_1$: the accuracy is $\pm 0.01\ d_1$.[12] The accuracy of the atom coordinate determination is thus substantially better than the actual resolution. This can easily be explained by the fact that for very small changes of the phases different waves may result in an observable change in the intensity distribution and in the location of the intensity maxima during formation of a phase contrast image.[1,2]

The relative displacement of perfect lattices against each other (e.g., in the neighborhood of phase boundaries or even at grain boundaries) can be analyzed from HREM micrographs. The values obtained by HREM are often sufficiently accurate for a differentiation between different grain boundary models. The resolution d_2 = 0.16nm (400kV HREM) is not sufficient to determine small displacements in the vicinity of ordered grain boundaries (e.g., twins) in metals.

Figure 5 shows a flow chart for quantitative HREM.

Complementary Diffraction Investigations

Selected area diffraction investigations provide important information on the periodicity along the common axis in the case of periodic defects (e.g., periodic twin boundaries). This is the only possible technique which permits any type of reconstruction along the atom columns with regard to the common axis. This method was recently used to complement high-resolution electron microscopy for determining the structure of grain boundaries.[1,2,8]

THE POTENTIAL OF HIGH-RESOLUTION ELECTRON MICROSCOPY IN THE SCIENCE OF CERAMICS

HREM studies can be used for identifying the structure of defects which possess exact periodicity parallel to the direction of the incoming electron beam. Such defects encompass domain boundaries and twins, tilt grain boundaries and dislocations.

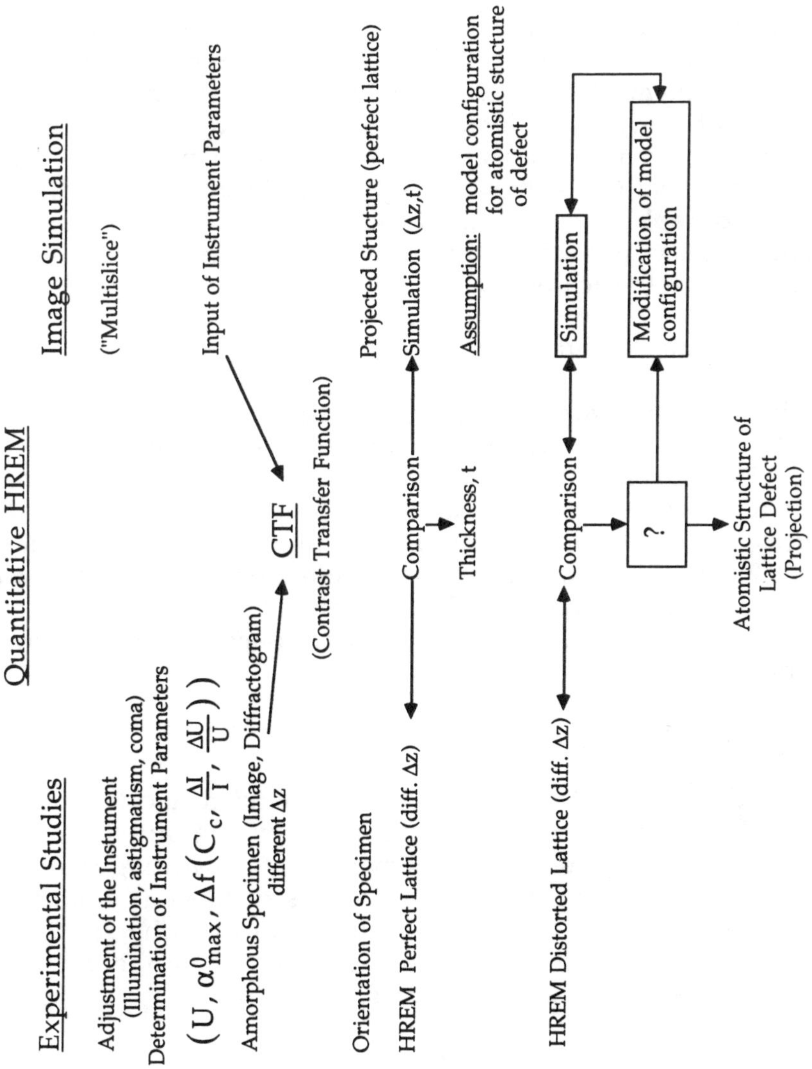

Fig. 5. Quantitative HREM — a flow chart

Domain Boundaries and Twins

Domain boundaries and various types of twins are two-dimensional lattice defects which possess a perfect lattice adjacent to the boundaries. The defects are formed by simple symmetry operations (displacement, mirror image). The atomic structure of the defects and lateral displacement of the two crystals can be determined by HREM. An example of a twin boundary in a periodic lattice is given in Fig. 6, which represents an incoherent twin boundary in monoclinic ZrO_2. The lattice constant of the monoclinic ZrO_2 is 0.54nm. The light spots in this micrograph represent open channels and the dark spots are superimpositions of various Zr atoms which have been slightly displaced relative to each other.

Structure of Grain Boundaries

The atomistic structure of grain boundaries in simple oxides was recently predicted by molecular static computer simulations.[12–13] Merkle[11,15] analyzed tilt boundaries in NiO. He found a reasonable agreement between observed and the predicted grain boundary structure. The volume increase at the boundary resulted in $\Delta V/V \sim 0.2$. In addition, quantitative HREM studies revealed that the scattering intensity by individual atom columns next to the interface were reduced by ~ 30 %, indicating that the ion density is reduced by the same amount.

Atomis Structure of Dislocations

The displacement and stress fields around dislocations can be described by various approximations whereby the structure of the dislocation core is probably not properly simulated by using continuum theoretical models. The real dislocation core structures $d_1 < .13$nm can only be analyzed by HREM if the resolution in the Scherzer focus is $\alpha_1 < 13$nm. This resolution can only be achieved by the new high-resolution, high-voltage electron microscopes.

CASE STUDY: CdO/Ag INTERFACE

The system chosen for the present case study, Ag/CdO, is a useful model for metal/ceramic interfaces for two reasons. Firstly, Ag is a noble metal and thus, no chemical reaction layer is expected to form; secondly, both phases possess a f.c.c. crystal structure. Additionally, very clean interfaces are likely to be formed by the preparation procedure used in this study: internal oxidation[16,17] of Ag–Cd alloys, which results in the formation of CdO precipitates inside the Ag matrix. Since the precipitates have the freedom to form with any shape and relative orientation with respect to the metal matrix, it is expected that low-energy interfaces can be produced by this method.

Conventional transmission electron microscopy (CTEM) and HREM studies were performed in order to obtain answers to the following questions: (1) Are misfit dislocations present at Ag/CdO interfaces? (2) What is the structure and the chemical composition of the Ag/CdO interphase boundaries formed *in situ* by internal oxidation?

EXPERIMENTAL DETAILS AND RESULTS

Homogenized Ag–1.5at.%Cd alloys were rolled into a sheet with a thickness of ~150μm. After annealing in an argon atmosphere, the specimens were oxidized at 973–1123K for several hours in air. TEM specimens were prepared by electropolishing with an alcoholic perchloric acid solution followed by brief (5–10 min) ion-beam milling under a small-angle (12°) beam incidence to the specimen plane.

The following results were obtained by CTEM studies. As a result of rolling, the Ag matrix showed a <110> texture. Only the thermodynamically stable CdO phase with a NaCl structure was observed to form dispersed precipitates which are 70–150nm in diameter and exhibit an octahedral shape where all eight facets are parallel to the close-packed {111} lattice planes. Electron diffraction studies revealed a perfect cube-on-cube orientation relationship between Ag matrix and CdO precipitates, in agreement with earlier observations.[18] Therefore, the Ag/CdO interfaces are parallel to {111} lattice planes of CdO as well as of Ag.

In bright-field, dark-field and weak-beam images using the diffraction vectors **g** = {111}, **g** = {200} and **g** = {220} of Ag and CdO, contrast from misfit dislocations was not observed. Also moiré contrast fringes from inclined Ag/CdO interfaces are undisturbed within the same thickness contour indicating a very small strain field of possibly existing misfit dislocations.

A JEM 200CX electron microscope (operated at 200kV) was used to obtain lattice images of very thin (thickness $t \leq 10$nm) interface regions in the <110> zone axes. The microscope is equipped with a top-entry small-bore pole piece having a low spherical aberration coefficient (C_s = 1.0mm). In the course of the HREM work, it turned out that the first reverse passband of the phase-contrast transfer function is optimized for the simultaneous imaging of {111} lattice planes of Ag ($d(111)_{Ag}$ = 0.236nm) as well as of {111} planes of CdO ($d(111)_{CdO}$ = 0.271nm).[19]

An example of a lattice image of the Ag/CdO interface is shown in Fig. 7. This image was formed using an objective aperture including space frequencies up to $\alpha_1 = 1/0.18\text{nm}^{-1}$ and by focusing the illumination on the specimen with a beam divergence of 1.2mrad. The objective lens defocus value, Δf = 95nm (underfocus), could be determined by the evaluation of optical diffractograms taken from the image of the thin amorphous region present at the edge of the foil. In Fig. 7, the two sets of {111} lattice planes are visible in each crystal and the interface lying parallel to the $(11\bar{1})$ planes is atomically flat for a distance of more than 20nm. The strict periodicity of clearly resolved white dots in the image of both crystals indicates that the lattices of Ag as well as CdO remain undistorted even very close to the interface. The interface itself exhibits diffuse white dots which are elongated parallel to the lattice planes; occasionally, a white-dot (200) contrast can be observed.

Interfacial steps could be commonly observed in lattice images close to precipitate edges where two {111} faces join. The HREM image clearly shows that the step line is exactly parallel to the [011] direction of projection and that the step height corresponds to the spacing of the close-packed {111} lattice planes. Closer inspection of the contrast indicates a slight bending of the $(11\bar{1})_{Ag}$ lattice planes near the step. This observation can only be explained by the fact that the step height amounts to the spacing of the $(11\bar{1})_{CdO}$ lattice planes forcing the $(11\bar{1})_{Ag}$ planes with the smaller spacing to a slight bending.

In order to obtain detailed information on the atom locations in the perfect lattices as well as in the interfacial region, image simulations were performed using the multislice method (EMS program package).[4] Because {111} planes of CdO are polar planes formed either by O^{2-} or Cd^{2+} ions, two different structural models of unrelaxed Ag/CdO interfaces were used: in model I (Fig. 8a), an oxygen layer forms the boundary on the oxide side; in model II (Fig. 8b), a Cd layer faces the Ag matrix. The distance between the two crystals in both

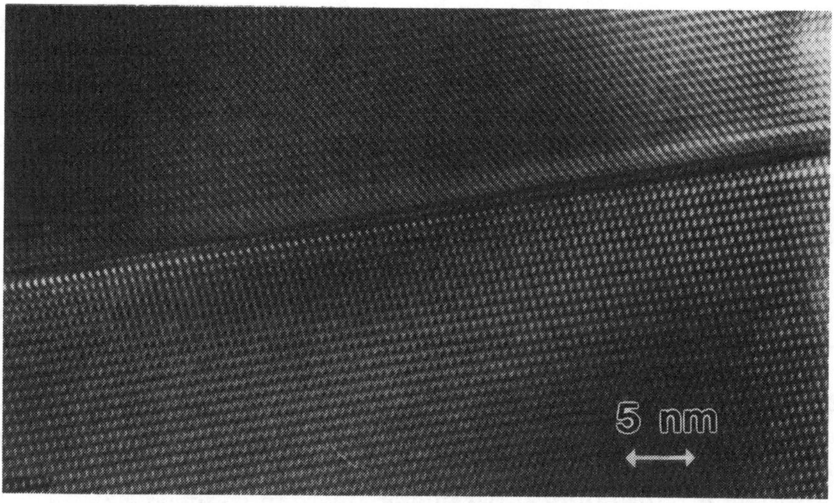

Fig. 6. Twin boundary in monoclinic ZrO_2. Two perfect lattices match at the interfaces.

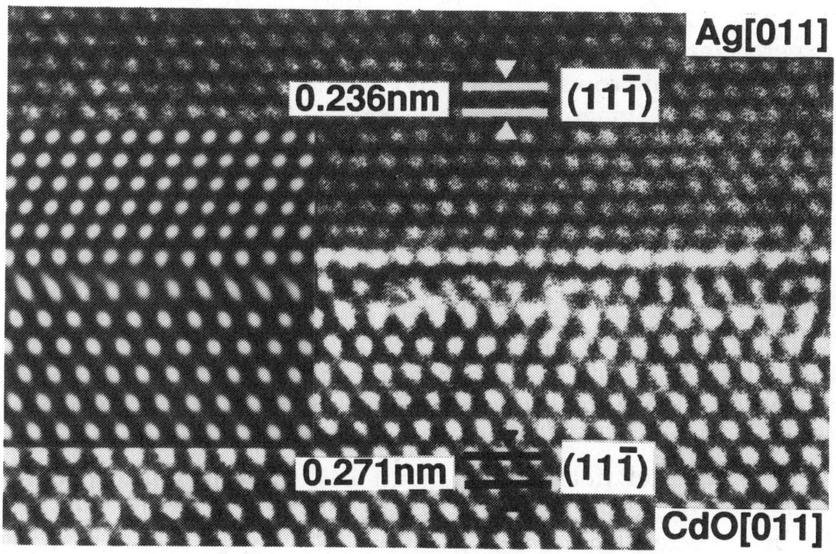

Fig. 7. Lattice image of interface between Ag matrix and CdO precipitate. Inset represents the computer simulated image for the unrelaxed interface model I when the terminating plane of the CdO is an oxide layer.

models was assumed to be determined by the exactly opposite atoms/ions of Ag and CdO at the interface (see Figs. 8a and 8b). The distance between opposite Ag atoms and O(Cd) ions has been chosen to be equal to the shortest interatomic distance, d_{A-B}, in Ag_2O (d_{Ag-O} = 0.205nm) and is AgCd (d_{Ag-Cd} = 0.288nm), respectively. A three-dimensional atomic model of the bicrystal was built up by using a large 4.0 x 4.7nm) simulation cell and by stacking the (022) planes of Ag and CdO in the correct sequence and at the corresponding distances parallel to the direction [011] of the electron beam. Within the periodicity length of 2.3nm parallel to [011], the bicrystal contains sixteen planes of $(022)_{Ag}$ and fourteen planes of $(022)_{CdO}$.

Multislice computations with dynamical scattering arrays of 512 x 512 reciprocal lattice points were performed on these layers until the correct foil thickness was achieved. The electron-wave-field computation was then followed by image calculations where the appropriate electron optical parameters were introduced.

The thickness t of the Ag crystal at the interface imaged in Fig. 7 could be evaluated to be $t \cong$ 7nm (in Ag, no contrast reversal was observed along the path from the very thin region at the edge of the foil to the imaged area). Assuming the same value of t for Ag and CdO, lattice images for both interface models were calculated. Images with the experimentally determined defocus value are shown in Figs. 8c and d where the atomic rows of the heavy elements (Ag, Cd) are imaged as white dots. The image of the interface for model I (Fig. 8c) appears broad because the terminating O layer together with the first adjacent Cd layer cause somewhat less intense and streaked dots; whereas, for model II, the image of the interface (Fig. 8d) exhibits a direct transition from Ag to CdO without any disturbances. In Fig. 7, an excellent matching of the observed and the simulated image for model I can be observed from the contrast width and other finer details, like streaked dots, pointing in the same direction. This comparison strongly indicates that O forms the terminating layer of the CdO precipitate facing the Ag matrix. A contrast similar to the one in Fig. 7d was never observed.

A hexagonal network of misfit dislocations with spacings of 2.2nm and with Burgers vectors of a/2 <110> and <211> line vectors would be expected from geometrical models[20] of an Ag/CdO interface lying parallel to {111} lattice planes. The observations in this study and especially the appearance of perfect lattice images show no evidence for the existence of misfit dislocations: The core region and the strain field

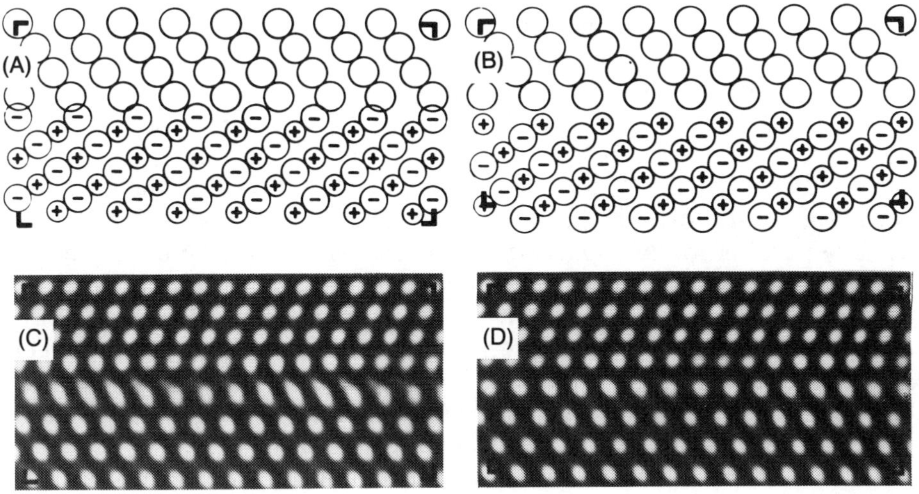

Fig. 8. a, b) Model structure of unrelaxed Ag/CdO interfaces where O or Cd forms the terminating layer of the oxide. One (022) plane of either crystal is drawn with corresponding atomic/ionic radii valid within metal and oxide causing virtual overlaps or standoffs at the interface.
c, d) Calculated lattice images for Ag/CdO bicrystals with interface model structures shown in (a) and (b) and with the same display size and area (see marks) as in (a) and (b). Foil thickness t = 7nm, objective lens defocus value Δf = 95nm (underfocus).

of inclined dislocations (in a foil with a normal parallel to [011], two sets of misfit dislocations would be inclined to the foil normal by 30°) will necessarily influence the lattice image contrast; however, undisturbed lattice images of both Ag and CdO were observed up to the interface. Therefore, contrary to statements by Pedder[18], it must be concluded that the Ag/CdO interface is incoherent despite the excellent crystallographic orientation relationship and the atomically flat interface parallel to close-packed lattice planes. This may be explained by (1) the large lattice misfit of 14 percent in this system where the formation of a necessarily high dislocation density is energetically unfavorable and (2) by the presumably weak adhesive forces between Ag and O. At Nb/Al_2O_3 interfaces, also produced by internal oxidation, misfit dislocations were observed[21–23]; however, in this system with a misfit of 3 percent, Nb exhibits a strong affinity for O as well as for Al.

The comparison of the observed and simulated images demonstrates that the (unrelaxed) structure model I with oxygen ions forming the outermost plane of CdO is very close to the structure of the Ag/CdO interface produced by internal oxidation. It should be noted that the simulated images in Fig. 8 cannot be distinguished from images where Ag is replaced by Cd and *vice versa* because of the very similar electron scattering potentials of Ag and Cd; that is, from these observations it could be possible that the terminating plane on the Ag side consists of Cd atoms located at positions of Ag atoms. However, such a structure will not be stable because the strong O–Cd bond will force the Cd atoms to occupy positions of the oxide lattice and produce an interface with structure equivalent to the one in model II which was not observed.

The chemical composition of the Ag/CdO phase boundary observed in this study appears to be reasonable because (1) the oxide precipitates were formed in an environment of oxygen excess (high oxygen activity) and because (2) the Cd–O bond is by far the strongest of the A–B bonds in this system so that essentially all free Cd bonds and hence, also the Cd bonds at the Ag/CdO interface will be saturated with oxygen.[24] Obviously, the existence of an interfacial oxygen layer common to CdO as well as Ag results in a low-energy configuration of the Ag/CdO phase boundary with close similarities to the case of free surfaces exposed to an oxygen-containing atmosphere where the segregation of oxygen reduces the surface energy. The chemical composition is not changed on either side of an interfacial step since the step heights are equal to or multiples of the spacing of $\{111\}_{CdO}$ planes containing a layer of Cd and O. These steps can be interpreted unequivocally as growth steps with step lines parallel to close-packed directions and equivalent to growth steps at {111} surfaces of f.c.c. crystals. CdO crystals grown from the vapor phase preferentially exhibit surface facets parallel to {100} and {110}.[24] In the Ag matrix, the formation of facets parallel to the polar {111} planes again indicate a low interfacial energy of {111} Ag/CdO phase boundaries. Finally, it is worth noting that the O–Ag bond seems to be strong enough to cause a bending of Ag lattice planes adjacent to interfacial steps.

The observations in this study are complemented by results of hydrogen trapping experiments in Pd and Al_2O_3 precipitates produced by internal oxidation.[19] Strong irreversible traps for hydrogen atoms could be related to the formation of O-H bonds at the metal/oxide phase boundary. This result confirms suggestions that a limiting oxygen layer is always present at metal/oxide interfaces produced in oxygen excess environments, even in the case of strong metal-metal bonds as in the example described above where a strong Pd–Al bond can be assumed from the heat of solution of Al in Pd.

SUMMARY

This paper summarizes the possible applications of high-resolution electron microscopy of ceramic materials. It is important to have an instruments with the best possible resolution (small values of d_1). Reducing d_1 (below 0.2nm) gives a better insight into the atomic structure of lattice defects. With the instruments available today, in particular the middle high-voltage instruments (400kV), and with the use of precision pole pieces, it is expected to improve the resolution of instruments to about 0.15μm. With this resolution and by avoiding all interference effects of the microscope, important information on the atomic structure of dislocations, grain boundaries and phase boundaries can be obtained for the first time. This information can be utilized to explain various material properties.

REFERENCES

[1] J. C. H. Spence, Experimental High-Resolution Electron Microscopy (2nd Ed.), Oxford University Press, New York and Oxford (1988).

[2] P. Buseck, J. Cowley, L. Eyring (eds.), High-Resolution Transmission Electron Microscopy, Oxford University Press, New York and Oxford (1988).

[3] M. A. O'Keefe, Electron Image Simulation: A Complementary Processing Technique, Electron Optical Systems, SEM, Inc., AFM O'Hare (Chicago) IL (1985) p. 209–220.

[4] P. A. Stadelman, Simulation of High Resolution Images, Ultramicroscopy 21 (1987) 138–147.

[5] J. J. Hren, J. I. Goldstein and D. C. Joy (eds.), Introduction to Analytical Electron Microscopy, Plenum Press, New York and London (1979).

[6] D. C. Joy, A. D. Romig, Jr. and J. I. Goldstein (eds.), Principles of Analytical Electron Microscopy, Plenum Press, New York and London (1986).

[7] R. F. Egerton, Electron Energy-Loss Spectroscopy, Plenum Press, New York and London (1986).

[8] P. R. Buseck (ed.), High Resolution Electron Microscopy, Ultramicroscopy 18 (1985) Nos. 1–4.

[9] T. E. Mitchell and P. K. Davies (eds.), Electron Microscopy of Ceramic Superconductors, J. Electron. Micr. Tech. 8 (1988) No. 3.

[10] J. Barbier, K. Hiraga, L. C. Otero-Diaz, T. J. White, T. B. Williams and B. G. Hyde, Electron Microscope Studies of Some Inorganic and Mineral Oxide and Sulphide Systems, Ultramicroscopy 18 (1985) 211–234.

[11] K. L. Merkle, HREM of Grain Boundaries in Oxides, in Mat. Res. Soc. Symp. 82 (1987) 383–402.

[12] A. Bourret, Atomic Structure of <011> and <001> Pure Tilt Grain Boundaries in Germanium and Silicon, J. de Physique 46 (1985) C4–27—C4–38.

[13] D. Wolf, On the Stability of (001) Twist Boundaries in MgO: A Theoretical Study, J. de Physique 43 (1982) C6–45—C6–64.

[14] D. M. Duffy, Grain Boundaries in Ionic Crystals, in: Condensed Matter Structure (A. M. Stonham, ed.) Adam Hilger, Bristol (1987) p. 143–162.

[15] K. L. Merkle and D. J. Smith, Atomic Structure of a Symmetric Tilt Boundary in NiO, Phys. Rev. Letters 59 (1987) 2887–2890.

[16] C. Wagner, Theoretical Analysis of the Diffusion Processes Determining the Oxidation Rate of Alloys, Z. Elektrochemie 63 (1959) 772–791.

[17] J. L. Meijering, Internal Oxidation of Metals, in Adv. in Materials Research (H. Herman, ed.) Vol. 5 (1971) p. 1–81.

[18] D. J. Pedder, The Thermal Stability of Cadmium Oxide in Internally Oxidized Silver-Cadmium Alloys, Met. Trans (AIME) 9A (1987) 659–670.

[19] W. Krakow, J. T. Wetzel and D. A. Smith, Observation and Interpretation of the Atomic Structure of a $\Sigma 19/[110]$ $(3\bar{3}1)$ Tilt Boundary in Au Thin Films, Phil. Mag. A53 (1986) 739.

[20] X. Y. Huang, W. Mader, J. A. Eastman and R. Kirchheim, Modulating the Copmpositionof a Metal/Oxide Interface and Using Hydrogen as a Probe, Scripta Metall. 22 (1988) 1105.

[21] W. Mader, TEM Analysis of Interfaces Between Nb and Al_2O_3 Precipitates, Mat. Res. Soc. Sym. Proc. 82 (1987) 403–408.

[22] M. Kuwabara, J. C. H. Spence and M. Rühle, HREM Studies of Al_2O_3 Precipitates in Nb, Proc. 46th EMSA Meeting (G. W. Bailey, ed.) San Francisco Press (1988) 732–733.

[23] M. Kuwabara, J. C. H. Spence and M. Rühle, Growth Processes of Al_2O_3 Precipitates in Nb, J. Mat. Science, in press.

[24] V. Kovák, P. Höschl, J. Dillinger and V. Prosser, CdO Particles Grown Out of the Vapour Phase, in: Crystal Growth, Supplement to Physics and Chemistry of Solids (H. S. Peiser, ed.) Pergamon Press, Oxford (1967) 341–348.

ULTRAMICROTOMY OF POROUS CERAMICS

Helen M. Kerch and Rosario A. Gerhardt
Department of Ceramics
P. O. Box 909
Rutgers University
Piscataway, N. J. 08855-1179

ABSTRACT

The method of ultramicrotomy was explored for potential use in the microstructural characterization of porous ceramics. The microstructure of 10 wt %: 90 wt % colloidal silica: potassium silicate gels were examined in the transmission electron microscope. Gel monoliths ranging from 14.5% to 40.5% theoretical density were infiltrated with Spurr epoxy [1]. All thin sectioning was done with a biological grade diamond knife. Electron micrographs reveal the internal structure of the silica particles. Two types of pores are evident, a set of large pores which connect loosely aggregated clusters of silica particles within which a smaller set of pores reside (viz. interdomain and intradomain pores). Changes in microstructure as a function of density are also clearly discerned.

INTRODUCTION

Ultramicrotomy is a sample preparation technique for transmision electron microscopy whereby the sample itself is rendered electron transparent (<100 nm thick) by the cutting action of a diamond knife against a resin embedded specimen[2]. Utilized primarily by the biological community, this preparation technique can find further applications in the field of ceramics. Ultramicrotomy has previously been employed to reveal the structure of zeolite catalysts [3] as well as the microstructure of TiO_2 and γ- alumina agglomerates [4]. The current investigators chose ultramicrotomy as a way to examine the highly porous structure of silica gels sintered to various densities.

Porosity is a common constituent of most ceramic materials at some stage in their fabrication. There is a great interest in their behavior and subsequent elimination during sintering. Conversely, for other ceramic materials the porosity is a deliberate and final element of the finished piece. Examples of the latter case include catalytic converters, filter materials, and porous fiber optic sensors. For either circumstance, it is necessary to characterize the pores according to their size, distribution, connectivity, and shape.

Two conventional procedures which are employed to examine the pore structure of materials are mercury penetration porosimetry (pore sizes> 50 nm) and nitrogen sorption technique (pore sizes < 50 nm). Interpretation of the data obtained by these techniques require assuming a model for the pore structure. On the other hand, microscopy provides direct observation of the porous media . The largest obstacle in microscopic techniques lies in sample preparation, image interpretation, and statistical quantification of the data.

A variety of microscopic sample preparation methods for porous materials have been explored. They include ultramicrotomy [3, 4, 5], powder dispersion [6], and flake formation [5]. Other workers [7, 8, 9] have utilized the technique of Pt/C replication. Ultramicrotomy was chosen for this investigation because it offered several advantages. First, the internal structure of the particles and the pores contained within can be observed without disruption. The entire sample may be viewed in contrast to other thinning techniques where only portions made electron transparent can be examined. Because the sections produced with ultramicrotomy are uniformly thin, density gradients in the sample are clearly discerned. If serial sections are collected, 3-dimensional reconstruction of the microstructure is possible.

In order for the technique of ultramicrotomy to be successful, a number of specimen and/or epoxy requirements must be fulfilled. The sample must possess open porosity so that the impregnant can infiltrate the pore space. Epoxy properties should include low viscosity and low curing temperature, and a long setting time. The mechanical properties of the cured epoxy must include a balance between rigidity and ductility in order that a coherent thin section can be produced during cutting. The microstructure of the infiltrant must not obscure that of the specimen. Finally, a strong bond must be formed along the epoxy/specimen interface in order to prevent ripping or tearing [2].

PROCEDURE

Gel Preparation

The material chosen for the investigation was a colloidal silica*: potassium silicate** gel as first described by Shoup [10]. The gel may be processed with varying ratios of colloidal silica (A) to the potassium silicate (B) depending on the properties desired. The compositional ratio of 10 wt % A: 90 wt % B was selected because conventional methods

* LUDOX HS-40 **Kasil #1

indicate that it possessed a narrow pore size distribution and a large average pore size (250 nm). A detailed description of the gel processing can be found elsewhere [11].

In order to insure representative microstructures, vertical and horizontal X-ray tomographic analysis [12] was performed on the unsintered gel. No detectable density gradients were found within the 1μm resolution limit of the technique.

The gels prepared for this study were sintered isochronally in air for 1 hour. The holding temperatures ranged between 1150 C - 1350 C with 50 C intervals. A list of the samples with their respective heat treatments is given in Table I. Density of the monoliths is also shown.

Table 1. Sintering schedule and density data for 10 wt% A: 90 wt% B gels.

Temperature (C)	Density (g/cc)*	% Theoretical Density
unsintered	.320	14.5
1150	.368	16.7
1200	.368	16.7
1250	.404	18.4
1300	.486	22.1
1350	.890	40.5

* Bulk density obtained from geometrical measurements.

Preparation for Ultramicrotomy

The gel monoliths were cut into equivalent sized disks with a diamond blade using a Buehler Ltd. Isomet Low Speed

Saw. A portion for microscopic analysis was then taken from the center of the disk which itself was taken from the median of the monolith. Spurr epoxy [1] was chosen as the infiltrant.

Before the sample is impregnated, it must first be dehydrated as Spurr epoxy is not miscible with water. Dehydration was accomplished by subjecting the gel to 3 acetone baths each with a 2 hour duration. Following this, the sample is placed in 4 Spurr epoxy baths each with a higher viscosity than the preceeding bath. The viscosity is varied with propylene oxide. The epoxy:propylene oxide mixture is kept fluid by the motion of a rotator. After the final bath, the sample and fresh 100% epoxy was placed in a size 00 BEEM capsule. Curing was done in a vacuum oven at 60 C for 48 hours. After the specimen block is hardened, its tip must be trimmed down to a trapezoidal shape with approximate dimensions of 1 mm. This size and particular shape aid in cutting coherent sections.

All thin sections were made with an R.M.C. MT-2 Ultramicrotome used in conjunction with a D.D.K. diamond knife. The wetting agent employed was distilled water. The sections were collected directly onto carbon coated T.E.M. grids. Micrographs were taken in bright field mode using a JEOL 100CX microscope.

RESULTS AND DISCUSSION

The micrographs shown in Figure 1 detail the evolution of microstructure with increasing sintering temperature. The lighter regions correspond to the epoxy phase while the darker regions represent the silica gel. There are no discontinuities in the sections indicating complete infiltration which is a direct result of the interconnected porosity of the material.

As can be seen in Figure 1, the ultramicrotomed thin sections reveal the internal structure of the aggregated silica particles. The silica gel appears to possess two species of pores; those with larger dimension located between the silica gel clusters and smaller pores situated between the silica particles themselves. These shall be designated as interdomain and intradomain pores, respectively. Such a domain model was previously proposed by Cao et. al. [11] based on a combination of Hg porosimetry and nitrogen desorption measurements. It is interesting to note that the pore diameters taken from T. E. M. micrographs show the interdomain and intradomain pores to have values of approximately 1000 nm and 100 nm respectively in the unsintered state. These values are substantially higher than those obtained from Hg porosimetry and nitrogen desorption measurements. The reasons for such a discrepancy require further investigation.

The microstructural features do not appear to change substantially between the unsintered samples and those fired up to 1200 C, although some particle rearrangement has occurred. The silica aggregates exhibit a fine texture in this temperature range which is shown in Figure 1 (a) for the unsintered case. This fine structure no longer appears in gels sintered to 1250 C (Figure 1 (b)), and their elimination coincides with an increase in density (see Table I).

The silica phase as a whole retains its interconnected structure during densification; however, the particles themselves become larger and more spherical in shape. Between 1300 C (Figure 1 (c)) and 1350 C (Figure 1 (d)) the silica gel experiences an almost two-fold increase in density which seems to be associated with particle growth and closing off of intradomain pores. Nevertheless, the microstructure still shows two distinct types of pores up to 40.5% theoretical density. SANS measurements [13] on similar materials show that a wide pore size distribution exists even when samples have achieved 70 % theoretical density. This supports the contention that a second generation of intradomain pores are being created as sintering proceeds.

Further evaluation of the microstructure of porous silica gels and the method for statistical quantification of the microstructural changes related to thermal processing will be discussed in subsequent publications.

CONCLUSIONS

Ultramicrotomy is a viable technique for the investigation of the microstructure of porous materials. This has been successfully demonstrated with a series of porous silica gels sintered to various densities. Micrographs reveal the nature of the internal structure of the silica gel, and how particle and pore morphology change with heat treatment.

ACKNOWLEDGEMENTS

The authors are indebted to John Grazul, Dept. of Biology, Rutgers University for his advise and expertise in the area of ultramicrotomy. Special thanks to Michelle Levin for her help in sample preparation. The support of the Fiber Optic Materials Research Program is gratefully acknowledged.

REFERENCES

[1]A.R. Spurr, "A Low Viscosity Epoxy Resin Embedding Medium for Electron Microscopy," *J. Ultrastruct. Res.,* **26** [1] 31- 43 (1969).

[2]P. Smart and N. K. Tovey, *Electron Microscopy of Soils and Sediments: Techniques.* Clarendon Press, Oxford, 1982.

[3]R. Csencsits and R. Gronsky, "Preparation of Zeolites for TEM using Microtomy," Specimen Preparation for Transmission Electron Microscopy of Materials , Vol. 115. Edited by J. C. Bravman, R. N. Anderson, and M. L. McDonald. M.R.S., Pittsburgh, PA. , 1988. pp. 1-6.

[4]D. G. Pickles and E. Lilley, "Ultramicrotoming of Ceramic Powders for Electron Microscopy," *J. Am. Ceram. Soc.* , **68** [9] C-222 - C-223 (1985).

[5]P. H. Tewari, A. J. Hunt, K. D. Lofftus, and J. G. Lieber, "Microstructural Studies of Transparent Silica Gels and Aerogels," Better Ceramics Through Chemistry II, Vol. 73. Edited by C. J. Brinker, D. E. Clark, D. R. Ulrich. M. R. S., Pittsburgh, PA., 1986. pp. 195-205.

[6]J. H. Mazur and C. M. Lampert, "High Resolution Electron63 Microscopy Study of Silica Aerogel Transparent Insulation," Optical Materials Technology for Energy Efficiency and Solar Energy Conversion III, Vol. 502, S.P.I.E., 1984. pp. 123-127.

[7]G. Ruben and M. W. Shafer, "Stereo-TEM Imaging of Sol-Gel Glass Surfaces," Better Ceramics Through Chemistry II, Vol. 73. Edited by C. J.Brinker, D. E. Clark, D. R. Ulrich. M. R. S., Pittsburgh, PA., 1986. pp. 207-212.

[8]M. W. Shafer, V. Castano, W. Krakow, R. A. Figat, and G. C. Ruben, "Structural Observations of Porous Silica Gels," Better Ceramics Through Chemistry II, Vol. 73. Edited by C. J. Brinker, D. E. Clark, D. R. Ulrich. M. R. S., Pittsburgh, PA., 1986. pp. 331-336.

[9]M. W. Shafer, D. D. Awshalom, J. Warrick, and G. Ruben, "The Chemistry of and Physics with Porous Sol-Gel Glasses," *J. Appl. Phys.*, **61** [12] 5438-5446 (1987).

[10]R. D. Shoup, "Controlled Pore Silica Bodies Gelled from Silica Sol-Alkali Silicate Mixtures," *Colloid and Interface Science*, **3** 63-69 (1976).

[11]W. Cao, R. Gerhardt, and J. B. Wachtman, Jr., "Preparation and Sintering of Colloidal Silica-Potassium Silicate Gels," *J. Am. Ceram. Soc.*, **71** [12] in press (1988).

[12]W. J. Glantschnig and A. Holliday, "Mass fraction profiling based on x-ray tomography and its application to characterizing porous silica boules," *Applied Optics,* **26** [6] 983-989 (1987).

[13]G. G. Long, S. T. Krueger, R. A. Gerhardt, in this proceedings.

Figure 1. Ultramicrotomed sections of 10 wt% A: 90 wt% B gels: (a) unsintered sample; (b) 1250 C/ 1 hr.; (c) 1300 C/ 1 hr.; (d) 1350 C/ 1hr.

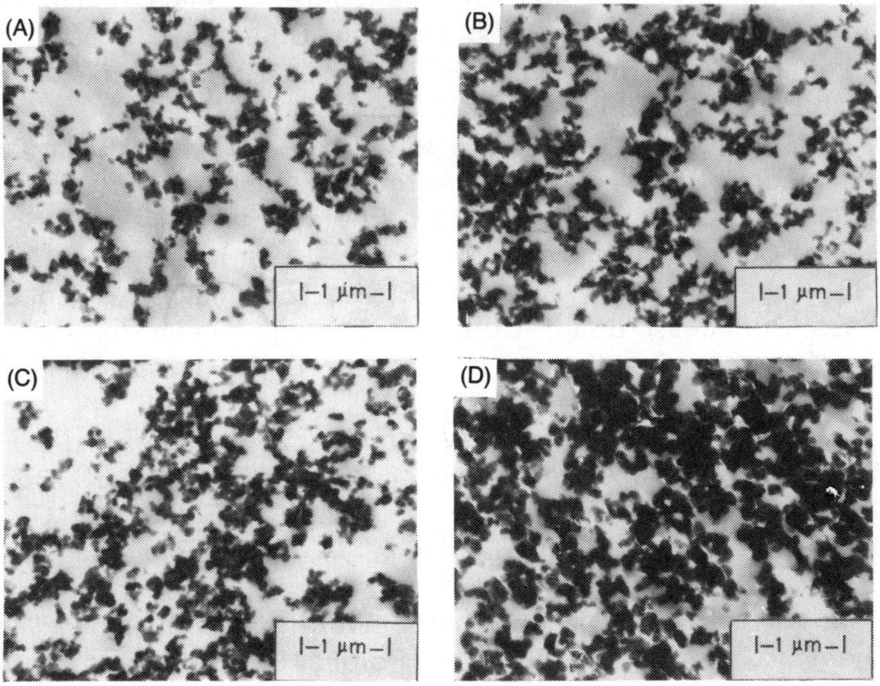

Section V. Optical Techniques

PHOTODEFLECTION AND PHOTOCONDUCTIVITY STUDIES IN CERAMICS

C. H. Seager
Sandia National Laboratories, Albuquerque, N.M. 87185

ABSTRACT

Sensitive calorimetric technques such as photothermal deflection spectroscopy (PDS) offer many advantages over conventional transmission spectroscopy for measuring the optical properties of ceramics. Insensitivity to light scattering is of particular importance for optical studies of these materials. The physics of the PDS method will be briefly reviewed as well as the experimental arrangement necessary to carry out sensitive (3 -5 ppm) determinations of absorptance on ceramic specimens over a wide range of photon energies. Examples will be given of PDS studies on PLZT and doped ZnO ceramics. In the latter case coupled PDS/photoconductivity measurements have given new insights into the role of impurities added to optimize the electronic properties of ZnO varistors.

I. INTRODUCTION

In this review we will discuss the application of photothermal deflection (PDS) and photocurrent spectroscopies to studying the properties of several ceramic materials used in electronic applications. Our purpose is to provide a brief tutorial on the use of the PDS technique and to emphasize the advantages that it has in the determination of the optical properties of weakly absorbing thin films and /or highly scattering polycrystalline bulk specimens which are difficult to polish and measure by traditional transmission techniques. These same circumstances occur frequently in the study of many types of ceramics; indeed, it can be argued that PDS is the most appropriate optical technique available to the ceramicist. The materials discussed here were chosen from among those studied by the author and his coworkers while developing a working PDS system at Sandia National Laboratories. While they represent investigations primarily aimed at extracting information about electronic properties, these methods are equally useful for obtaining knowledge about the chemical or compositional makeup of ceramic specimens.

II. THE PHYSICS AND OPERATIONAL DETAILS OF PHOTOTHERMAL DEFLECTION SPECTROSCOPY

The PDS method is the most sensitve of the common calorimetric spectroscopies. Others in this group include photoacoustic spectroscopy and thermal lensing techniques. It was developed in the early 1980's by Boccara et al.[1] and by Jackson and co-workers[2] at the University of California at Berkeley. A calorimetric spectroscopy relies on the fact that, in most cases, the majority of optical energy absorbed by a solid appears as heat rather than being re-emitted as optical energy (luminescence). Thus detection of photon absorption is accomplished by monitoring the temperature of the sample, rather than comparing intensities of an incident and a transmitted light beam as is done in more traditional transmission spectroscopies. When the transmitted beam intensity is greater than, say, .999 of the incident beam minus the reflected beam, system noise, scattering of the beam as it passes through the sample, and the difficulty of accurately measuring the reflection coefficient make the transmission technique a poor choice for accurately estimating absorption. Even when the transmitted intensity is considerably less than the incident beam, scattering due, for instance, to polycrystallinity can result in large errors. So, for weakly absorbing, roughly polished, or polycrystalline specimens, calorimetric spectroscopies are often the preferred techniques.

Figure 1 is a schematic diagram of the author's apparatus for generating and analyzing PDS signals. The sample of interest is suspended in a high purity quartz cuvette filled with an organic liquid such as carbon tetrachloride or pentane. This liquid has been carefully filtered to remove particulates. A low power, high stability HeNe laser is focussed to produce a narrow (30-50 micrometer waist) beam which is parallel to the sample surface and just touches it. A partially masked photocell is positioned further along the beam path to detect deflections of the beam as it passes through the layer of liquid near the sample surface. Chopped monochromatic light, if absorbed by the sample, produces periodic variations in the surface temperature of the sample and results in an oscillatory index of refraction gradient in the organic liquid due to the small, heat induced density fluctuations. The resulting angle of deflection of the probe (HeNe) laser is given by:

$$\theta = w/N \, (dN/dT) \, (dT/dk)$$

where N is the refractive index of the liquid, w is the sample width, and k is the direction normal to the sample surface. In general, θ is proportional to the temperature rise at the sample surface. In measuring θ with a lock-in amplifier the issue of the phase of this ac deflection becomes important. It is possible to define a quantity called the thermal length in both liquid and the sample. This length is just a measure of the distance a heat pulse will propagate in a time period equal to one over the frequency of the chopped heating light. If the heating light is absorbed in a sample thin compared to its thermal length or, for a thick sample, within a fraction of the thermal length from the front surface, θ will always be in phase with the chopped light. More generally, absorption may occur at depths smaller or larger than a thermal length, depending on photon energy. In this case θ will show a variable phase which must be measured to obtain the correct optical properties of the specimen. In these cases knowlege of the thermal properties of the specimen are required and solution of the full differential heat flow equation must be employed. The detailed properties of these solutions have been explored by Rosenczwaig et al[3]. Fortunately, the vast majority of specimens can be measured by simply comparing their θ (hν) spectra at a single phase with the θ(hν) response of a suitable calibration specimen. Such a calibration specimen can be made by applying a flat black coating to a piece of sample material, itself, or to a quartz blank. By taking the ratio of these spectra, the thermal properties of the sample as well as the spectral density function of the lamp/monochromator/focussing lens combination are automatically taken into account. In the following sections we will explain how the absorption coefficient was extracted from sample and reference spectra for each of the materials discussed.

In a practical sense, the difficulty in making PDS measurements lies in the present day unavailability of commercially available, turnkey systems. The individual components necessary for a high quality apparatus are available, however , and the few components that are not easily purchased can be fabricated in a modestly equipped workshop. The first requirement is a temperature stabilized HeNe laser (preferably single mode) having minimal amplitude fluctuations and pointing noise. Since the detected signal is an angular deviation of the laser beam, the latter is of tantamount importance. Some of the amplitude and pointing fluctuations can be removed by the split beam path, dual detector scheme shown in figure 1. This method uses a differential amplifier to cancel the amplitude and pointing noise seen by the main and split beam detectors. Not all of this noise is removed however, because of some beam occlusion by the sample. Major noise reductions are acheived with the entire apparatus being isolated from room vibrations on an air table and from air currents (which bend the probe laser beam) by enclosing the major components in an airtight box. Of equal importance is extensive filtration of the deflecting liquid and the procurement of high brightness light sources and high throughput monochromators. On-line monitoring of the heating beam intensity is also desirable to ensure that the reference and sample runs are corrected for any change in the heating light intensity due to dust, lamp ageing,etc.. Extensive computerized processing of the sample and reference signals is, of course, useful in enhancing sample throughput.

III. OPTICAL ABSORPTION IN LEAD LANTHANUM ZIRCONATE TITANATE CERAMICS

Lead lanthanum zirconate titanate (PLZT) ceramics hold considerable promise for use in optical image storage devices. While image storage in unmodified PLZT devices requires illumination with near band gap light (3.35eV) and is rather inefficient, it has been recently been demonstrated[4] that implanting the surface of these materials with various species of ions permits the use of imaging light in the visible part of the spectrum and greatly enhances the imaging efficiency. Peercy and Land[5] have demonstrated that these effects can be explained by assuming that much of the voltage applied across the PLZT during the imaging process is dropped across the highly resistive implanted layer. Illumination and subsequent photoconductivity in this region "switches" a large fraction of the total applied voltage onto the bulk of the material, thereby causing rapid field reorientation in the ferroelectric grains. The difference in domain ordering between illuminated and dark regions scatters "viewing light" subsequently passed through the PLZT plate to reveal the stored image. In the studies to be discussed here the author and C. E. Land used the PDS technique to make the first measurements of the absorption induced by implanting PLZT plates with Al and Ne ions.

Hot pressed PLZT ceramic plates of several different grain sizes having thicknesses around 0.025 cm were obtained from a commercial supplier. The nominal chemical composition was $Pb_{0.93}La_{0.07}(Zr_{0.65}Ti_{0.35})_{0.98}O_3$. Implants of Al ($5\times10^{14}$450 keV Al/cm^2), Ne ($10^{15}$500 keV Ne/cm^2), and co-implants of Al and Ne at these fluences and energies were performed on several samples. The implants were followed by a 200 C air anneal for 1 hr in air to simulate the thermal cycling experienced by actual PLZT devices, which have a layer of indium-tin oxide (ITO) deposited on each surface as an electrical contact. No ITO was present on any of the samples measured in this study.

By measuring the ratio of the PDS signals with the laser beam on the same and on the opposite side of the sample wafers as the heating light, the thermal length was found to be about 0.01 cm at the 19 Hz chopping frequency. Because sub-band gap absorption of unimplanted PLZT is fairly weak, virgin PLZT plates showed a PDS signal whose phase varied considerably over the 2 to 4 eV photon energy range. As mentioned before this is expected when the absorption depth is comparable to the thermal length. Under these conditions evaluation of the full solution of the heat flow equations is necessary to deduce the bulk absorption coefficient. Seager and Land[6] discuss this procedure in detail; in this review we will concentrate only on deducing the additional absorptance (unitless) created by the implantation process. Because our PLZT plates are several thermal lengths thick, only a small fraction of the heat absorbed on the side facing the heating beam propagates to the reverse side of the PLZT plate. So, by measuring the PDS signal (at the phase appropriate to absorptance at the surface) originating from an implanted sample, first with the probe beam on the front and then on the back side of the plate, we can use the following expression to deduce the absorptance of the thin, implanted layer[6]:

$$A = \log_e \left[\frac{T_I T_U - I T_U}{T_I T_U \quad U T_I} \right] \qquad (2)$$

where I and U are the PDS deflection signals measured on the implanted and unimplanted surfaces and T_U and T_I are the respective transmissivities of the unimplanted and implanted surfaces. These latter quantities were determined by measuring the above band gap (>3.4 eV) PDS response on each side of an implanted plate and comparing it to a reference sample made by depositing a flat black coating on a virgin PLZT plate.

The surface layer absorptance calculated from equation 2 is shown in figure 2 for PLZT plate implanted with Ne. The surface absorptance is independent of the average grain size while we have previously shown[6] that the bulk absorption coefficient varies considerably with this parameter (not shown). Samples implanted with Al and co-implanted with both species show essentially the same surface absorptance, suggesting a common defect may exist in large numbers in all the implanted samples. Since the titanates can tolerate substantial variations in the oxygen/metal ratio, defects related to oxygen loss might be good candidates for this effect. It is also possible that amorphization of a fraction of the surface region has taken place; the rather featureless, Urbach-like form of the data in figure 2 is frequently observed in amorphous materials.[7]

It is interesting to examine how this measured surface absorptance correlates with the enhanced imaging efficiency observed in implanted PLZT devices. We have previously observed, from optical studies on unimplanted PLZT, that the absorptance of virgin PLZT plates is very small (less than .0025) at photon energies less than 3 eV. This means that imaging light in this energy range is essentially uniformly absorbed, producing little photo-induced field rearrangement and a correspondingly small imaging efficiency. However as figure 2 illustrates, the thin (< 1 micron thick) implanted layer absorbs 1/10 of the incident light in an implanted PLZT plate; if this layer drops a substantial fraction of the applied voltage in the dark, considerable voltage rearrangement might be produced by illumination. In fact the imaging efficiency data of Peercy and Land[8] (shown as circles in figure 3) exhibits a spectral dependence which is quite similar to the absorptance in the implanted layer. This agreement appears to be strong supporting evidence for the photoconductive imaging model of Peercy and Land and suggests that exploring alternative implantation processes which enhance this surface absorptance might lead to improved PLZT device efficiencies.

V. ABSORPTION AND PHOTOCONDUCTIVITY STUDIES IN ZnO VARISTORS

ZnO varistors are polycrystalline, n-type, semiconducting ceramics which are electrically insulating at low applied voltages, but become good conductors with a very non-linear current/voltage relationship at high voltages. Their electrical properties are reminiscent of back-to-back Zener diodes, except that typical switching voltages range from 1 to 100 kV per centimeter of material. It is well established that their electrical characteristics are controlled by the electrostatic potential barrier, double depletion layer structure which exists at the grain boundaries.[9,10] Empirically it is found that percent levels of two types of impurities must be added to the polycrystalline ZnO to produce devices with good switching characteristics. First a large atom, such as Bi, is required. Bismuth is located in the grain boundaries and apparently increases the density of deep, grain boundary trap states and hence raises the equilibrium potential barrier height. A transition metal additive, typically Co or Mn, is also needed to raise the current/voltage non-linearity at high voltages, but the mechanism for this is not known. Electron spin resonance measurements have established that these metals are fairly soluble and substitute for the Zn in the ZnO lattice.[11,12]

In this section we will discuss the way that combined PDS and photoconductivity data have been used to suggest a novel mechanism for the effect of these transition additives on varistor properties. While measurements have been carried out on ZnO doped with several transtion metals[13], only the data on Co doped ZnO will be reviewed here.

The varistors in this study were prepared from ZnO powder containing dissolved Co. The preparation method involved precipitation of varistor constituents from a chloride solution as oxalates which were then converted to oxides by heating at 600 C. The powders were subsequently sintered with 0.5 mol% Bi_2O_3 at temperatures near 725 C.

PDS measurements were made by abrasively thinning the specimens to about 50 microns after they were bonded to a quartz substrate with a clear cyanoacrylate adhesive. Because this thickness is less than the 19 Hz thermal length in ceramic ZnO, PDS signals at all wavelengths were at a single phase relative to the chopper. By measuring the PDS signals at photon energies above the ZnO energy gap (3.2 eV), where all the incident light was absorbed, the thermal respose of the bonded specimens is established; the absorption coefficient is then obtained by comparing the response at an arbitrary wavelength with that seen above 3.2 eV.

In figure 3 the absorption coefficient is shown for a varistor doped with 1 atomic % Co. Also shown for comparison are the curves for polycrystalline ZnO and ZnO doped only with Bi. The varistor absorption is very similar to that measured in single crystal ZnO doped with Co, and the small peaks in the absorption spectra have been labelled as in the work of Pappalardo et al.[14] Each of these peaks is due to a transition from the ground state of Co_{2+} to a localized excited state. The photoexcited electron on the Co_{2+} can either decay to a lower energy state or it can escape to the conduction band. There is evidence for both effects. When the incident photon energy is 1.96 eV (near peak D), two photoluminescent peaks have been observed.[15] The peak energies are 1.77 and 1.44 eV and result from Stokes shifted transitions from the D state to the ground and the S state, respectively.

Ionization of the Co_{2+} may be inferred from the

photoconductivity spectrum shown in figure 4. This spectrum was taken with a constant dc voltage applied to the varistor sufficient to bias it into its nonlinear regime. Also shown in figure 4 is the absorption data plotted as the fraction of photons absorbed in the 50 micron thick specimens (assuming none reflected). There is a clear coincidence of the photoconductivity rise with the absorption edge for peak D. Thus, although this peak is associated with transitions to the bound state of Co_{2+}, this state must be easily ionized. Its energy is probably resonant with the conduction band states which implies that the ground state is less than 1.9 eV below the conduction band edge.

This data suggests a mechanism for the large varistor I-V nonlinearities which has not been considered before. If photons of sufficent energy can ionize transition metal additives, then hot electrons in the grain boundary depletion region could also. Indeed, sub-bandgap electroluminescence peaks in both Co and Mn doped varistors have recently[15,16] been reported, showing that electrons are exciting these additives. The effect of ionizing just a small fraction of these impurities can be a significant alteration of the net charge density in the depletion region, since Co is present at the percent level while the backround space charge density (from oxygen vacancies which act as shallow donors) is a few ppm. Recent calculations by Pike et al.[13] show that this increased space charge density can greatly enhance the rate of collapse of the space charge barrier height which occurs at large fields as the grain boundary traps are filled. Calculations have also shown that the ordinary trap filling mechanism (without the presence of extra ionized impurities in the depletion region) can only lead to rather small non-linearity coefficients and this is in accord with measurements on undoped ZnO varistors.

It is thus fair to say that the use of PDS and photoconductivity measurements has given new insights into the effect of transition metals in varistors and may point the way to the design of better devices in the future.

ACKNOWLEDGEMENTS

The author would like to acknowledge the

contributions of C. E. Land, G. E. Pike, R. G. Dosch, and W. R. Bayless in carrying out the research reviewed here.

contributions of C. E. Land, G. E. Pike, R. G. Dosch, and W. R. Bayless in carrying out the research reviewed here.

REFERENCES

1. A. C. Boccara, D. Fournier, and J. Badoz, Appl. Phys. Lett. 36 , 130(1980).
2. W. B. Jackson, N. M. Amer, A. C. Boccara, and D. Fournier, Appl. Opt. 20 , 1333(1981).
3. A. Rosencwaig and A. Gersho, J. Appl. Phys. 47 , 64(1976).
4. C. E. Land and P. S. Peercy, Ferroelectrics 38 , 947(1981).
5. P. S. Peercy and C. E. Land, Appl. Phys. Lett. 37 , 815(1980).
6. C. H. Seager and C. E. Land, Appl. Phys. Lett. 45 , 395(1984).
7. See for example the discussion in Electronic Processes in Non-Crystalline Materials ,edited by N. F. Mott and E. A. Davis (Clarendon, Oxford),p. 273.
8. P. S. Peercy and C. E. Land, Nucl. Instrum. Methods 209/210 , 1167(1983).
9. G. E. Pike, Mat. Res. Soc. Proc. 5 , 369(1982).
10. G. D. Mahan, L. M. Levinson, and H. R. Phillip, J. Appl. Phys. 50 , 2799(1979).
11. P. B. Dorain, Phys. Rev. 112 , 1058(1958).
12. A. Hausman, Phys. Stat. Sol. 31 , K131(1969).
13. G. E. Pike, C. H. Seager, and R. G. Dosch, Mat. Res. Soc. Symp. Proc. 46 , 409(1984).
14. R. Pappalardo, D. L. Wood, and R. C. Linares, J. Chem. Phys. 35 , 2041(1961).
15. G. E. Pike, S. R. Kurtz, P. L. Gourley, H. R. Phillip, and L. M. Levinson, J. Appl. Phys. 57 , 5512(19785).
16. M. Rossinelli, G. Blatter, and F. Greuter, in Electrical Ceramics , edited by B. C. H. Steele (Institute of Ceramics, Shelton, U.K., 1985), p.1.

PHOTOTHERMAL DEFLECTION APPARATUS

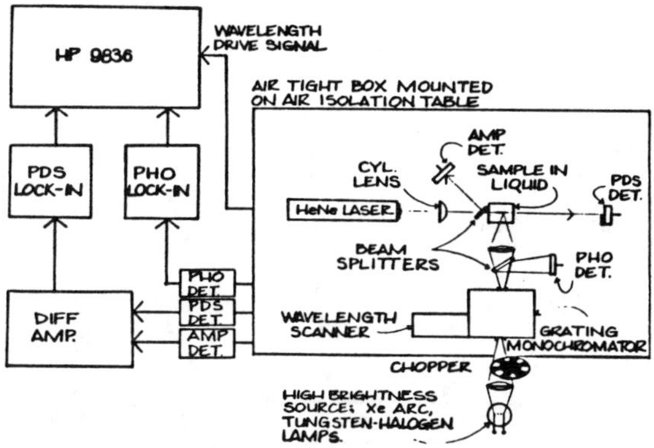

Figure 1. A schematic of the photothermal deflection apparatus. The "PHO" detector monitors the intensity of the heating light, while the "AMP" and "PDS" detectors monitor the positon and intensity of the laser beam before and after being deflected by the oscillatory temperature wave near the sample surface.

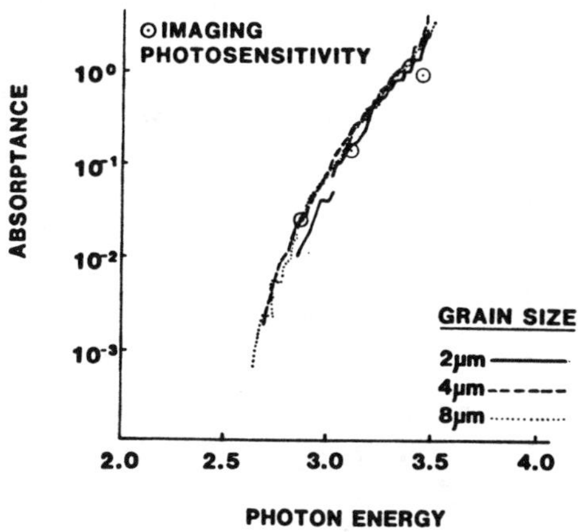

Figure 2. Surface absorptance in several plates of Ne-implanted PLZT deduced using Eq. 2. Open circles are the imaging photosensitivity data (arbitrary scale) obtained for Ne-implanted PLZT by Peercy and Land (Ref. 8).

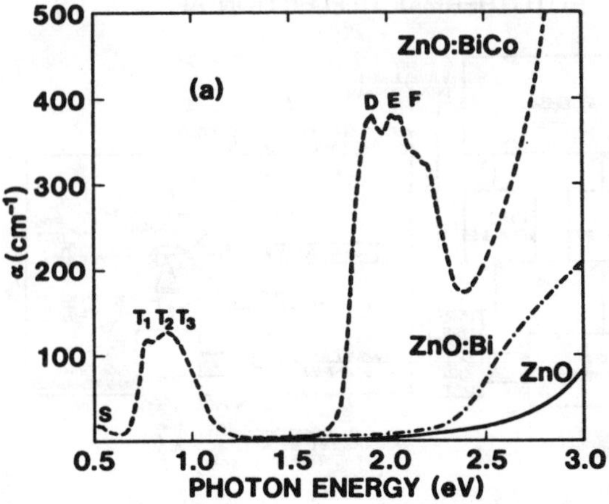

Figure 3. Optical attenuation coefficient for varistors containing pure ZnO, ZnO with 0.5 at % Bi, and ZnO with 0.5 at % Bi and 1 at % Co.

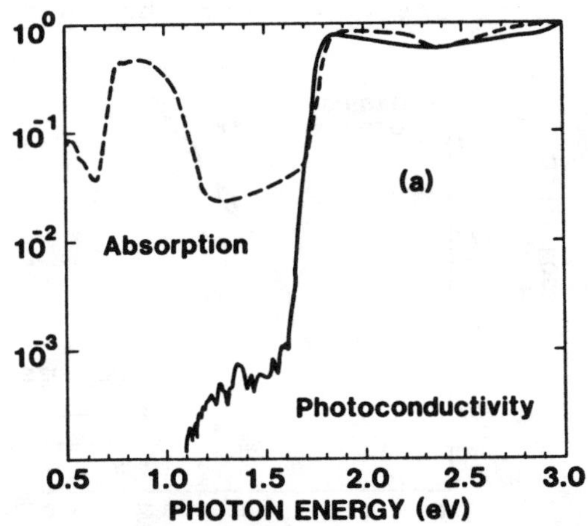

Figure 4. Photoconductivity and absorption data for varistors containing Co. While the vertical scale for the photoconductivity data is arbitrary, the absorption data is plotted as the fraction of photons absorbed in the actual sample thickness.

RAMAN STUDIES OF CHEMICAL BONDING AND LOCALIZED INTERACTIONS

Gregory J. Exarhos
Pacific Northwest Laboratory
Richland, Washington 99352

ABSTRACT

The availability of high intensity laser systems, intensified gated array detectors, and micro-optical imaging devices has led to a resurgence of laser-based molecular spectroscopic techniques for rapid non-destructive analysis of materials. The utility of Raman spectroscopy as an *in situ* probe of chemical bonding in bulk materials, dielectric films, and at interfaces will be emphasized in this review. Recent work in this laboratory concerning the response of materials to extreme conditions of pressure and temperature based upon changes in the vibrational Raman spectrum will be discussed.

INTRODUCTION

Laser Raman spectroscopy is a particularly attractive method for the characterization of materials owing to the adaptability of the measurement as an *in situ* probe and to the amount of chemical bonding information that can be extracted from measured spectra. A Raman spectrum displays the intensity of inelastically scattered light from a material as a function of frequency shift from the exciting probe laser frequency. This vibrational density of states can be separated into a low frequency region where the vibrational excitations are delocalized and strongly coupled to the lattice, and a high frequency region characteristic of more localized vibrational modes. Localized modes are relatively insensitive to structural changes in a material, while the delocalized modes can exhibit marked changes in frequency and intensity during a structural phase transformation. The temperature and pressure dependence of these "soft" modes is an important subject in spectroscopic studies of condensed phase materials.

Vibrational bands in measured spectra are characterized by a number of parameters including peak frequency, bandwidth at half maximum, peak asymmetry, and relative intensity. The degree of polarization of the scattered light contributing to a particular vibrational mode is necessary for determining the symmetry of that mode. Polarization information has been successfully used to establish molecular orientation in glass fibers [1], and to establish the degree of grain orientation in sputter-deposited titania films [2]. Based on the number of modes observed and their symmetry, the structure of a free molecule or crystal may be deduced [3,4,5]. Therefore, Raman spectra can be used for the identification of crystalline phases in a material. Relative phase composition in mixed phase systems can also be determined from analysis of band intensities intrinsic to each phase. For example, deposition of carbon films by chemical vapor or reactive sputtering methods often yields a mixture of phases. Raman measurements of such films are used to discern the diamond phase, characterized by a strong mode at 1334 cm^{-1}, from the graphitic phase (1580 cm^{-1}), and from amorphous carbon (broad features near 1350 cm^{-1} and 1600 cm^{-1}).

A recent review emphasizes the application of Raman spectroscopy to the characterization of thin film dielectrics and semiconductor materials [6]. Phase identification, homogeneity, and thickness have been obtained from the number and intensity of observed vibrational modes [7]. Such measurements are important in the characterization of thin films and film/substrate interfaces. Allred, et. al. have identified binary alloy interfacial layers in silicon-germanium multilayer stacks [8]. The average thickness of these interfacial layers was derived from deconvolution of measured Raman spectra and found to depend upon deposition method. Crystalline phase alteration near an interface in metal oxide films on silica has also been observed. Sputter-deposited ZrO_2 films on silica exhibit a transformation from a monoclinic phase to a boundary tetragonal phase near the silica/zirconia interface. The presence of interfacial phases or chemical reaction products can markedly influence the optical and electrical properties particularly of a multilayered structure. Therefore, a non-destructive molecular spectroscopic probe, such as Raman spectroscopy, is paramount for characterization of the film or multilayer structure.

The behavior of materials subjected to high temperature or high pressure or chemically reactive environments can be studied _in situ_ using Raman scattering methods. Chemical changes or phase changes resulting from the applied stress can be identified from analysis of the vibrational spectrum. In addition, variations in vibrational frequency with applied temperature or pressure can be related to bulk properties of the material. For example, the dependence of vibrational frequency on applied pressure (temperature) is proportional to the bulk compressibility (thermal expansion coefficient). The proportionality constant, Gruneisen parameter, is related to the anharmonicity of the associated vibrational mode [9]. Conversely, measurement of a vibrational frequency shift can be related to the magnitude of localized strain in a material or the sample temperature. Strain heterogeneity measurements in thin films using Raman spectroscopy has been demonstrated in metal oxide structures as well as semiconductors [10,11].

Results are presented in this review which demonstrate the utility of Raman spectroscopy for in situ materials characterization as a function of applied temperature or pressure. Topics to be discussed include temperature and pressure induced phase transformations in dielectric films. An experimental arrangement is also described for Raman measurements of materials at temperatures in excess of 2000°C. Raman spectra of boron nitride and molten silica in this temperature regime will be presented.

RAMAN SCATTERING PHENOMENA

A brief introduction to the theory of the Raman effect is presented here. More detailed discussions appear in the literature [5,12].

Interaction between incident electromagnetic radiation and bound electrons results in a scattering process because the time-varying electric field of the radiation can induce a dipole moment (**M**) in the material. The induced moment varies with the frequency of the incident field and generates elastic scattering (Rayleigh) at the same frequency as the incident radiation, but scattered over 360°. An inelastic scattering event (Raman) can occur when energy is "removed" from the oscillating dipole moment by a lattice vibrational excitation (phonon). This excitation is characterized by a time-varying dipole moment during vibration of the associated normal mode, Q_k. As with an isolated molecule, this vibrational mode must exhibit a change in polarizability during the vibration to exhibit Raman activity. Following a Raman scattering event, the system will be excited to the next higher vibrational energy level (Stokes process) accompanied by scattering of red-shifted light from the wavelength of the incident excitation. Anti-Stokes scattering is also allowed in which a transition to a lower vibrational energy level occurs removing energy from the lattice and resulting in blue-shifted light. Figure 1 illustrates these two processes in the regime where the energy of the incident excitation is much lower than the bandgap of the material.

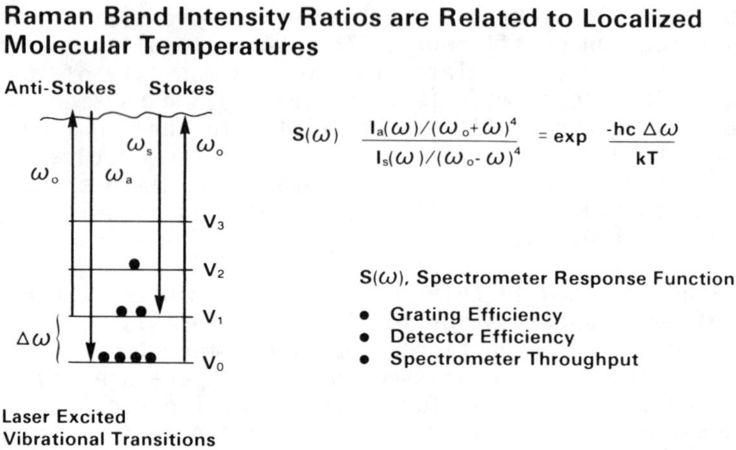

Figure 1. Vibrational level energy diagram and relationship between band intensities and temperature.

In general, the Raman intensity is proportional to the fourth power of the scattered frequency and the transition probability which involves the vibrational level populations and transition moment. (The probability for a Stokes process depends on the vibrational populations in all levels while only those levels at V_1 and above contribute to the probability for an anti-Stokes process.) The fraction of oscillators in each vibrational level is determined from a statistical mechanics treatment of the quantum harmonic oscillator. An expression results from this formalism which relates the Stokes/anti-Stokes band intensity ratio to the average temperature of the vibrational system as shown in Figure 1. This model provides a means for determining sample temperatures from Raman intensity measurements and has been evaluated recently by Malyj and Griffiths [13].

In addition to spontaneous Raman scattering processes which are usually quite weak, a number of enhanced scattering techniques are available for improving the sensitivity of the method. These include Resonance Raman scattering, Interference Enhanced Raman scattering, and waveguiding techniques. While not discussed here, these techniques have been of great importance in thin film characterization work where the limited amount of material often results in weak or non-detectable signals. A summary of signal enhancement techniques is discussed in a recent review of thin film characterization using Raman spectroscopy [6].

INSTRUMENTATION

Visible laser excitation is the most efficient manner for excitation of Raman scattering due to the relatively weak nature of the Raman effect. Near infrared and ultraviolet excitation is also used for samples that fluoresce strongly under visible excitation and to enhance Raman scattering intensities through resonance phenomena. A 90° or 180° sampling geometry relative to the incident laser direction is usually preferred to minimize Rayleigh scattering in the first case and for microprobe applications in the second case. The polarization direction of the probe laser can be varied, and scattered light can be analyzed in one of two directions normal to the propagation direction of the scattered light. Thus the 180° backscattering geometry may be defined according to Damen, et. al. by Z(XY)Z where the incident laser, propagating in the Z direction, is polarized along the X-direction and light polarized along the Y-direction is analyzed by means of a polarizing filter [14]. The scattered light, which consists of a major elastically scattered component (Rayleigh), is dispersed in a monochromator having good stray light rejection capability. Typically double or triple monochromators (0.25 - 1.00 meter focal length) equipped with holographic gratings are used. Scanning monochromators equipped with high sensitivity photomultiplier detectors that operate in a photon-counting mode generate the usual intensity versus frequency spectra. Silicon array detectors offer the capability of time resolved measurements and markedly decrease spectrum acquisition time for a given signal to noise ratio over single channel detectors.

By interfacing an optical microscope to the entrance slits of a monochromator, spatially resolved Raman measurements are possible with a diffraction limited resolution of about one micrometer. The probe laser is imaged onto the sample using the microscope optics which also act to collect the scattered light for analysis. Surface profiling of dielectric films and heterogeneous minerals and ceramic composites has been a major application of this surface sensitive technique.

Time resolved Raman spectroscopy (TRRS) can be performed using multichannel detectors. The availability of and improvement in intensified gated array detectors during the past several years has allowed new applications of Raman spectroscopy for materials analysis. One application involves high temperature spectroscopy. A schematic diagram for high temperature measurements is depicted in Figure 2. In this system, an ancillary CO_2 laser (10.6 micrometer emission) collinear with the Raman probe laser is used for sample heating. Localized

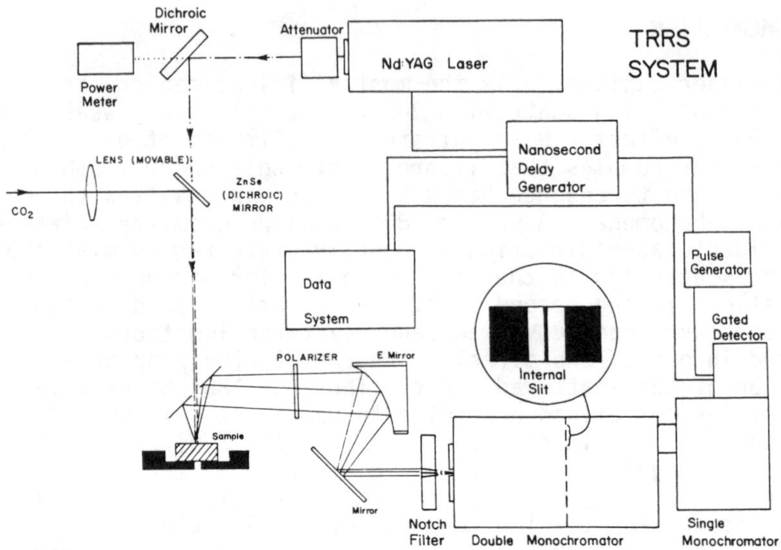

Figure 2. Time resolved Raman spectroscopy system for measurements at high temperatures. An ancillary CO_2 laser provides localized sample heating. The pulse-excitation gated-detection system is built around a silicon array detector.

temperatures of ceramic substrates in excess of 2500°C. have been achieved using CW laser powers of about 100 watts. Under these conditions the sample emits intense visible blackbody radiation which is many orders of magnitude stronger than the Raman scattered light from the sample. In order to suppress the strong visible emission, a synchronous detection scheme is used. Raman scattering is excited by pulsed radiation from a doubled Nd:YAG laser (532 nm, 8 ns bandwidth) operating at a frequency of 30 Hz. An intensified gated array detector (8 ns gatewidth) is triggered by the laser pulse after a suitable time delay. In this manner, spontaneous Raman scattering is detected and the CW blackbody emission from the hot sample is rejected. This system has been used to measure vibrational spectra of materials at very high temperatures and to determine sample temperatures from intensity measurements using the Stokes/anti-Stokes formalism presented earlier. For such measurements, an internal filter is located in the double monochromator (subtractive mode) to reject the Rayleigh component of the scattered light.

TEMPERATURE DEPENDENCE OF VIBRATIONAL MODES

The pulsed-excitation gated-detection technique was used to measure Raman scattering from the hexagonal phase of boron nitride at temperatures up to 2000°C. A boron nitride coating was deposited on the surface of a hollow cylindrical graphite furnace (3 cm in length and 1 cm in diameter) which served as a resistive heater. This heating arrangement is identical to that used for analysis by atomic absorption spectroscopy. High electrical currents applied to the furnace give rise to rapid heating and subsequent volatilization of a deposited analyte. The ramp time to maximum temperature occurs over a two second time interval. However, equilibrium measurements could also be made at a fixed temperature determined by the current input to the furnace.

Figure 3 shows the time evolution of the Stokes/anti-Stokes spectrum of boron nitride as a function of time in fractions of a second. The primary Raman line assigned to the intralayer

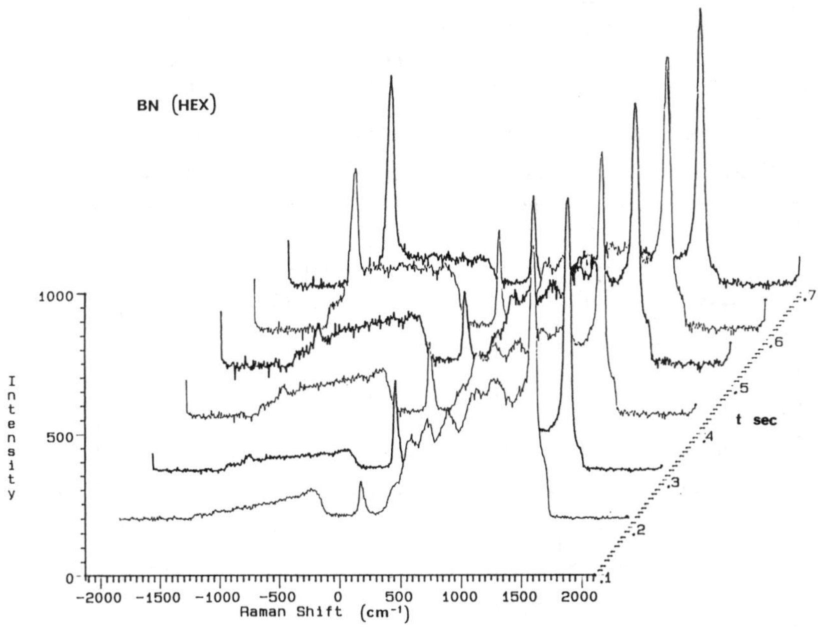

Figure 3. Simultaneous measurement of Stokes and anti-Stokes Raman scattering from the hexagonal phase of boron nitride. Spectra are recorded every tenth of a second.

vibrational stretching mode between boron and nitrogen occurs at 1365 cm^{-1} and is evident in the spectrum. The anti-Stokes line is not discernable on this scale at room temperature. As the material heats, the intensity of the anti-Stokes line increases relative to the Stokes line indicating an increase in molecular temperature. The laser exciting frequency appears as an attenuated line at 0 cm^{-1} due to the monochromator filter and the presence of a narrow band rejection filter tuned to the probe laser wavelength at the entrance slits to the monochromator. The filter also gives rise to transmission fringes which can be removed following digital processing of the spectra. Sample temperatures were determined from line intensities according to the formalism discussed by Malyj and Griffiths [13]. Sample temperatures were verified in equilibrium heating experiments using a two-color optical pyrometer imaged on the hot furnace.

In addition to band intensity changes as a function of temperature, the mode frequency also decreases. A correlation between temperature and the vibrational frequency of the strong BN mode is presented in Figure 4. At temperatures above those presented in the figure, boron nitride begins to volatilize and the line intensity decreases markedly. Sample temperatures can be correlated directly with mode frequency shifts which significantly improves measurement sensitivity since the

TEMPERATURE DEPENDENCE OF THE HIGH FREQUENCY E_g MODE IN BORON NITRIDE (HEXAGONAL)

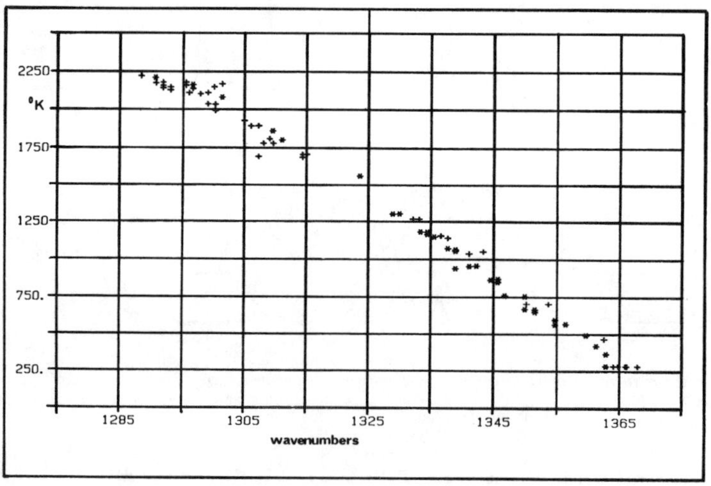

Figure 4. The variation in mode frequency for BN as a function of temperature.

uncertainty for two intensity measurements exceeds that for one frequency measurement. Successive single laser shot Raman spectra have been acquired from boron nitride using a pulsed laser operating at 30 Hz placing a time resolution of 33 milliseconds on the temperature measurement.

High temperature measurements of molten silica were also obtained using the same technique. However, sample heating was achieved by means of a CO_2 laser collinear with the Raman probe laser. To prevent disproportionation of the silica at these extreme temperatures, the sample was flooded with oxygen. Raman spectra of the glass at room temperature and a high temperature melt are presented in Figure 5. The two high frequency modes broaden and shift to lower frequency in the melt. A broad feature near 900 cm^{-1} appears and the 803 cm^{-1} mode broadens and shifts to lower frequency. The low frequency modes at 603 cm^{-1} and 490 cm^{-1} persist in the melt at greater intensity than the glass and exhibit little change in frequency. These features have been

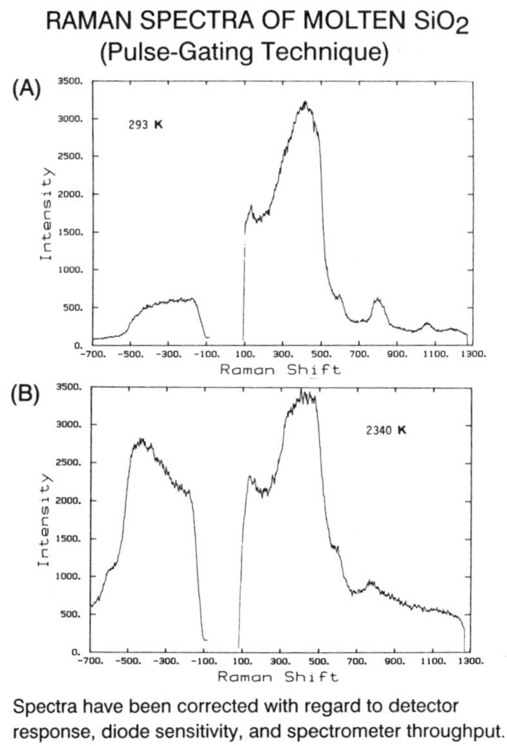

Figure 5. Raman spectra of fused silica at room temperature and the melt.

assigned to 3- and 4-member ring vibrations. Molecular dynamics simulations of the structure of molten silica also predict a preponderance of these species and are in agreement with spectroscopic results [15]. Temperature-induced phase transformations can be followed in real-time using Raman spectroscopy. Previous work successfully demonstrated that Raman measurements could be used to study the kinetics of the glass-crystal second order phase transformation in alkali metaphosphate systems [16]. The rate constant for the transformation was found to correlate inversely with the length of time the sample was annealed. An analogous measurement is reported here for temperature induced crystallization phenomena in sol-gel deposited films.

Titania films having an average thickness of 200 nm were spin cast on silica substrates from acidified (HCl) alcoholic solutions of titanium ethoxide. Coated substrates were placed in a heated furnace at a fixed temperature and Raman spectra were recorded as a function of time as the initial amorphous coating transformed to the crystalline anatase phase. Figure 6 illustrates the time evolution of the Raman spectrum during isothermal heating. The principal features ascribed to the

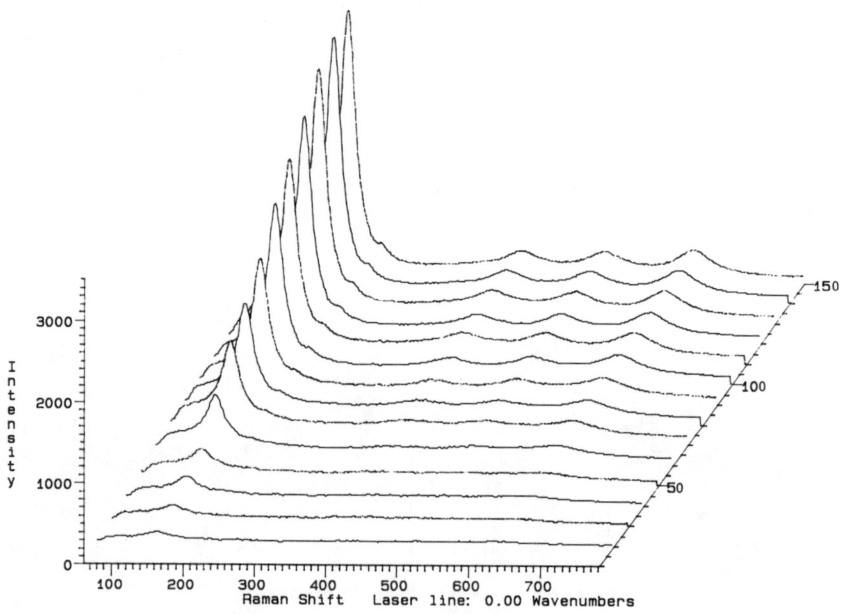

Figure 6. Transient Raman spectra of an amorphous sol-gel film transforming to the crystalline anatase phase.

anatase crystalline phase at 143, 197, 397, 517, and 638 cm^{-1} are evident. The intensity increase of the 143 cm^{-1} mode is shown in Figure 7 as a function of time at 350°C. Results at other processing temperatures yield a similar sigmoidal dependence. Isothermal rate constants are derived from this data based upon a formalism developed by Avrami and used in previous glass crystallization studies [16]. From initial measurements at several temperatures an activation energy for the phase transformation process of 17 kcal/mole is indicated. Further measurements are in progress to establish the dependence of rate constant and activation energy on sol-gel processing parameters.

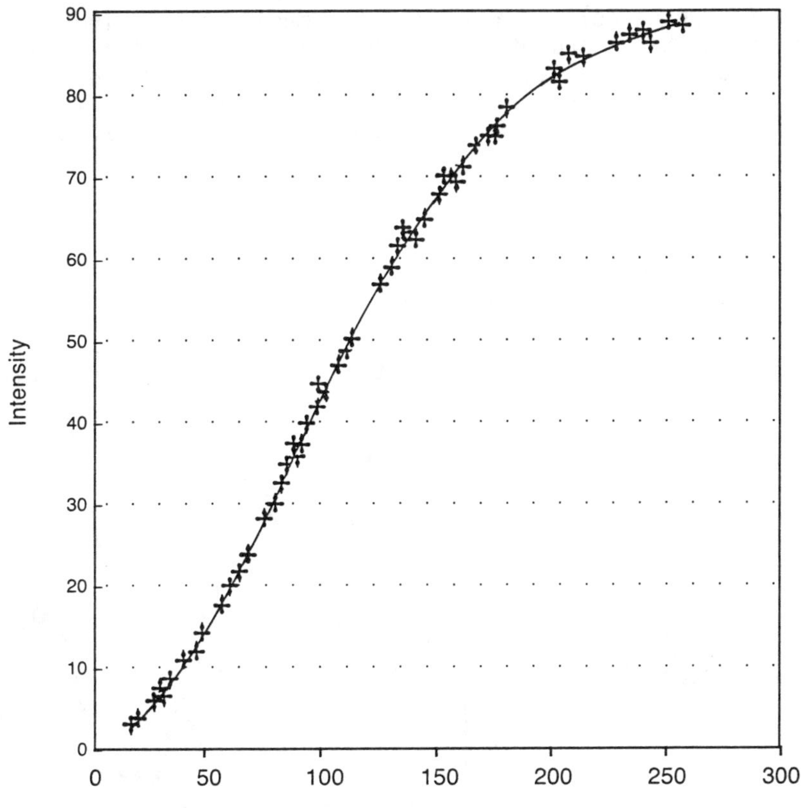

Figure 7. Degree of crystallinity as a function of time during isothermal crystallization of a sol-gel deposited titania film at 350°C.

PRESSURE DEPENDENCE OF VIBRATIONAL MODES

Applied pressure can also induce structural phase transformations in crystalline materials. Pressures at which such transformations can occur lie in the kbar regime making diamond anvil cell (DAC) techniques necessary for spectroscopic measurements. The response to applied pressure of anatase sol-gel deposited films in comparison with the bulk crystalline material formed the basis for current work in this area. Experimental details along with a review of high pressure spectroscopy are discussed in a recent article [17].
A 0.25 micrometer sol-gel deposited film was mounted in a DAC of Merrill-Basset design along with several ruby chips (used as a pressure sensor through pressure-induced shifts in the laser-excited fluorescence) and a hydrostatic medium consisting of a mixture of methanol and ethanol. The sample is constrained between the diamond anvils in a circular area having a diameter of ca. 200 micrometers. Figure 8 depicts Raman spectra from the thin film and a bulk sample at an applied pressure of 70 kbar.

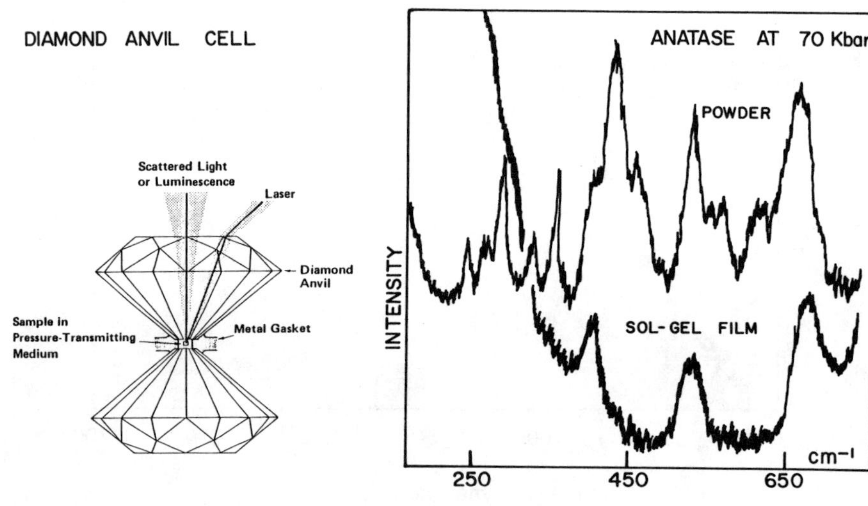

Figure 8. Schematic diagram of a diamond anvil cell. The anomalous pressure response of a sol-gel anatase film in comparison with a crystalline sample of anatase.

The powdered material has undergone a phase transformation to the TiO_2-II phase while the film exhibits Raman lines characteristic of the anatase phase. Marked splitting in the vibrational features accompanied by band intensity changes signify formation of a new phase. The pressure dependence of mode frequencies also exhibits anomalous behavior as seen in Figure 9. Results suggest

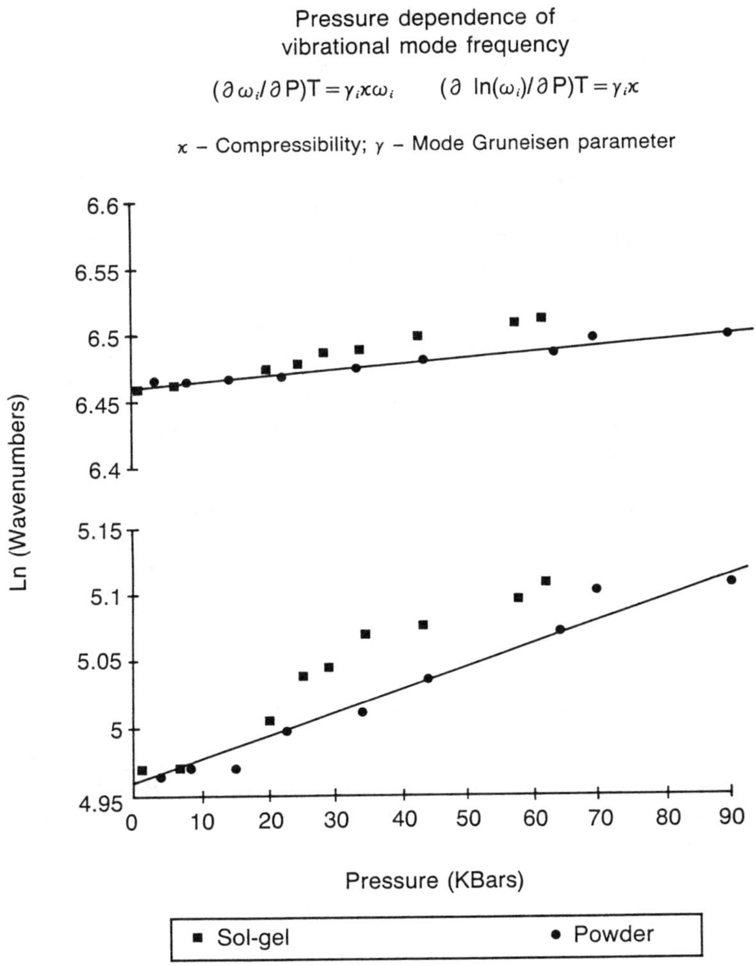

Figure 9. The logarithmic variation in mode frequency with applied pressure for crystalline anatase and a sol-gel anatase film.

that the sol-gel film can be characterized by a higher compressibility when compared with the bulk powder. The logarithmic variation in mode frequency with applied pressure is directly proportional to the mode Gruneisen parameter and the compressibility. For both materials the Gruneisen parameters are identical implying a higher compressibility for the film. TEM cross-sections of sol-gel deposited anatase films reveal a unique spheroidal microstructure consisting of crystalline grains embedded in an amorphous titania matrix with localized voids encompassing the grains. The anomalous pressure response of titania sol-gel films is attributed to this morphology [18].

SUMMARY

Ongoing work in this laboratory involves vibrational studies of structural phase transformations and the response of materials to applied pressure and temperature. Raman spectroscopy has been used successfully in many applications to elucidate molecular changes which result from an applied stress to a material. These changes may involve atomic rearrangement and often are accompanied by alteration of chemical bonding. In tribological processes the combination of applied pressure and resulting temperature may act in concert to promote chemical reactivity and thereby enhance materials degradation. Recent Raman investigations in this area are helping to develop a fundamental understanding of friction and wear at the molecular level [19,20]. The application of this technique to other materials-related areas will allow direct probing of molecular-level interactions which could lead to the design of improved materials.

ACKNOWLEDGMENTS

This work has been supported by the Materials Sciences Division of the Office of Basic Energy Sciences, U.S. Department of Energy under contract DE-AC06-76RLO-1830. Experimental assistance by Professor Joel W. Schaaf (Lower Columbia College), Ms. Nancy J. Hess (University of Washington), and Ms. MaryJo Aloi (Florida Atlantic University) is gratefully acknowledged.

REFERENCES

1. P.J. Miller, G.J. Exarhos, and W.M. Risen, Jr., "Interionic Vibrations and Glass Transitions in Ionic Oxide Metaphosphate Glasses," J. Chem. Phys. 60(11), 4145 (1974).

2. G.J. Exarhos, "Raman Characterization of Molecular Bonding and Phase Homogeneity in Thin Optical Films," Proc. Soc. Photo-Opt. Instrument. Eng. 678, 51 (1986).

3. E.B. Wilson, Jr., J.C. Decius, and P.C. Cross, Molecular Vibrations: The theory of Infrared and Raman Vibrational Spectra, McGraw-Hill, New York (1955).

4. G. Herzberg, Molecular Spectra and Molecular Structure. II. Infrared and Raman Spectra of Polyatomic Molecules, Van Nostrand Reinhold, New York (1945).

5. P.M.A. Sherwood, Vibrational Spectroscopy of Solids, Cambridge University Press, Cambridge (1972).

6. G.J. Exarhos, "Molecular Characterization of Structure and Bonding in Thin Film Materials," Proc. Soc. Photo-Opt. Instrument. Eng. 895, 194 (1988).

7. G.J. Exarhos, "Substrate Signal Suppression in Raman Spectra of Sputter Deposited TiO_2 Films," J. Chem. Phys. 81(11), 5211 (1984).

8. D.D. Allred, J. Gonzalez-Hernandez, O.U. Nguyen, D. Martin, and D. Pawlik, "Raman Scattering and X-Ray Diffraction Characterization of Amorphous Semiconductor Multilayer Interfaces," J. Mater. Res. 1(3), 468 (1986).

9. G.A. Samara and P.S. Peercy, "Pressure and Temperature Dependence of the Static Dielectric Constants and Raman Spectra of TiO_2 (Rutile)," Phys. Rev. B 7(3), 1131 (1973).

10. A.C. deWilton and M. Simard-Mormandin, "Raman Spectroscopy for Non-Destructive Depth Profile Studies of Ion Implantation in Silicon," J. Electrochem. Soc.: Solid State Science and Tech. 133(5), 988 (1986).

11. K. Yamazaki, M. Yamada, K. Yamamoto, and K. Abe, "Raman Scattering Characterization of Residual Stresses in Ge and GaAs/Ge on SiO_2-Coated Si Substrates," Jpn. J. Appl. Phys. 23(6), 681 (1984).

12. W.A. Guillory, Introduction to Molecular Structure and Spectroscopy, Allyn and Bacon, Boston (1977).

13. M. Malyj and J.E. Griffiths, "Stokes/Anti-Stokes Raman Vibrational Temperatures: Reference Materials, Standard Lamps and Spectrophotometric Calibrations," Appl. Spec. 37(4), 315 (1983).

14. T.C. Damen, B. Tell, and S.P.S. Porto, "Raman Effect in Zinc Oxide," Phys. Rev. 142, 570 (1966).

15. G.J. Exarhos, "Vibrational Spectra of Silica Near 2400 K: Measurement and Molecular Dynamics Simulation," in Proceedings of the Eleventh International Conference on Raman Spectroscopy, eds. R.J.H. Clarke and D.A. Long, John Wiley, Chichester, pp. 503-504 (1988).

16. G.J. Exarhos and W.M. Risen, Jr., "Raman Study of Isothermal Devitrification Kinetics of $NaPO_3$ Glass," J. Am. Ceram. Soc. 57(9), 401 (1974).

17. R.J. Hemley, P.M. Bell, and H.K. Mao, "Laser Techniques in High-Pressure Geophysics," Science 237(4815), 605 (1987).

18. W.S. Frydrych, G.J. Exarhos, K.F. Ferris, and N.J. Hess, "Chemical Control of Stress in Sol-Gel Derived Titania Films and Their Pressure-Dependent Optical Properties," In Proc. Mat. Res. Soc. Better Ceramics Through Chemistry 3, 121, 343-348 (1988).

19. G.J. Exarhos and M.S. Donley, "Real-Time Raman Detection of Molecular Changes in Ceramics Undergoing Sliding Friction," Microbeam Analysis 1987, 125.

20. B.E. Hegeman, S. Jahanmir, and S.M. Hsu, "Microspectroscopy Applications in Tribology," Microbeam Analysis 1988, 193.

CATHODOLUMINESCENCE AND PHOTOLUMINESCENCE IN ALUMINUM NITRIDE

R.A. Youngman, J.H. Harris and D.A. Chernoff
BP Research International
Research Center Warrensville
Cleveland, Ohio 44124

ABSTRACT

Aluminum nitride is a known luminescent material, emitting a range of wavelengths from U.V. through infrared. The luminescence spectrum discloses the presence of impurities and defects in the AlN lattice, which is important because of the connection between lattice defects and thermal conductivity in this material. In this work, we present data relating various photoluminescence spectra to impurity and defect concentrations in AlN powders and single crystals, including defects formed when oxygen substitutes for nitogen in the AlN lattice. Using an SEM modified for cathodoluminescence imaging, these data are utilized to provide a defect map of sintered polycrystalline samples.

INTRODUCTION

Aluminum Nitride is a known luminescent material [1,2], emitting a range of wavelengths from the ultra-violet to infrared. Two major peaks are present in the luminescence spectra- a U.V.-blue peak (c.a. 375nm) and a red peak (c.a. 600 nm). Both of these peaks are extremely broad at room temperature (1-1.5eV). The red peak sharpens significantly at low temperature and exhibits the presence of phonon side-bands, the origin of which has been attributed to the presence of Mn^{4+} in the AlN lattice[3]. The U.V.-blue peak does not sharpen as the temperature is lowered and has been associated with oxygen impurities in the AlN lattice[4]. A model has been proposed to explain the broadness of this peak invoking the presence of donor-acceptor pairs in the material[5].

Aluminum nitride has a unique ability to dissolve oxygen into the lattice up to values of 10-12 w/o. The mechanisms for this large accomodation of oxygen are of technological and scientific interest due to the established deleterious effects of oxygen on the thermal conductivity in this material[6]. We present in this paper the results of our own photoluminescence study on single crystal and polycrystalline AlN prepared under a range of conditions and with varous dopants, including oxygen. The results of this work can be used to: (i) corroborate the results of previous studies concerning the identity of the oxygen luminescence peak at 375nm; (ii) disclose the observation of new peaks at 320 and 460nm; (iii) suggest some initial ideas to explain the extremely broad linewidths observed; and (iv) illustrate how cathodoluminescence can be a useful tool for imaging impurity distributions in polycrystalline AlN.

EXPERIMENTAL

Our photoluminescence experiment utilizes a mode-locked argon ion laser which synchronously pumps a R6G dye laser tuned to a wavelength of 536 nm. This is followed by a second harmonic generation crystal which frequency doubles the dye laser pulses to produce the final excitation wavelength of 268 nm. The excitation beam is loosely focused on the sample which is mounted on the cold-finger of a variable temperature cryostat for low temperature measurements (down to 10K). The photoluminescence is focused on the entrance slit of a 0.5 meter monochrometer via a quartz optical fiber. The exit slit of the monochrometer is coupled to the detector of an optical multi-channel analyzer (EG&G OMA-3), where all data aquisition and signal averaging operations are performed. For cathodoluminescence, a quartz optical fiber is focused onto the specimen area of a SEM and interfaced to the detection system in the same manner as for the photoluminescence experiment.

RESULTS AND DISCUSSION

Figure 1(a) presents the photoluminescence spectrum from a powder of AlN whose oxygen content has been determined by neutron activation studies to be 1.53 $\pm$ 0.05 w/o. Consistent with other work in this field[4,5], we assign the large dominant line in this spectra to be related to the defect created when oxygen substitutes for nitrogen in the AlN lattice. In order to check this assignment, we fabricated a set of AlN samples by reacting ultra-pure aluminum metal (99.9999) with gettered N_2 gas ($PO_2 < 10^{-14}$ atm) mixed with a small, controllable quantity of O_2. The resulting luminescence spectra is virtually identical with that of the 1.53 w/o oxygen-containing powder. We label this peak as related to oxygen content of the powder, but not necessarily directly due to it, because we do not observe a linear scaling of this peak intensity with the oxygen flow in our powder fabrication process. We will comment more on this question at the end of this section.

The luminescence spectrum from a single crystal of AlN, grown under nitrogen deficient conditions by Slack et al.[7], is given as Figure 2. The oxygen content of this crystal is extremely low, measured at about 380 ppm- thus accounting for the absence of a peak at 375 nm. As is evident from Figure 2, the single crystal spectrum is characterized by broad peaks centered at 320 and 460nm; peaks which may be related to nitrogen vacancies in the sample. To further support this suggestion, we prepared highly nitrogen deficient AlN powders by reactively evaporating pure Al metal in a nitrogen atmosphere and quenching the AlN product before full nitridation was allowed to occur. In these highly N-deficient, dark grey, AlN samples, the same 320 and 460nm peaks are observed. These peaks diminish, and eventually disappear, when the samples are annealed in 1 atmosphere N_2 at 1800°C for various lengths of time up to 1 hour.

A remaining challenge in understanding the luminescence spectra of doped AlN is to explain the extremely broad linewidths observed for a wide range of impurities in this material. As is illustrated in Figure 1(a) and (b), where the 300K and 10K spectra for the O-related impurity are presented respectively, the broad line at 375nm is approximately temperature independent. The same is observed for the 320 and 460nm lines. This is very different behavior than is observed for AlN spectral lines in the red, where significant phonon structure emerges at low temperature.

Clearly the magnitude of the linewidths observed for these impurities (1-1.5eV) are much too large to be explained by strain broadening alone. Using an average (as opposed to local) strain approximation, we estimate this broadening to be a maximum of about 100meV; more than an order of magnitude less than the observed width. These simple considerations are a strong indication that some form of electronic interaction is responsible for the line broadening in this system. For example, as originally proposed by Thomas et al.[8] and similar to the work of Pasternak et al for AlN[5] and recent work in ZnS[9], one could explain these linewidths by constructing a donor-acceptor pair model where the coulomb interaction between ionized donors and acceptors in the ground-state would provide an electronic interaction in the correct energy range. For the case of AlN, the O_N and V_N defects would act as deep donors with V_{Al} as an acceptor state. In the ideal case, these donor-acceptor pairs manifest themselves as discrete lines, but as recently shown for ZnO, in the presence of other interactions (strain, disorder, lattice polarization) and depending on the particular lattice sites occupied, a continuum is often observed, even at low temperatures. Incidently, a donor-acceptor pair model would account for the non-linear relationship between O-content and the magnitude of the O-related spectral line, since in this picture two different (but possibly linked) impurity populations would determine each transition. A number of experiments are presently underway, including time-resolved measurements, to determine the validity of this approach.

Independent of the origin of the luminescence signal in AlN, it is clear that one can correlate specific spectral lines with the presence of well defined impurities in the sample. As a result of this correlation, cathodoluminescence becomes an extremely useful tool in spatially resolving such defect populations within the AlN microstructure. Figure 3a presents a backscattered electron micrograph of a typical microstructure in sintered polycrystalline AlN. In this image, the contrast arises from atomic number differences in the sample, and in this case the light grey areas are AlN and the bright areas at triple points and grain boundaries are Y-Al-O phases resulting from the additions of Y_2O_3 as a sintering aid. Since it is established that the 375nm luminescence peak is related to oxygen content in AlN, one can utilize the emitted light from this peak to produce an image of the oxygen distribution in the sample. Figure 4 is the cathodoluminescence spectrum obtained from the same area as shown in Figure 3a where two peaks, one at 375nm and one at 600nm, are evident. The cathodoluminescence spectrum is very similar to the photoluminescence spectrum in AlN. An image formed with the 375nm peak represents the oxygen distribution in the lattice. Such an image is given in Figure 3b encompassing the same area as that shown in Figure 3a. Significant intensity distribution is seen within the AlN grains, and such imaging can allow elucidation of the oxygen profile during sintering of AlN. These data represent a unique and powerful method for spatially resolving defects in AlN, heretofore an extremely difficult problem in microstructural analysis.

CONCLUSIONS

Initial observations made in this study indicate that: (i) the broad temperature independent luminescence peak in single crystal, polycrystalline, and sintered AlN centered on 375 nm is related to oxygen dissolved in the lattice and that one possible origin of the broadness is coulombic interaction between donor (O_N) and acceptor (V_{Al}) pairs though more detailed time-resolved measurements will be necessary to confirm this picture; (ii) two new peaks centered at 320 and 460nm have been observed and are possibly due to the presence of nitrogen vacancies in the AlN lattice (it may also be due to a donor (V_N)- acceptor (V_{Al}) transition as for the oxygen peak above) and (iii) it is possible to utilize cathodoluminescence imaging of the appropriate impurity-related peak to establish the distribution of this impurity in the microstructure of polycrystalline or sintered material.

ACKNOWLEDGEMENT

The authors wish to acknowledge useful and stimulating discussions with Drs. Leonid Turkevich and R.C. Enck, and to thank Dr. G.A. Slack of General Electric Corporate Research Laboratories for providing the large single crystal of AlN for our measurements and for helpful comments concerning AlN crystal growth.

REFERENCES

1. G.A. Wolff, I. Adams, and J.W. Mellichamp, Phys. Rev. 114, 1262 (1959)
2. I. Adams, J.W. Mellichamp, and G.A. Wolff, Proc. of Ultrapurif. Semicond. Mater., MIT Press, Boston, MA (1962) p.1262
3. F. Karel and J. Pasternak, Czech. J. Phys. B19, 78 (1969)
4. J. Passternak, S. Pacesora, J. Sanda and J. Rose, Bull. Acad. Sci. USSR, Phys. Ser. 31, 123 (1973)
5. J. Pasternak, S. Pocesova and L. Roskovcova, Czech. J. Phys. B24, 1149 (1974)
6. G.A. Slack, J. Phys. Chem. Solids, 34. 321 (1973)
7. G.A. Slack and T.F. McNelly, J. or Cryst. Growth 42, 560 (1977)
8. D.G. Thomas, J.J. Hopfield and W.M. Augustyniak, Phys. Rev. 140, A202 (1965)
9. A. Abdel-Kader and H.I. Farag, J. Mat. Sci., 23, 2382 (1988)

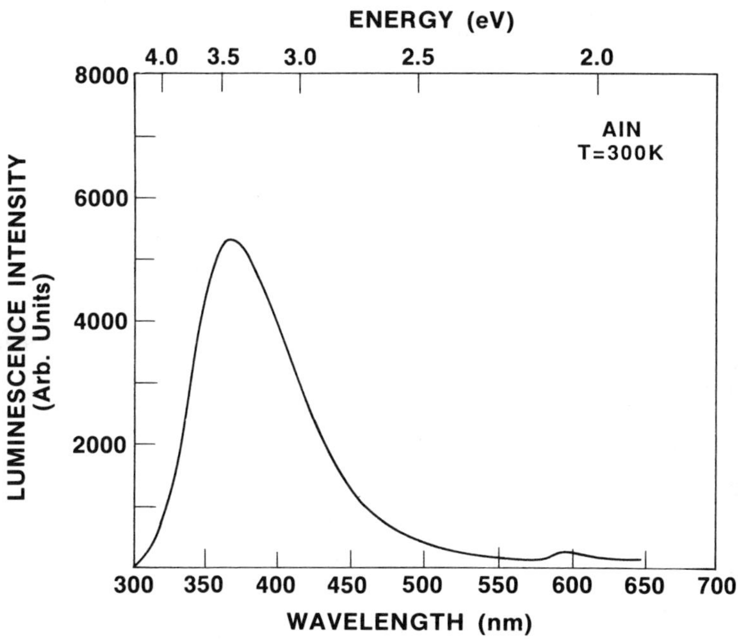

Figure 1a. Photoluminesence spectrum at 300K from a powder of AlN with an oxygen content of 1.53 wt/o.

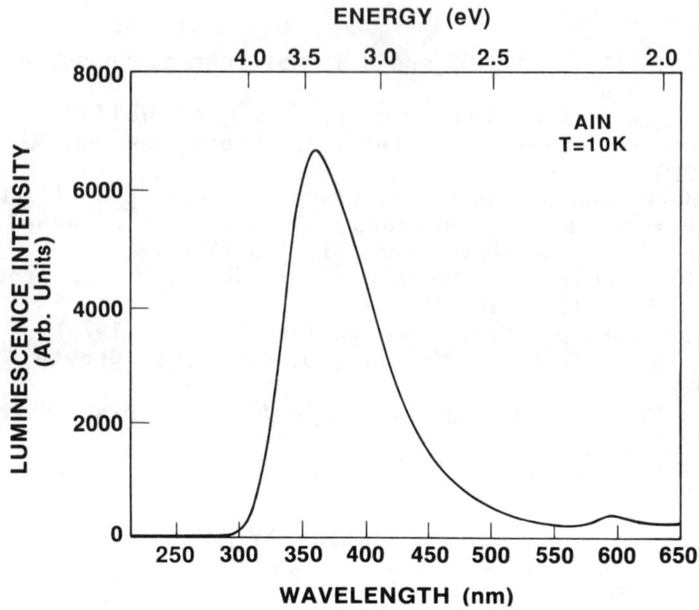

Figure 1b. Photoluminescence spectrum at 10K from a powder of AlN with an oxygen content of 1.53 wt/o.

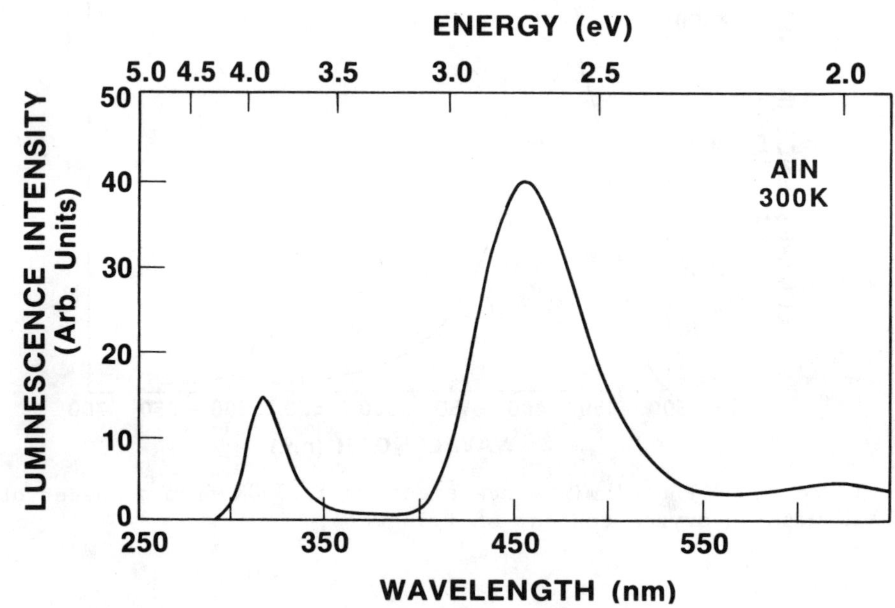

Figure 2. Photoluminescence spectrum at 300K from a large single crystal of AlN grown under nitrogen defficient conditions.

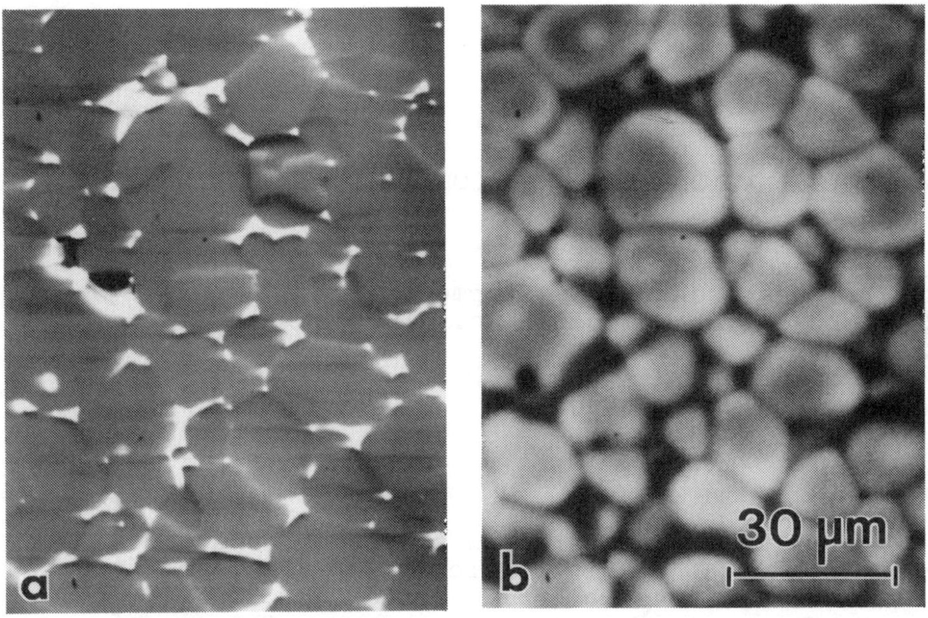

Figure 3. Micrographs of sintered AlN utilizing the BE signal (a), the CL signal from an 80 nm bandpass filter on the 375 nm peak (b).

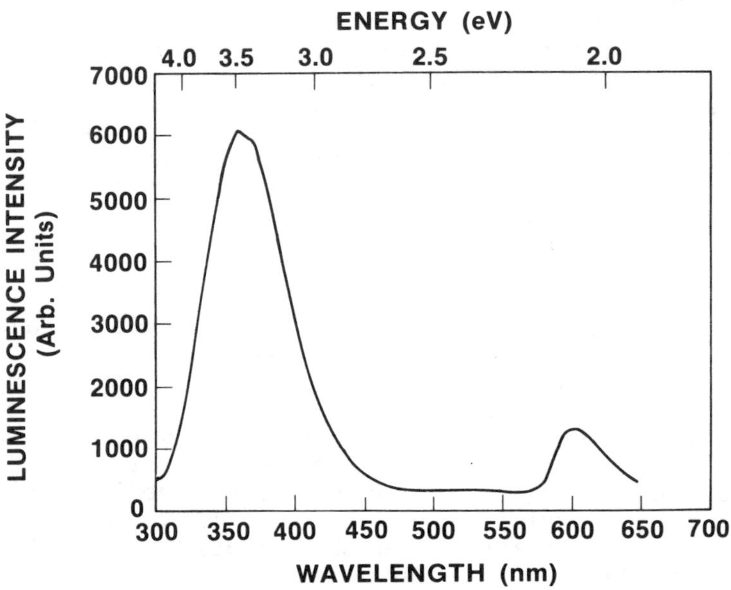

Figure 4. Cathodoluminesence spectrum at 300K from same area of sintered AlN as is shown in Figure 3.

VACUUM ULTRAVIOLET SPECTROSCOPY OF CERAMICS

R. H. French
Central Research and Development
E. I. DuPont de Nemours & Co., Inc.
Experimental Station
P. O. Box 80356
Wilmington, DE 19880-0356

ABSTRACT

Many important ceramics are large band gap materials with both intrinsic and extrinsic electronic structure features which lie above 6 eV, outside the range of normal optical spectroscopy. The use of vacuum ultraviolet (VUV) techniques can extend the range of reflection, transmission, absorption and flourescence measurements up to 25 eV (50 nm), thereby covering the complete electronic structure of ceramic materials. We have developed temperature dependent VUV methods to study the electronic structure as a function of temperature from 4.2K (cryostat) to greater than 1500 K (CO_2 laser heating). Examples will be given of high temperature and strain induced effects, and variations in the electronic structure associated with compositon and phase transitions.

INTRODUCTION

The electronic structure of ceramic materials is of increasing importance due to the many new applications of ceramics. The electronic structure of a ceramic material is the basis of its optical and electronic properties and is also the foundation of its structural properties. Our knowledge of the electronic structure of ceramics is not presently well developed because many ceramic materials have large band gaps (>6 eV) which lie above the "vacuum cutoff" of normal optical spectroscopy techniques. Even for ceramics with smaller band gaps, many of their intrinsic electronic structure features appear at higher energies. Optical methods when extended to the higher energies appropriate for electronic structure studies of these materials, have many advantages over the electron photoemission processes which have been applied very successfully to more conductive materials such as metals and semiconductors. Optical methods do not induce charging problems in insulating ceramics, the energy resolution of optical measurements can be very high (5 meV) and both polycrystalline and single crystal materials can be studied. Also it is possible to perform in situ optical measurements on ceramic materials at the temperatures (1500 K) of importance for both processing and applications. One need not study quenched samples assuming the high temperature structure is preserved, instead the actual effects of high thermal energies, processing and phase transitions can be determined.

In this paper we will discuss basic vacuum ultraviolet spectroscopy techniques, and then present three examples of VUV spectrscopy studies of ceramics; the temperature dependence and high temperature electronic structure of Al_2O_3, surface stresses and strains in Al_2O_3 and MgO and their effects on the electronic structure and the variation of the electronic structure with composition and phase transformations in AlN-AlON-Al_2O_3 thin film materials.

VACUUM ULTRAVIOLET SPECTROSCOPY

The vacuum ultraviolet is traditionally the region above 6 eV and below the 25 eV energy where grazing incidence methods are required. A good review of VUV experimental techniques is given by Samson.[1] This region is characterized by two important experimental limitations. Spectroscopy above 6 eV (<200nm) requires that all measurements be performed in high vacuum due to the strong absorption of air. In addition cleanliness of both the vacuum and samples is important because hydrocarbons are also strongly absorbing in this region. The optical elements used are all reflection optics, such as mirrors and concave focusing gratings; the dispersion of transmissive optics is very strong in this region. Windows of MgF_2 can be used to separate regions of different vacuum, for example between gas discharge lamps and sample chambers. Typically deuterium lamps and photomultiplier detectors with MgF_2 windows are used up to 11 eV (the window cut-off).

The second experimental cutoff, the 11 eV window cutoff (<115 nm) arises due to the lack of useful transparent solids above this energy. Windows can not be used to isolate light sources or detectors. Therefore differential pumping methods must be used to isolate discharge lamps. Order sorting of light is performed with gas phase differentially pumped filters, and either vacuum photodiodes, electron multipliers or sodium salicylate phosphors are used for detection.

The DuPont Laser Plasma Sourced (LPS) VUV spectrophotometer covers the optical region from the visible (1.7 eV, 700 nm) to the VUV (25 eV, 50 nm). The light source is a Nd:YAG laser produced plasma, which is a high intensity, synchrotron brightness, broad band, time resolved source. The 800 mj/pulse, 10 nanosecond output of a Q-switched YAG laser is focused onto a samarium metal target. The resultant 150,000 K plasma's light is collected, focused off of the sample and on to the entrance slit of a 1 meter monochromator. This dispersive analysis, coupled with a 1024 element time-resolved photon-counting detector allows the collection of a 20 or 80 nm spectrum in a single pulse. The resolution is 5 to 20 meV over the complete spectral region. Reflection, transmission, absorption and luminescence measurements can be performed. The samples can be heated using a CO_2 laser up to 1500 K or cooled down to 10 K using a flowing He cryostat, with spectroscopy performed at temperature in this range.

HIGH TEMPERATURE ELECTRONIC STRUCTURE OF Al_2O_3

Al_2O_3 and MgO are mixed ionic and electronic conductors at high temperatures. Their high temperature (1800 K) electrical conductivity behavior has been studied extensively to elucidate the defect chemistry with a goal of understanding the mass transport associated with sintering. Since both ionic and electronic carriers play an important role in the conductivity a clear understanding of the electronic conductivity of Al_2O_3 at high temperature is important. By observing the electronic structure as a function of temperature, the band gap temperature dependence, the effects of lattice expansion, and the electron-phonon interaction can be determined.[2,3]

The room temperature VUV transmission and reflection of Al_2O_3 are shown in figure 1. The 8.8 eV room temperature band gap can be seen in the transmission cut off while the 9 eV exciton peak associated with the band gap is seen in the reflectance. The absorption edge arises from transitions between the O 2p states in the valence band and the Al 3s states at the bottom of the conduction band, while the 13 eV peak in reflectance is due to transitions to Al 3p states. When the crystal is heated, and the reflectance measured in situ, (figure 2) the band gap absorption edge, the exciton and the 13 eV peak shift to lower energies. The band gap temperature dependence can be determined from these measurements and is found to be 1.1 meV/K. The band gap shifts linearly with temperature and the band gap changes from 9.1 eV at 0K to 7.2 eV at 1763K.

The temperature dependence of the band gap leads to a 10^5 increase in the conductivity product $\sigma_e\sigma_h$ at 1500 K from that expected for a 8.8 eV band gap material. Through a study of the high temperature electronic structure it is possible to predict from first principles the intrinsic electronic conductivity of Al_2O_3 at 1800 K.[4] This analysis, along with a comparison to the compensation point conductivity, shows that Al_2O_3 is a predominantly electronic conductor at high temperature, and allows the evaluation of reported conductivity results and defect chemistry models.

SURFACE STRESS DETERMINATION IN Al_2O_3 AND MgO

The temperature dependence of the band gap can be modeled by considering the effects of thermal lattice expansion and the electron-phonon interaction. The temperature dependences observed in Al_2O_3 and MgO are 3 to 6 times larger than observed for Si of Ge. By modeling the lattice expansion contribution to the experimentally observed temperature dependence one can also derive the band gap strain coefficients. These are the basis of a VUV surface stress determination method; the surface stresses and the associated variation in bond lengths will change the band gap energy, leading to increases in the band gap for compressive stresses, and decreases for tensile stresses. Since this is a VUV reflection measurement at the band gap energy, the high absorption coefficient at the gap means the measurement is inherently surface sensitive, information derives from the top 1 to 10 microns of the sample.

Surface stresses play an important role in the mechanical properties of ceramics. Tempering, or induced compressive stresses can strengthen ceramics by effectively closing the griffith cracks with no applied stress. We studied polishing induced surface stresses which arise due to far field stresses of the polishing induced dislocations in the surface layer.[5] The shift in the band gap associated exciton peak caused by surface stresses in chemically and diamond polished 10 micron grain size Al_2O_3 are shown in figure 3a. The polishing damage produces a 0.26 ev shift arising from a compressive stress. In addition this is unaffected by the single or polycrystalline nature of the samples, figure 3b compares the surface stresses in a variety of single and polycrystalline samples, the damage due to diamond polishing is the same in all samples. By successive chemical polishing the surface stresses can be profiled, and for Al_2O_3 are found to penetrate to a depth of 1 micron.

Similar results have been found for MgO. Figure 4 compares chemically polished and cleaved MgO, both of which are relatively stress free. No observable shift of the 7.6 eV exciton peak is seen, but diamond polishing the MgO single crystal exhibits a 0.16 eV shift. By successive profiling, the stressed layer is found to penetrate 40 microns into the bulk.

COMPOSITION AND STRUCTURE IN AlN, AlON AND Al_2O_3 THIN FILMS

Changes in composition and crystal structure can lead to dramatic changes in the electronic structure, which can be studied using VUV techniques. In this example thin film samples of AlN, AlON and Al_2O_3, were fabricated using reactive sputter deposition[6] and the band gap and absorption edge were used to study the effects of varying the anion progressively from nitrogen to oxygen.[7] The samples were fabricated on single crystal sapphire, which has a larger band gap and transparency region than the Al-N-O materials, and reflection and transmission measurements were analyzed using thin film absorption edge analysis. These results were coupled with X-ray Photoemission measurements of the valence band region to confirm the changes in the valence band structure.

The thin film absorption coefficients are shown in figure 5, plotted for direct band gap determination. In AlN thin films doped with 2 or 8 % oxygen, the band gap is found to be 5.8 eV, and no noticeable change is observed in the absorption edge with oxygen doping. The films are one to 3 microns thick with 300 Å columnar grains. The oxygen in the film is most probably present in the grain boundary regions of the film, thereby not profoundly effecting the electronic structure of the AlN grains. At higher oxygen contents (33 % O) an AlON phase is formed causing the band gap to shift slightly to 6.2 eV and, in addition, the absorption edge slope is seen to decrease. At high oxygen contents (66% O), a γ Al_2O_3 phase is formed and the band gap increases rapidly to 7.2 eV.

Even though all three of these materials have conduction bands arising from Al, the change in the valence band states from nitrogen to oxygen character changes the band gap and absorption edge of the material. N 2p valence bands reside above (at lower binding energy) O 2p valence bands. These results can be understood by considering a schematic electronic structure model (figure 6). The 5.8 ev band gap of AlN arises from N 2p to Al 3s. When half of these N 2p states are removed from the top of the valence band of AlON, the number of valence to conduction band transitions at the band gap energy is reduced, leading to a decrease in the absorption edge slope, but no dramatic change in the band gap energy. In addition the valence band for AlON has a 4 peak nature consisting of N 2p, O 2p, and Al-N and Al-O hybridized orbitals. With the complete removal of the N 2p states in the γ Al_2O_3 material the band gap opens up dramatically to 7.2 eV, the energy between the O 2p and Al 3s bands.

The change in the character of the top of the valence band from nitrogen to oxygen has strong implications for the physical properties of these two important ceramics. We have found for the case of metal ceramic bonding involving copper,[8-10] that the more covalent bonding of AlN leads to strong metallic bond formation between Cu and the nitrogen valence band, with preferential surface sites near the Al atom. In the more ionic material Al_2O_3 bond formation is weak and ionic in nature, with Cu forming bonds at the surface oxygen sites.

The 7.2 eV band gap of the γ Al_2O_3 thin film sample is much smaller than that observed for α Al_2O_3 (8.8 eV). This can be demonstrated in situ by studying the γ Al_2O_3 sample and then using laser heating to transform the thin film to the more stable α phase. This is shown in figure 7 where one sees the shift of the transmission cutoff from 7.2 to 8.8 eV at room temperature, before and after heating to 1500 K. γ Al_2O_3 crystalizes in the spinel structure dictating that Al ions are found in both octahedral (as in α Al_2O_3) and tetrahedral coordination. One can therefore suggest that the conduction bands of Al in tetrahedral coordination lie below the bands arising from Al in octahedral coordination. We are studying the effects of this phase transition at present.

CONCLUSIONS

Vacuum ultraviolet spectroscopy is a powerful electronic structure probe, which is especially suited to ceramic materails due to their large band gaps. The ability to optically probe the intrinsic and extrinsic electronic strucutre, without the experimental limitations of the vacuum cutoff or sample charging, allows one to better understand the optical, electrical and structural properties of ceramics. These electronic structure studies, when coupled with low and high temperature capabilities, permit the study of ceramics at the temperatures of interest for processing and application. We have discussed examples of the high temperature electronic structure, the use of VUV measuremtns to determine surface stresses and strains and the effects of both composional variations and phase transformations and their effects on bonding. This demonstrates the wide variety of problems to which experimental determination of the electronic structure of ceramics can be applied.

ACKNOWLEDGEMENTS

Much of this work has been done in collaboration with F. S. Ohuchi and R. V. Kasowski of Dupont Co., and R. L. Coble and H. Song of M. I. T. The spectrophotometer was designed in conjunction with R. D. Fancy of Acton Research Corp.

REFERENCES

1. J. A. R. Samson, Techniques of Vacuum Ultraviolet Spectroscopy, John Wiley and Sons Inc., New York, 1967.

2. R. H. French, R. L. Coble, R. V. Kasowski, F. S. Ohuchi, "Vacuum Ultraviolet, Photoemission and Theoretical Studies of the Electronic Structure of Al_2O_3 up to 1000°C", Physica B, 150, 1988, 47-49.

3. R. H. French, R. L. Coble, R. V. Kasowski, "Temperature-dependent Vacuum Ultraviolet Optical Spectra and Electronic Structure of Al_2O_3", to be published.

4. R. H. French, J. B. Blum, "Electronic Structure and Conductivity of Al_2O_3", to be published in Advances in Ceramics, Ed. by C. A. Handwerker, J. B. Blendell, W. A. Kaysser, American Ceramic Society, 1989.

5. H. Song, R. H. French, R. L. Coble, "Effect of Residual Strain on the Electronic Structure of Al_2O_3 and MgO", to appear in the J. Amer. Ceram. Soc., 1989.

6. F. S. Ohuchi, P. E. Russell, "AlN Thin Films with Controlled Crystallographic Orientations and their Microstructure", J. Vac. Sci. Technol. A, 5,4, 1987, 1630-4.

7. F. S. Ohuchi, R. H. French, "Summary Abstract: Effect of Oxygen Incorporation in AlN Thin Films", J. Vac. Sci. Technol. A, 6, 3, 1988, 1695-6.

8. F. S. Ohuchi, R. H. French, "Cu Deposition on Al_2O_3 and AlN Surfaces: Electronic Structure and Bonding", J. Appl. Phys., 62, 6, 1987, 2286-9.

9. F. S. Ohuchi, R. H. French, R. V. Kasowski, "Summary Abstract: A Study of Room-Temperature Cu-Al_2O_3 and Cu-AlN Interfacial Interactions", J. Vac. Sci. Technol. A, 5, 4, 1987, 1175-7.

10. R. V. Kasowski, F. S. Ohuchi, R. H. French, "Theoretical and Experimental Studies of Cu Metallization of Al_2O_3", Physica B, 150, 1988, 44-6.

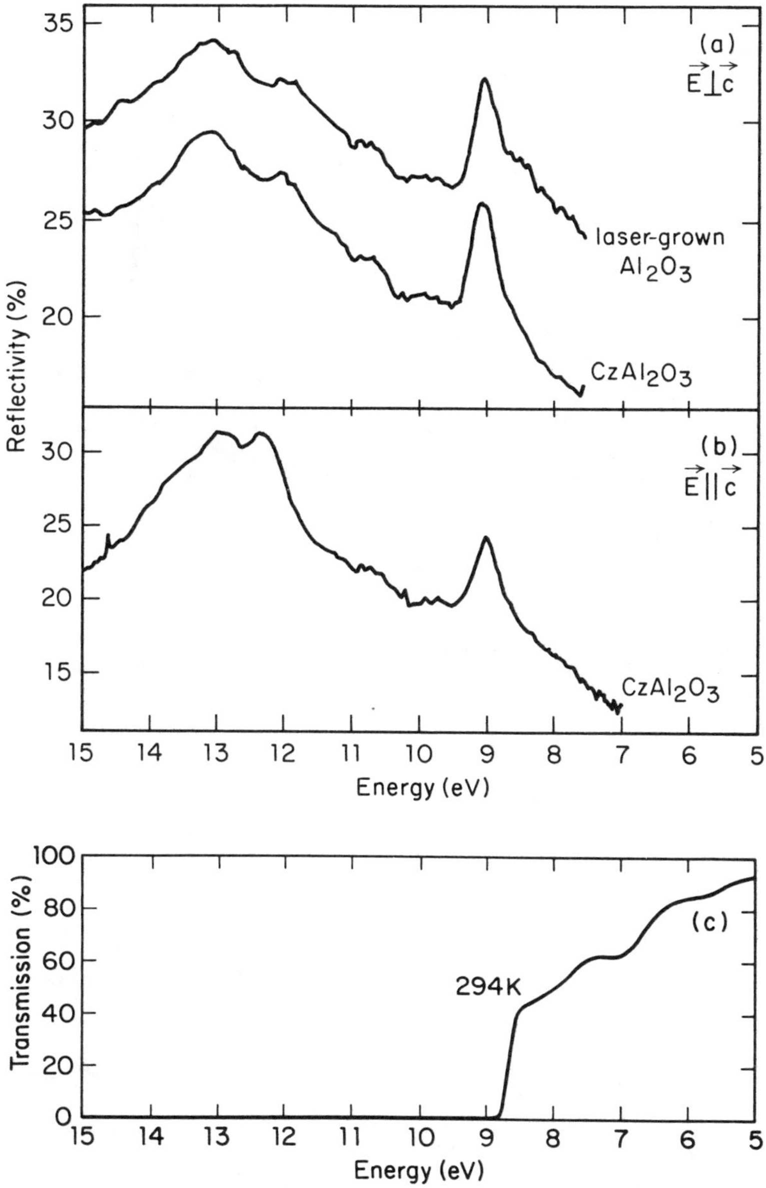

1. Room temperature VUV reflectance of single crystal Al_2O_3 for E polarized a) perpendicular and b) parallel to the c-axis, and c) room temperature transmission of single crystal Al_2O_3.[3]

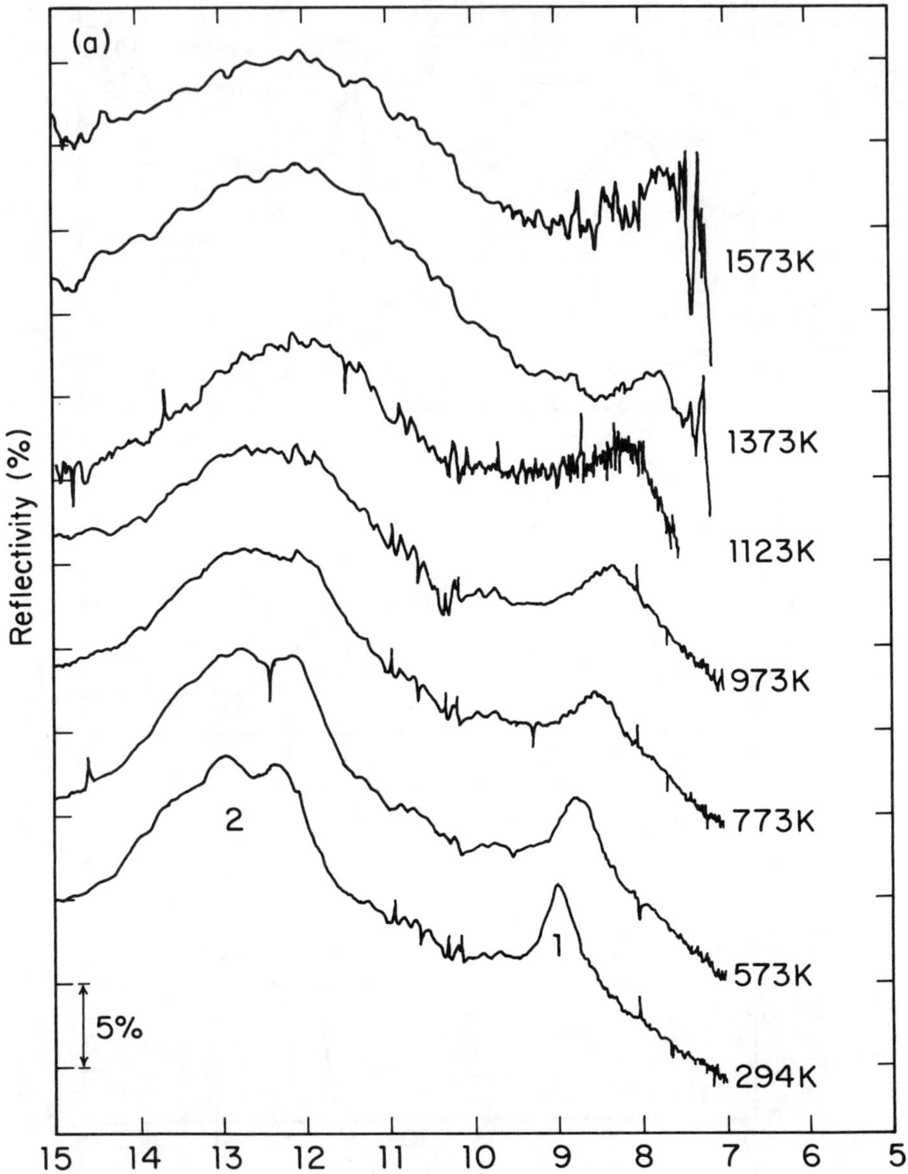

2. Temperature dependent VUV reflectance of single crystal Al_2O_3 for E polarized parallel to the c-axis.[3]

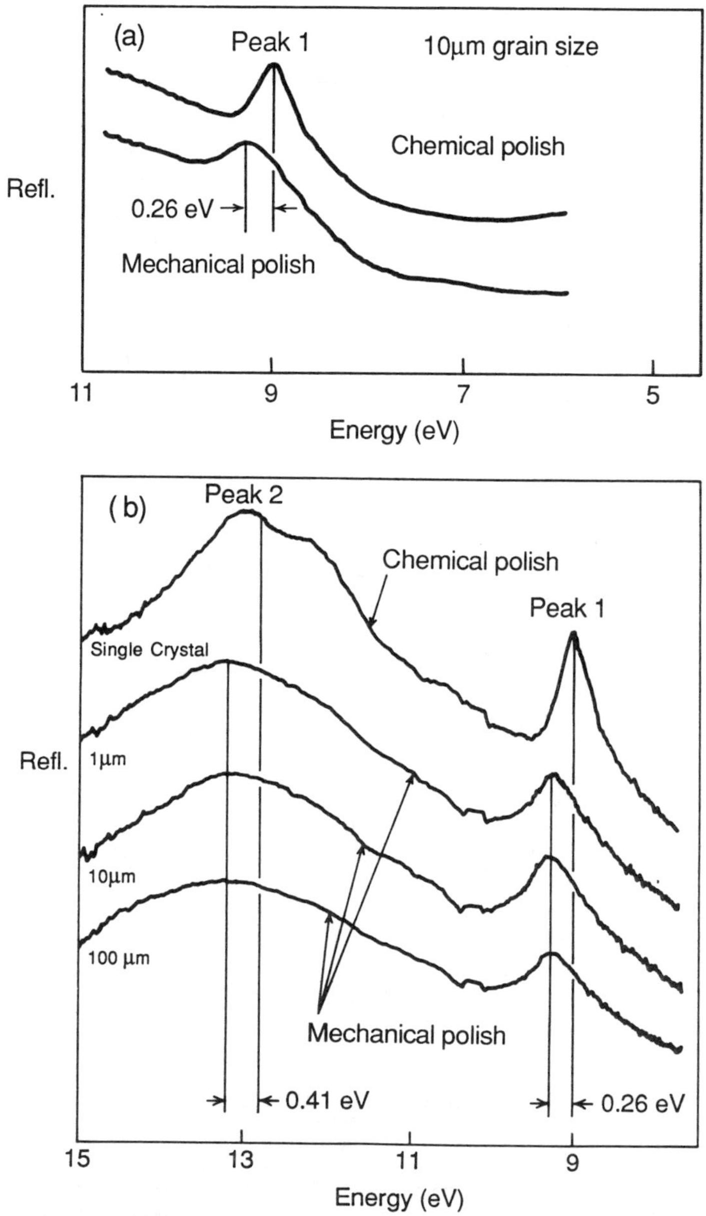

3. The VUV reflectivity of a) 10 micron grain size Al_2O_3 chemically and mechanically polished and b) chemically polished single crystal Al_2O_3 compared to mechanically polished polycrystalline Al_2O_3.[5]

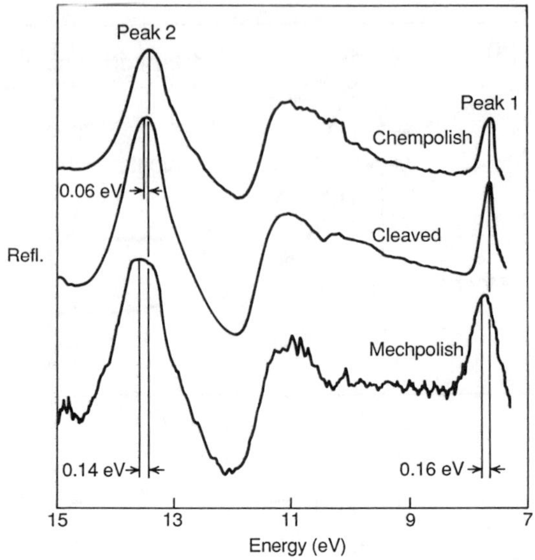

4. The VUV reflectivity of single crystal MgO which has been chemically polished, cleaved, or mechanically polished.[5]

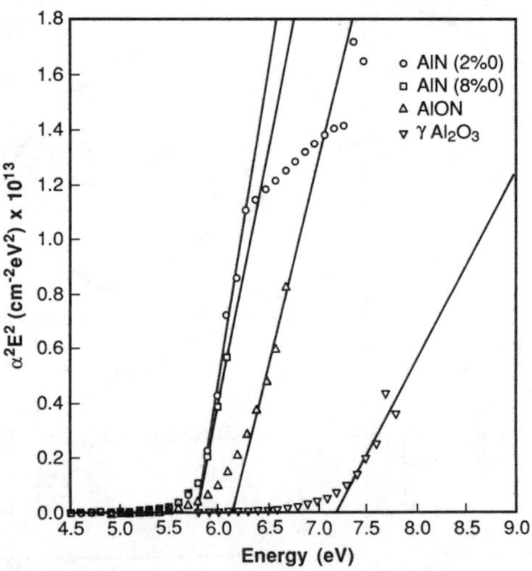

5. Absorption coefficients plotted for direct absorption edge analysis of AlN doped with 2 % O (E_g = 5.8 eV), AlN doped with 8 % O (E_g = 5.8 eV), AlON (E_g = 6.2 eV), and γ Al_2O_3 (E_g = 7.2 eV) thin films, showing the changes in absorption edge shape and band gap energy with changing N to O ratio.

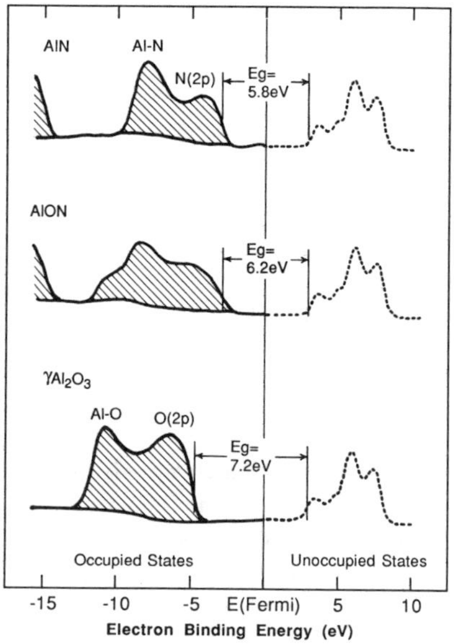

6. Schematic density of states for the electronic structure of AlN, AlON and γ Al_2O_3.

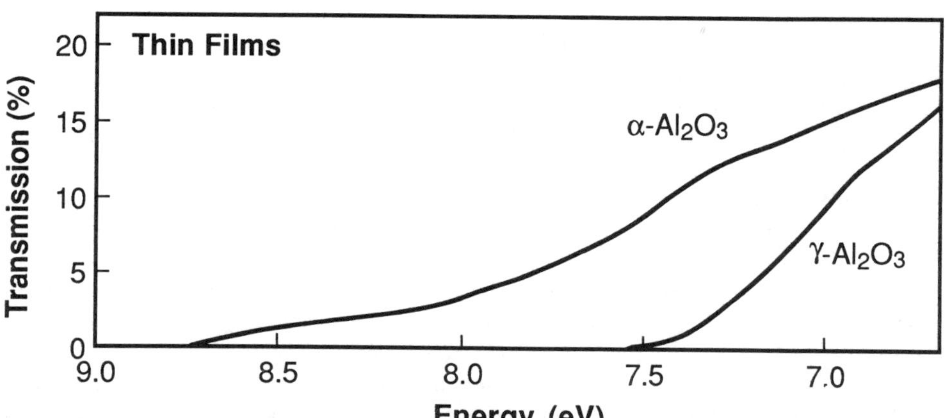

7. Room temperature transmittance cutoff for γ Al_2O_3 thin film, and the same film after the phase transition to α Al_2O_3 due to in situ heating to 1500 K, note the shift of the absorption edge from 7.3 eV to 8.8 eV.

FTIR REFLECTANCE SPECTRA OF ZIRCONIUM TITANATE BASED DIELECTRIC CERAMICS

R. Kudesia and A. E. McHale
NYS College of Ceramics at Alfred University, Alfred, NY 14802

ABSTRACT

A series of tin doped zirconium titanate compositions was analysed for dielectric characteristics using Far IR reflectance data. The trends in quality factor data were found to be as expected. For comparative studies or for the development of new microwave ceramic materials, the method is promising.

INTRODUCTION

Zirconium titanate based dielectric ceramics show very interesting properties at microwave frequencies. These ceramics have low dielectric loss and a unique composition dependence of the temperature coefficient.[1,2] Solid solution compositions in ZrO_2-TiO_2-SnO_2 system are commonly used as resonator material in microwave devices.[3] Solid solutions in the ZrO_2-TiO_2-CeO_2-SiO_2 system are also of interest.[4]

A dielectric post resonator technique is the conventional way of measuring the dielectric characteristics of these materials at microwave frequencies.[5] Other methods include measurement using a network analyser and the analysis of the Far Infrared reflectance spectra by the Kramers-Kronig relation and/or classical dispersion theory.[6] In the reflectance method, dielectric functions are obtained from the fundamental properties of the material, making the method quite useful for exploratory measurements on materials. This method can also be used to determine the critical importance of ceramic preparation and microstructural control for a particular microwave material. This method does not require any special sample preparation while, for resonance methods, a standardized sample geometry is necessary. Wakino, et al.[6,7] have calculated the dielectric constant and quality factor for TiO_2 and Ba(Zn,Ta)O_3-$BaZrO_3$ from the dispersion parameters. However, the calculated values were found to be larger than the directly measured values. This discrepancy could be understood on the basis that not all normal modes of vibrations are infrared-active. Non-active modes do not contribute to spectra. Considering all the modes, we can correlate the data and predict correct values.

The dielectric resonator materials based on zirconium titanate solid solution with SnO_2 or CeO_2/SiO_2 have a structure which is that of high temperature phase of pure $ZrTiO_4$.[1] The crystal structure of the high temperature phase $ZrTiO_4$ is orthorhombic and isomorphous with α-PbO_2 (orthorhombic space group D_{14}^{2h} = Pbcn).[8] The unit cell consists of two formula units. This space group implies a random distribution of the cations in available octahedral cation sites. Neutron diffraction studies have shown that the coordination octahedra of the Zr^{4+} ions are highly distorted.[9]

In the present experiments, infrared reflectivity measurements have been obtained on $Zr_xTi_ySn_zO_4$ ($x + y + z = 2$) compositions. The reflection spectra of the system are analyzed by the Kramers-Kronig (K-K) analysis. The K-K relation used in the evaluation of the reflectivity R may be written as

$$\theta(\nu_m) = \frac{2\nu_m}{\pi} \int_0^\infty \frac{\ln r(\nu)}{\nu^2 - \nu_m^2} d\nu, \tag{1}$$

where the reflectivity amplitude is $re^{\iota\theta}$ with $r = R^{0.5}$ and ν is frequency. The real and imaginary parts of the complex dielectric constant (ϵ', ϵ'') can be obtained from the following relations:

$$\epsilon' = n^2 - k^2, \tag{2}$$

$$\epsilon'' = 2nk, \tag{3}$$

$$re^{\iota\theta} = [(n-1) - \iota k]/[(n+1) - \iota k], \tag{4}$$

where n is index of refraction and k is extinction coefficient.

EXPERIMENTAL PROCEDURE

Sample Preparation

Powders containing the desired oxide compositions were prepared by a coprecipitation method from metal alkoxide solution. The method has been described elsewhere.[10] All specimens were fired at 1350 °C for 24 hours, and were then furnace cooled. Figure 1 shows the composition of different samples used in the study and the single phase range found in an earlier study here[10] which agrees with the work of Wolfram and Gobel.[1] Specimens were polished up to 400 grit (~ 23 μm) before collecting the spectra.

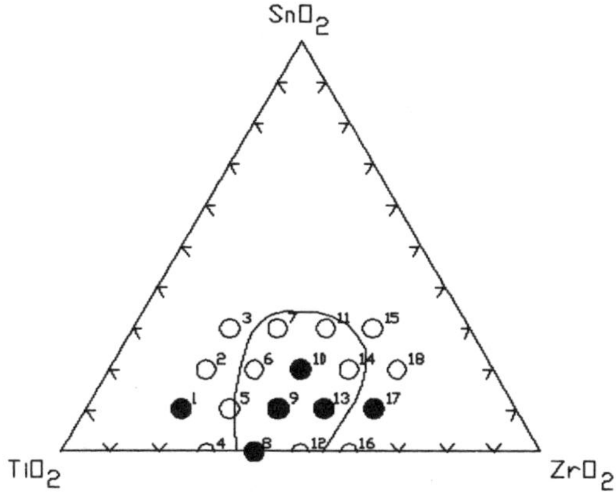

Figure 1: Composition of samples studied (full circles).

Experimental Measurement Technique

Nicolet 60 SXR and 20F FT-IR spectrometers with a Spectra-Tech Diffuse Reflectance (DRIFT) accessory were used to obtain the Mid and Far IR reflection spectra, respectively. A pure gas air drier purging device attached to the instrument maintained a moisture- and CO_2- free atmosphere. The spectra was recorded over the 5000 cm^{-1} to 50 cm^{-1} region at a resolution of 4 cm^{-1}. The polished ceramic sample was set in the sample cup of the DRIFT accessory. The reflectivity of the sample was obtained as the intensity relative to the reflectance from a deposited aluminium mirror.

ANALYSIS OF DATA

From the reflected intensity, R, and the phase angle, $\theta(\nu_m)$ (or θ_m), the optical parameters, n and k, may be determined from the following relationships:

$$n = (1 - R)/(1 + R - 2R^{0.5}\cos\theta_m), \quad (5)$$
$$k = (-2R^{0.5}\sin\theta_m)/(1 + R - 2R^{0.5}\cos\theta_m) \quad (6)$$

and hence the dielectric parameters. Since the phase angle is not experimentally determined it is generated from the reflected intensity measurements using the following K-K relationship:

$$\theta_m = \frac{2\nu_m}{\pi}\int_0^\infty \frac{\ln R^{0.5}(\nu) - \ln R^{0.5}(\nu_m)}{\nu^2 - \nu_m^2} d\nu. \quad (7)$$

The above equation has the following advantages over (1):

- The Integrand is always finite.
- There is minimum distortion in the term θ_m.

The integral was performed on a digital computer using a measured spectrum of R versus ν, the frequency of radiation. The range of integration of (7) was divided into three: a low frequency section, up to $\nu = \nu_l$, where $R(\nu)$ is approximately constant, the middle section of the range, and the final section starting from $\nu = \nu_h$, where $R(\nu)$ is again approximately constant up to high frequencies. The value of θ_m is then given by the expression:

$$\theta_m = L + H + \frac{2\nu_m}{\pi}\int_{\nu_l}^{\nu_h} \frac{\ln R^{0.5}(\nu) - \ln R^{0.5}(\nu_m)}{\nu^2 - \nu_m^2} d\nu, \quad (8)$$

where,

$$L = \frac{1}{2\pi}\ln(\frac{R_l}{R_m})\ln|\frac{\nu_m - \nu_l}{\nu_m + \nu_l}|, \quad (9)$$
$$H = -\frac{1}{2\pi}\ln(\frac{R_h}{R_m})\ln|\frac{\nu_h - \nu_m}{\nu_h + \nu_m}|. \quad (10)$$

Above relations are valid only for $\nu_l < \nu_m < \nu_h$.

EXPERIMENTAL RESULTS

Compositions selected for this study were along two series in the composition triangle: one along constant tin content and the other along constant zirconia content. Figure 2 shows a typical reflectance spectrum obtained on these ceramics. The Far IR reflectance data below 80 cm^{-1} was not taken into account as the noise level in this range was quite high. Instead, each spectrum was extrapolated from the available reflectance trend just above 80 cm^{-1}.

The resultant dielectric characteristics calculated by using the K-K analysis are listed in Table I. The calculated quality factor values are smaller than expected for ceramics of these compositions. Although the reason is not clear, these low values may be the effect of microstructure which was not optimized in these ceramics. This possibility is also supported by the fact that a similar analysis performed on a commercial zirconium tin-titanate sample gave satisfactory quality factor value.

The quality factor was found to be quite sensitive to the selection of the lower frequency limit ν_l. It was found that low value of ν_l gives a high value of the quality factor. However, this effect was insignificant in the case of dielectric constant.

CONCLUSIONS

It is concluded from this study that Far Infrared reflectance data can be used to determine dielectric functions at microwave frequencies. Dielectric functions can be calculated for the high GHz range where resonance methods do not give reliable values. It is not simple to calculate dielectric functions using the K-K relation. There are practical difficulties due to the range of integration. However, the use of modern computer software makes the analysis less time consuming. For the data reduction and K-K analysis, a sophisticated software is being developed.

There is an indication of the possibility that microstructure of these ceramics plays a significant role in the development of good microwave dielectric properties. Further studies are required to confirm this. Since the dielectric functions are obtained from the fundamental material chracteristics and the sample preparation is quite flexible, this technique could be very useful in the development of new microwave ceramic materials.

ACKNOWLEDGEMENTS

This research was funded by the Center of Advanced Ceramic Technology at NYS College of Ceramics, Alfred University, Alfred.

REFERENCES

1. G.Wolfram and H.E.Gobel, "Existence Range, Structural and Dielectric Properties of $Zr_xTi_ySn_zO_4$ Ceramics ($x + y + z = 2$)," *Mater Res Bull, 16*[11] 4155-63 (1981).

2. K.Wakino, K.Minai, and H.Tamara, "Microwave Characterstics of (Zr,Sn)TiO_4 and BaO-PbO-Nd_2O_3-TiO_2 Dielectric Resonators," *J Am Ceram Soc, 67*[4] 278-81 (1984).

3. M.Katsube, Y.Ishikawa, H.Tamura, and K.Tomono, "Dielectric Ceramic Composition for High Frequencies," U.S.Patent no. 4,102,696 (25th July, 1978).

4. K.Sasazawa and N.Yamaoka, "Dielectric Ceramic Composition," U.S.Patent No. 4,584,282 (22nd April, 1986).

5. W.E.Courtney, "Analysis and Evaluation of a Method of Measuring the Complex Permittivity and Permeability of Microwave Insulators," *IEEE Trans Microwave Theory Tech, 18*[8] 476-485 (1970).

6. K.Wakino, D.A.Sagala, and H.Tamura, "Far Infrared Reflection Spectra of Ba(ZnTa)O_3-$BaZrO_3$ Dielectric Resonator Material," *Jpn J Appl Phy, 24*, Suppl 24-2, 1042-1044 (1985).

7. K.Wakino and H.Tamura, "Microwave Characteristics and Far Infrared Reflection Spectra of TiO_2 Single Crystal," Abstract No. 24-E-87, 89th Ann. Meet. Am. Ceram. Soc., Pittsburgh, PA, 26-30 April, 1987 (Unpublished).

8. R.E.Newnham, "Crystal Structure of $ZrTiO_4$," *J Am Ceram Soc, 50*[4] 216 (1967).

9. P.Bourdet, A.E.McHale, A.Santoro, and R.S.Roth, "Powder Neutron Diffraction Study of $ZrTiO_4$, $Zr_5Ti_7O_{24}$, and $FeNb_2O_6$," *J Sol State Chem, 64*[1] 30-46 (1986).

10. A.H.Lee, "The Structure and Properties of Tin Doped Zirconium Titanate Made by Co-Precipitation," M.S. Thesis, Alfred University, Alfred, (1988) (Unpublished).

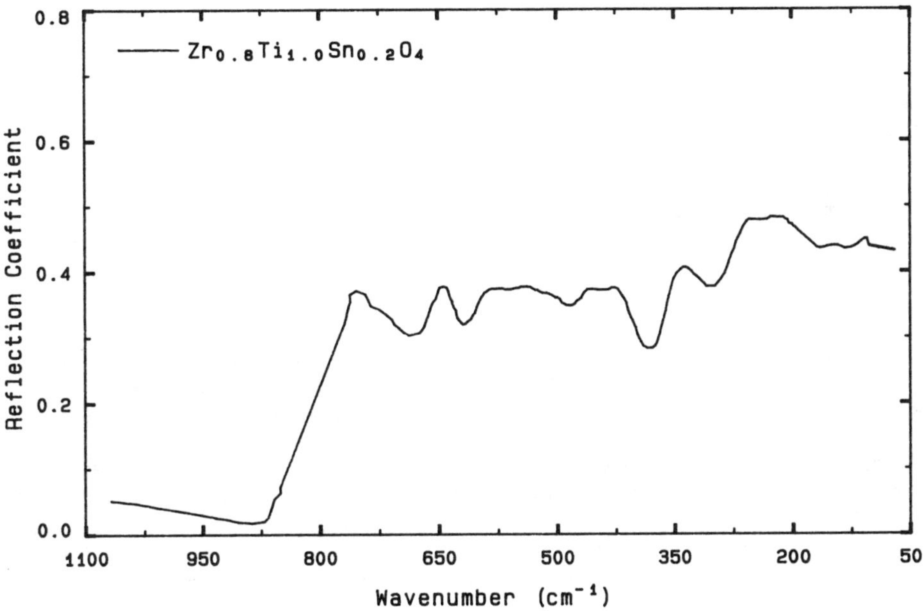

Figure 2: Infrared reflectivity of $Zr_{0.8}Ti_{1.0}Sn_{0.2}O_4$ (sample 9).

Table I: Calculated and Measured Dielectric Characteristics

Sample No.	Composition	Calculated at 300 GHz		Measured at 1 GHz[a]	
		ϵ	Q	ϵ	Q
1	$Zr_{0.4}Ti_{1.4}Sn_{0.2}O_4$	16.4	60	69.0	47
8	$Zr_{0.8}Ti_{1.2}Sn_{0.0}O_4$	21.4	820	-	-
9	$Zr_{0.8}Ti_{1.0}Sn_{0.2}O_4$	21.7	640	41.6	112
10	$Zr_{0.8}Ti_{0.8}Sn_{0.4}O_4$	43.2	35	35.8	79
13	$Zr_{1.0}Ti_{0.8}Sn_{0.2}O_4$	18.3	480	38.5	28
17	$Zr_{1.2}Ti_{0.6}Sn_{0.2}O_4$	9.7	420	-	-
Reference Sample[b]	$Zr_{0.6}Ti_{1.0}Sn_{0.4}O_4$	6.5	30 000	-	11 987[c]

[a]Measurements were done using a HP4191A impedance analyser.
[b]Nominal sample composition and measured dielectric characteristics were provided by Dr. T. Negas of TransTech Inc.
[c]Data obtained using a network analyser at approximately 10 GHz.

ADVANCED MICROSTRUCTURAL PREPARATION TECHNIQUES FOR CERAMICS

D.C. Zipperian and Mary Callahan
BUEHLER Southwest Research Center
Tucson, AZ 85715

ABSTRACT

Fracture mechanics of ceramic materials has been studied to differing degrees in the past; however, this application as applied to polishing and finishing operations has primarily been qualitative. This paper addresses the amount of subsurface damage that is produced by varying the diamond size and the support platen. It is quite evident that the amount of surface and subsurface damage for brittle ceramic materials is highly dependent upon the interaction of these parameters.

INTRODUCTION

The characterization techniques for advanced ceramics generally require special attention be given to the microstructural preparation techniques. The basic concepts and techniques for microstructural preparation also find application in machining and surface finishing techniques. Polishing and machining techniques, though somewhat tedious and time consuming, are a necessary requirement, primarily because the electrical as well as the structural properties of these ceramic materials vary dramatically with surface condition. In addition, analysis of the microstructural properties such as grain size and distribution, porosity, phase size and distribution provides valuable information about the physical properties of these materials.

In analyzing and discussing the mechanisms for machining and polishing advanced ceramics, a general description of the basic fracture mechanisms for brittle materials is required. The removal rates for brittle ceramic materials depends upon load, and abrasive size and shape. It has been suggested that a harder abrasive

grain can act similar to either a sharp indenter or a blunt indenter (1-4). The general sequence of events for both these types of phenomena includes the fracturing and cracking when the indenter (abrasive) is loaded or forced into the surface, followed by additional cracking upon unloading. Although the general series of events for both sharp and blunt indenters are similar, the actual crack formations are somewhat different. Figure 1 sketches the sequence of events that occur under an increasing load for a sharp indenter, followed by removal of the load. Initial loading creates a plastic zone, as can be witnessed by the permanent impression that is left in the surface. As the load is increased to a critical indentation load, a median crack is initiated below the deformation zone. Upon unloading the median cracks tend to close off. However further increases in the load above this critical indentation load causes the formation of lateral cracks to initiate. Upon unloading the relaxation of the irreversibly deformed material creates intense residual stresses that cause the lateral cracks to eventually reach the free surface (1-5). Figure 2 shows the cone crack formation that typifies a blunt indenter.
Upon fully developed conditions a ring crack forms just outside the contact circle and propagates downward to develop a cone crack (1,6).

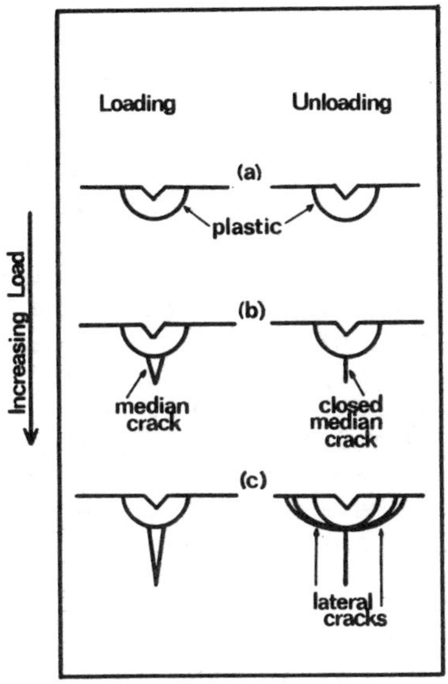

Figure 1. Fracture and growth for a sharp indenter.

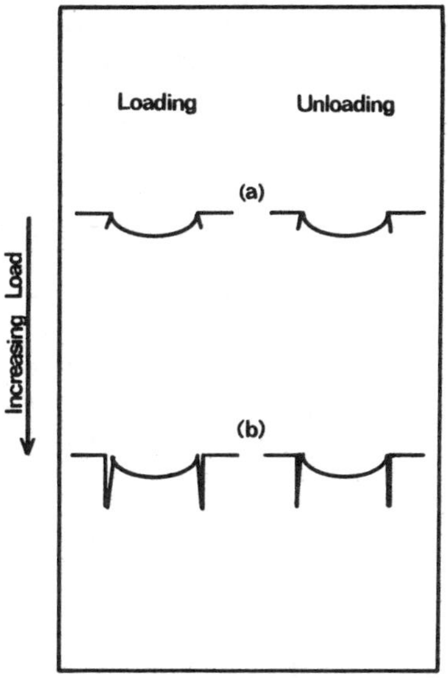

Figure 2. Cone cracking for a blunt indenter.

Typical sample preparation and machining operations can be encompassed into three major categories:

(i) planar grinding
(ii) sample integrity
(iii) surface finishing (polishing)

The planar grinding step involves surface grinding or lapping to produce the flat surfaces required for microstructural analysis, especially in the case where multiple samples are mounted. The sample integrity stage involves the removal of damage produced in the sectioning and planar grinding steps. These steps bring the surface back to a level which can be analyzed; however, to clean up the surface further a final polishing step is usually required.

Typical polishing and machining operations for ceramics utilize lapping techniques for the sample integrity stage. These lapping techniques can pertain to either a fixed or rolling abrasive. In general, the term lapping applies to the method by which the abrasive is applied to a flat platen surface, either in an aerosol or a slurry form. The abrasive grain then becomes either embedded or remains free. The literature describes both these mechanisms. With the general rule applying to platen hardness (i.e. abrasive becoming fixed in a soft platen and remaining free on a hard platen (7,8,9)). This paper focuses on the subsurface damage produced by the interaction of various lapping platen hardnesses with different grades of diamond.

EXPERIMENTAL

To study the interaction of a diamond abrasive with an alumina ceramic for various lapping platens, BUEHLER METLAP™ platens were utilized. These platens vary in their hardness, which is an important key in trying to determine the action of the diamond abrasive, i.e. fixing or rolling, which inherently affects the fracturing of brittle ceramic surfaces. The hardness of these lapping platens increase in increasing order from METLAPS™ 4,6,8, and 10. In addition, the diamond size and its interaction with the ceramic as well as the METLAPS™ is important for analyzing the polishing mechanisms for ceramics.

99% dry pressed alumina samples were planar ground with a 45 micron diamond on METLAP™ 10, using the BUEHLER DATAMET™ because of its ability to maintain specific forces, speeds, times, and directions. The degree and removal of damage was measured with the BUEHLER OMNIMET®II image analysis system. The lapping work with METLAP™ platens was done with the BUEHLER MINIMET® and BUEHLER diamond slurries.

RESULTS

The effect of diamond (abrasive) size is influenced by its particle shape and the way it interacts with the lapping platen (i.e. a loose abrasive acts as a coarser fixed abrasive). Figure 3 shows that the amount of alumina removal, for diamond sizes between 1 and 15 micron, is a function of the diamond size and not METLAP™ platen hard-ness. Figure 3, however does not take into account the degree of damage induced into the sample.

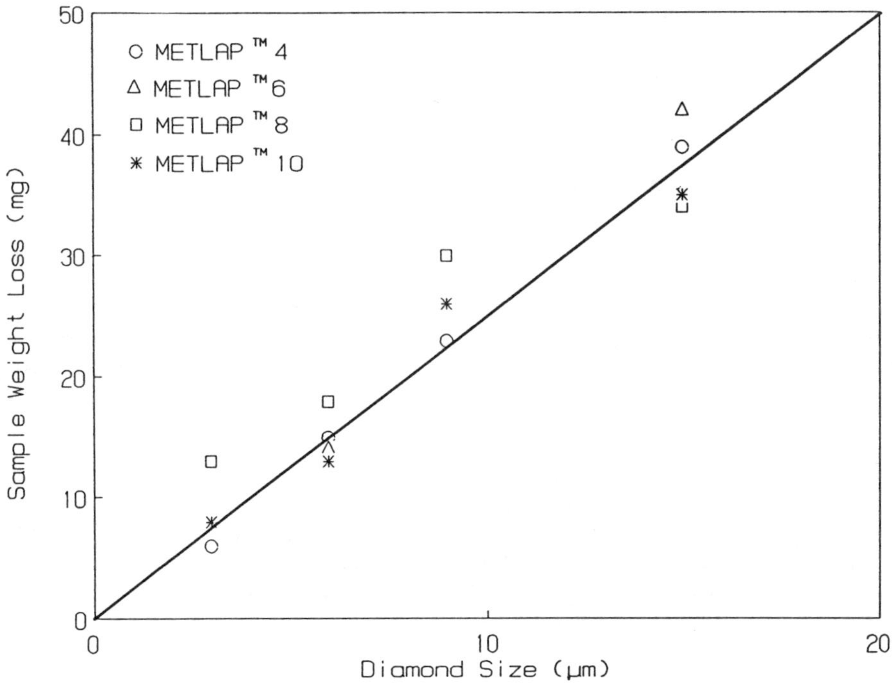

Figure 3. Sample weight loss vs diamond size after 30 min.

The amount of damage produced in the alumina subsurface layers is presented in Figure 4 for METLAP™ 4. From this figure we see that increasing diamond size up to 6 micron increases the rate of subsurface damage

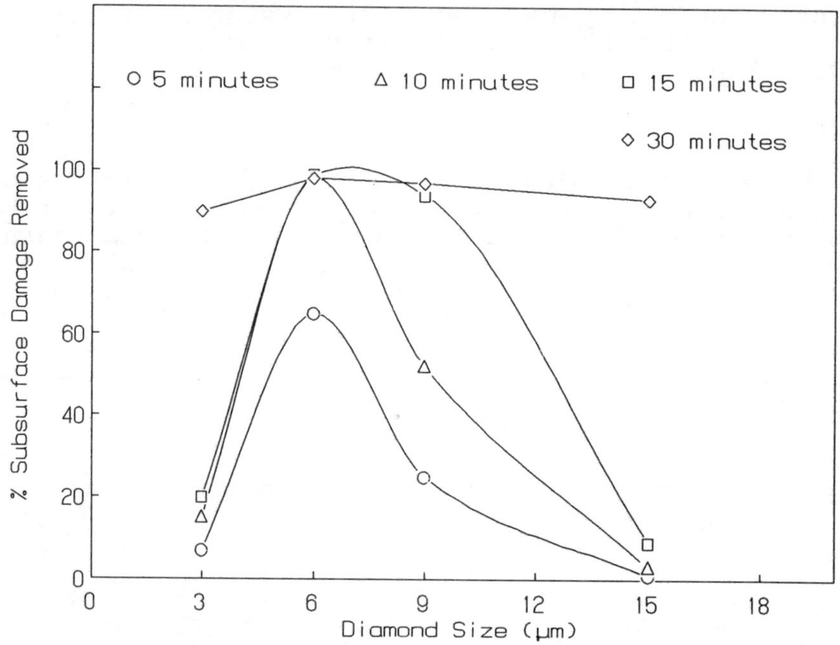

Figure 4. Percent damage removed vs diamond size.

removal. Above 6 micron diamond, increasing diamond size decreases the rate of subsurface damage removal. This along with the steadily increasing material removal rates seen in Figure 3, suggests that diamond sizes greater than 6 micron create their own additional damage (note that this damage is still removed after 30 minutes).

Figure 5 shows the role that the lapping platen has on subsurface damage removal. In the case of a hard brittle material such as alumina, METLAP™ 4 showed the highest rate of damage removal using a 6 micron diamond slurry. Note METLAP™ 4 is the softest and produced the least amount of additional damage.

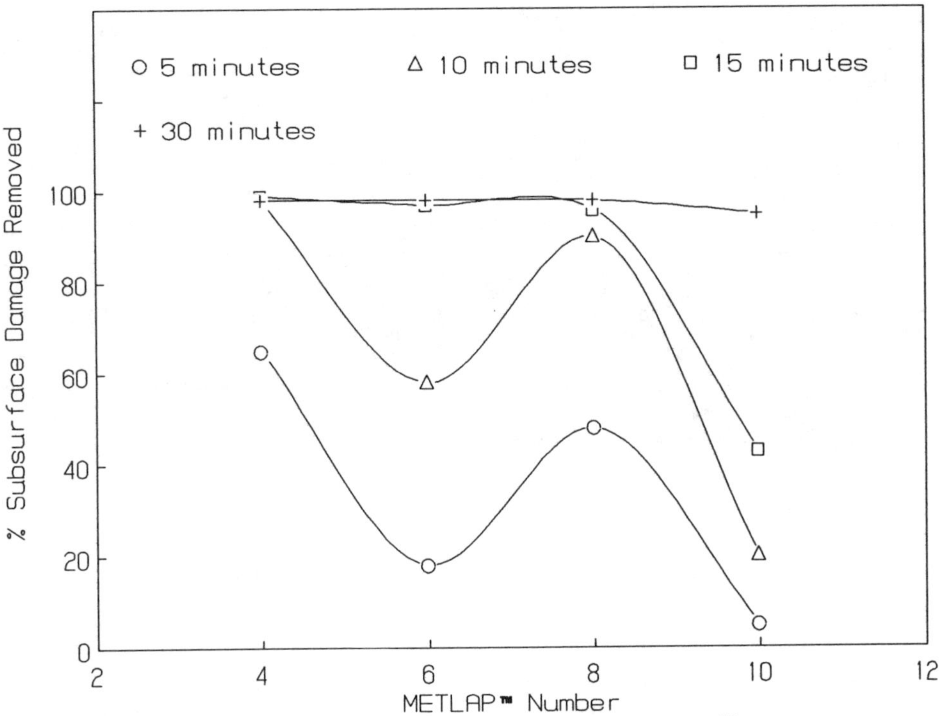

Figure 5. Percent damage removed vs METLAP™ No.

DISCUSSION

In the process of developing new techniques for handling the preparation of advanced ceramics, the fracture mechanics of an alumina ceramic was studied in significant detail. These include diamond particle size and lapping platen hardness.

Obviously from the data all three of these are important considerations in the preparation of alumina. The consideration of diamond size is important because of the different cracking mechanisms that occur. For smaller diamonds (3 micron) the cracks are localized and appear to correspond to sharp indenter damage (Figure 6a). Whereas for larger diamond sizes (>15 micron) the cracks have a shattered appearance (cone fracture) which is typical of blunt indenters (Figure 6b).

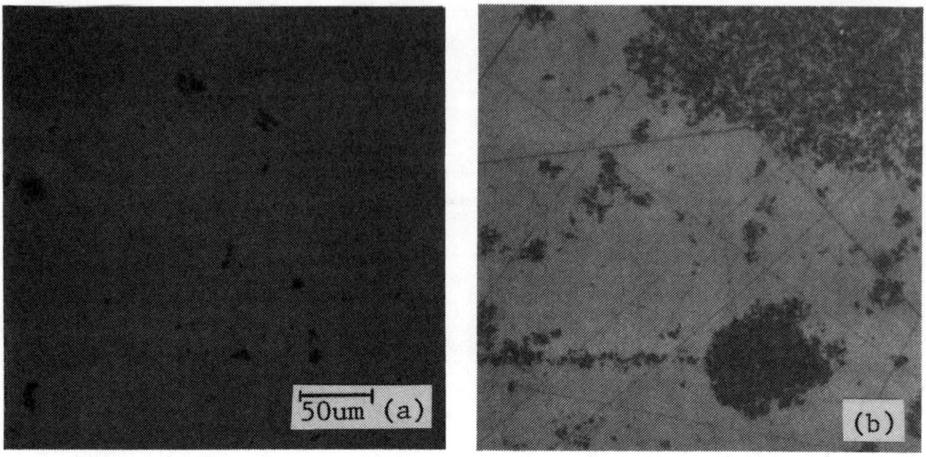

Figure 6. Micrograph of Surface Damage (a) 3μm (b) 15μm on METLAP™ 4.

Brian Lawn and Rodney Wilshaw (6) produced a general review paper on the effects of a "Fixed" indenting particle as opposed to a "Free" indenting particle. In their discussion they surmised that the specimen surface damage for "Fixed" indenting particles resembled a sliding indenter. They went on to describe this type of damage as being similar to a sharp indenter for sharp abrasives, where chipping occurred from lateral cracks intersecting the surface. For worn abrasive particles the damage compared more to blunt indenters, with removal rates arising from neighboring cone cracks combining.

For "Free" indenting particles Lawn and Wilshaw described similar mechanisms as discussed above. However, another noticeable difference can be seen in their review paper. This pertains to the occurrence of scratching in addition to the chipping phenomena. In the case of a "Fixed" indenting particle the scratches were quite deep. In contrast the "Free indenting particles showed no significant amount of scratching. In the preparation of alumina we see the occurrence of scratching for the coarser diamonds, indicating that some degree of diamond "Fixing" must be occurring.

CONCLUSIONS

The development of advanced preparation techniquesd for advanced ceramics is based upon the interaction and fracture mechanics of a sample with the lapping platen and the diamond size. These experiments focused on alumina, where both sharp and blunt indenter characteristics were seen for different diamond sizes. In addition, both "Fixed" and "Free" indenting particle characteristics have been noted. Thus, the use of a range of lapping platen hardnesses, with diamonds of varying sizes, produces maximum flexibility in microstructural preparation.

REFERENCES

1. L.E. Samuels, "Brittle Materials: Principles," Metallographic Polishing By Mechanical Methods, 3rd ed., American Society for Metals; Metals Park, Ohio, 203-209 (1982).
2. D.B. Marshall, A.G. Evans, B.T. Khuri Yakub, J.W. Tien, and G.S. Kino, "The Nature of Machining Damage in Brittle Materials," Proc. R. Soc. London, **A385**, 461-475 (1983).
3. D.W. Richerson, "Final Machining," Modern Ceramic Engineering Properties, Processing and Use in Design, Marcel Dekker, Inc.; New York, 260-273 (1982).
4. B.R. Lawn and D.B. Marshall, "Indentation Fracture and Strength Degradation in Ceramics" Fracture Mechanics of Ceramics, Vol.3, ed. R.C. Bradt, D.P.H Hasselman, F.F. Lange, Plenum Press, NY, pp. 205-229 (1978).
5. O.O. Ajayi and K.C. Ludema, "Surface Damage of Structural Ceramics: Implications for Wear Modeling," Wear, **124**, 237-257 (1988).
6. B.R. Lawn and R. Wilshaw, "Review Indentation fracture: principles and applications," J. Mat. Sci., **10**, 1049-1081 (1975).
7. E.J.A. Armarego and R.H. Brown, The Machining of Metals, Prentice-Hall; Englewood Cliffs, New Jersey, 375 (1969).
8. R. Bainbridge, "A Factorial Experiment in Lapping"; Thesis held at the University of Melbourne Engineering School Library).
9. F. Eugene, "Investigations of the Lapping Process," Industrial Diamond Review, **7**, pp. 35 and 67 (1947).

ACRONYM INDEX

AAS	Atomic absorption spectroscopy
AEM	Analytical electron microscopy
AES	Auger electron spectroscopy
AES	Atomic emission spectroscopy
AFM	Atomic force microscopy
AP	Atom probe (analyzer)
AREM	Atomic resolution electron microscopy
ARM	Atomic resolution microscope
BF	Bright field
CBED	Convergent beam electron diffraction
CCD	Charge-coupled device
CDW	Charge density wave
CI	Chemical ionization
CIS	Complex impedance spectroscopy
CITS	Current imaging tunneling spectroscopy
CMC	Ceramic matrix composites
CTEM	Conventional transmission electron microscopy
CTF	Contrast transfer function
CW	Continuous wave
DAC	Diamond anvil cell
DBM	Direct bombardment mode
DF	Dark field
DLTS	Deep-level transient spectroscopy
DSC	Differential scanning calorimetry
DTA	Differential thermal analysis
DXRD	Dynamic X-ray diffraction
EDAX	Energy dispersive X-ray analysis (see EDX)
EDS	Energy dispersive X-ray spectroscopy (see EDX)
EDX	Energy dispersive X-ray spectroscopy (also EDS or EDXS)
EELS	Electron energy loss spectroscopy
EI	Electron impact
ELNES	Energy-loss near-edge structure
EPMA	Electron probe microanalysis
EPR	Electron paramagnetic resonance
ESCA	Electron spectroscopy for chemical analysis (see XPS)
EXELFS	Extended energy-loss fine structure
FAB	Fast atom bombardment
FDMS	Field desorption mass spectroscopy
FEG	Field emission gun
FIM	Field ion microscopy
FIMS	Field ionization mass spectroscopy
FTICR	Fourier transform ion cyclotron resonance (also FTMS)
FTIR	Fourier transform infrared spectroscopy
FTMS	Fourier transform mass spectrometer (also FTICR)

HEIS	High energy ion scattering (see RBS)
HREM	High-resolution electron microscopy
ICP	Inductively coupled plasma
ICPMS	Inductively coupled plasma mass spectroscopy
IETS	Inelastic electron tunneling spectroscopy
IPS	Inverse photoemission
LAMMA	Laser microprobe mass analysis
LDMS	Laser desorption mass spectroscopy
LEED	Low energy electron diffraction
LIMS	Laser ionization mass spectroscopy
LPS	Laser plasma source
MMC	Metal matrix composites
MMR	Magnetically modulated microwave reflection
MS	Mass spectrometry
MSANS	Multiple small angle neutron scattering
NCEM	National Center for Electron Microscopy (Berkeley, CA)
NIST	National Institute of Standards and Technology (formerly National Bureau of Standards)
NMR	Nuclear magnetic resonance
PDS	Photothermal deflection spectroscopy
PIXE	Proton induced X-ray emission
PLZT	Lead lanthanum zirconate titanate
PSD	Position sensitive detector
PSZ	Partially stabilized zirconia
PTCR	Positive temperature coefficient of resistance
RAE	Resistive anode encoder
RBS	Rutherford backscattering spectroscopy (also HEIS)
REMPI	Resonantly-enhanced multi-photon ionization
RIR	Reference intensity ratios
RSF	Relative sensitivity factor
SAD	Selected area diffraction
SALI	Surface analysis by laser ionization
SAM	Scanning Auger microscopy
SANS	Small angle neutron scattering
SAXS	Small angle X-ray scattering
SBM	Separate bombardment mode
SEM	Scanning electron microscopy
SIMS	Secondary ion mass spectroscopy
SLAM	Scanning laser acoustic microscopy
SNMS	Sputtered neutral mass spectroscopy
STEM	Scanning transmission electron microscopy
STM	Scanning tunneling microscopy
STS	Scanning tunneling spectroscopy

TD	Theoretical density
TEM	Transmission electron microscopy
TGA	Thermogravimetric analysis
TMS	Tetramethylsilane
TOF	Time of flight
TRRS	Time-resolved Raman spectroscopy
TXRF	Total reflection X-ray fluorescence
UPS	Ultraviolet photoemission
UTW	Ultra thin window (detector)
VUV	Vacuum ultraviolet
WDS	See WDX
WDX	Wavelength dispersive X-ray spectrometry (also WDS)
XPS	X-ray photoelectron spectroscopy (See ESCA)
XRD	X-ray diffraction
XRF	X-ray fluorescence

INDEX